GENOMIC REGULATORY SYSTEMS

Genomic Regulatory Systems

Development and Evolution

Eric H. Davidson
Division of Biology
California Institute of Technology
Pasadena, California

ACADEMIC PRESS
A Harcourt Science and Technology Company

San Diego   San Francisco   New York   Boston   London   Sydney   Tokyo

Front cover photograph: Two images are represented on the cover. The larva of an indirectly developing hemichordate, *Ptychodera flava* is shown in blue [From Peterson *et al.* (1999). *Development* **126**, 85–95 and The Company of Biologists, Ltd.]. Overlying this in white is a functional model of a portion of the *cis*-regulatory system (double line) of the *endo16* gene of the sea urchin *Strongylocentrotus purpuratus*. The model indicates the logic transactions (circles) mediated by the target sites for DNA-binding factors (boxes) (from Yuh *et al.*, 2000). The type of embryogenesis the outcome of which is exemplified by the hemichordate larva, is discussed in Chapters 3 and 5 of this book, and the *endo16 cis*-regulatory system in Chapter 2.

This book is printed on acid-free paper.

Academic Press
A Harcourt Science and Technology Company
525 B Street, Suite 1900, San Diego, California 92101-4495, USA
http://www.academicpress.com

Academic Press
Harcourt Place, 32 Jamestown Road, London NW1 7BY, UK
http://www.academicpress.com

Library of Congress Catalog Card Number: 00-110671

International Standard Book Number: 0-12-205351-6

PRINTED IN THE UNITED STATES OF AMERICA
00 01 02 03 04 05   WP   9   8   7   6   5   4   3   2   1

CONTENTS

PREFACE

For about 40 years I have been engaged in the effort to perceive how development is encoded in the DNA, and thereby how animal evolution happened. These are the questions that lie at the heart of functional genomics, at least for those whose fascination is the academic one of learning how life works. But of course in the early days when I entered the field we didn't use terms such as "functional genomics," perhaps because the idea of genomics divorced from function had not yet been invented. I became a graduate student in the laboratory of Alfred Mirsky at Rockefeller in 1958, and in those days the major intellectual issue was whether differential gene expression, i.e., cell-by-cell regulatory control of the transcription of each gene, is indeed the fundamental basis of cell differentiation. We thought it must be so, but the evidence was weak, and it took about 15 years of increasingly incisive molecular biology to prove the point. But long before all that dust had settled, the logical consequences of this theory of the mechanism underlying differentiation had led me into the pathways of which this book is a current way station. Logic demanded that embryonic development is the outcome of a vast spatial and temporal series of differential gene expressions, and that the control of these must depend on a hardwired regulatory program built into the DNA sequence. You could get that far just by combining the three accessible major articles of what would become genomic bioscience: the 19th C. demonstration that the course of development is a heritable property of the organism, dependent on egg and sperm genomes; the demonstration, then almost new, that DNA is "it," the heritable genetic material; and the principle of differential gene regulation (or as we said then, "variable gene activity"). At the end of the 1960s Roy Britten and I explored the spray of logical inferences that led from this position out into the domains of development and evolution. But our field still had a universe to learn about: We should have to find out what the DNA of the animal genome is really like in complexity and organization; to measure the dimensions of gene expression in development; to absorb the stunning examples of regulatory genes that emerged from *Drosophila* developmental genetics; to learn how to really make use of the possibility of isolating and studying any individual gene.

Things did not hang together very well at the end of the 1980s: images of the network-level analyses that would tie it all together were still well in the future. The *Drosophila* examples were all "Chiefs," the "Indians" only implied. The enormous field of mammalian differentiation molecular biology was just the reverse. Some of us were busy learning how simple embryos work, sea urchins, ascidians, and then *C. elegans* embryos. But we discovered that how these embryos work illuminates mainly themselves, or at best each other, and not more elaborate developmental processes. We could study any gene we wanted at biochemical and cell biological levels. We could get answers to questions about intercellular signaling that had bedeviled embryologists since Spemann. But because we could, we did, and the foci of explication in our field seemed to be drifting in a hundred directions outward, further and further from the informational source of the whole program, the DNA.

Three things happened in the 1990s that are now creating the resolving power that we need to understand development and evolution in the informational terms that truly define these processes. The first, in which my laboratory along with a few others has been deeply involved, was the advent of sophisticated *cis*-regulatory analysis. At last, after all these years, we can bring regulatory hardwiring at the DNA nucleotide level into concise functional focus. The second major change was the sudden availability of genomic DNA sequence and sequence databases on the actual scale of animal genomes. The third was the revitalization of evolutionary arguments pointed at the kinds of genomic events that have caused change in body plans. Good minds from both evolution and development have converged on the explanatory power of genomic regulatory mechanisms, and meanwhile molecular phylogeny has washed away a lot of confusion about animal relationships. In the lines of thought that have led to this book, evolutionary considerations have provided crucial guides, something I might not have predicted a decade or two ago.

So this book could not have been written, by me at any rate, much before the present moment. It is the sum of many influences, insights, and inputs, that have served as prisms with which to refract the rush of incoming data. I shall single out a few of these influences, with the understanding that none bear responsibility for whatever nonsense shall be found within these covers. First, I think that the MBL Course in Developmental Biology, of which I was Director from about 1985 to 1995, provided the intellectual pot in which brewed many of the ideas and approaches that have become important for me. This was due to the tremendously stimulating give and take with a vocal and intellectually alive group of senior Faculty, and the lively students; and also to the deliberately arranged marriage between regulatory mechanism and comparative embryology that formed the overriding conceptual and experimental environment of the Course in those years. From that cross fertilization ultimately emerged the mode of parsing developmental regulatory networks that is laid out in Chapters 3 and 4. Second, there have been a small number of other scientists whose experiments, interpretations, and attitudes have been particularly important for me. Among the people who in these last few years have made especially interesting contributions to my own state of mind (and this will surprise some of them) are Mike Levine, André Adoutte, Ellen Rothenberg, Michael Akam, Joel Rothman, R. Andrew Cameron, Lee Hood, and Kevin Peterson. Third, partly by serendipity, I have been drawn into an edge of the genomics community as a consumer and an interlocutor, and have turned increasingly to the real details of the problem of how to obtain solutions to the great issues of evolution and development at the genome level.

It would all have been hot air instead of a book were it not for a number of people who have been specifically involved with this project in one way or another. James W. Posakony deserves special mention. This book was conceived in a series of conversations with him (on Nobska Beach at Woods Hole, and at the Crab Shack in Newport Bay). He is the one who convinced me that such a book should be constructed. Many of my colleagues at Caltech in and outside of my own group have read drafts, argued details, beat me over the head, and encouraged me. Practically every page of this manuscript has profited from the insights and critical judgment of Ellen Rothenberg. Mike Levine, Marianne Bronner-Fraser, Kevin Peterson, R. Andrew Cameron, Jonathan Rast, Paola Oliveri, and Hamid Bolouri each reviewed some or all of the chapters: tough critics and true intellectuals all.

There is a long way from my piles of barely legible handwritten drafts to the book that you are holding, a very long way. A few wonderful people accomplished this transformation. Jane Rigg, who has worked with me for about a third of a century and put two of my three previous books (*Gene Activity in Early Development*, 1968, 1976, and 1986) together, provided the essential organizing principle, and also a keen editorial eye and bibliographic wizardry. Deanna Thomas endlessly dug out references and on a daily basis handled a huge number of "get this for me yesterday" requests, always with amazingly good humor and intelligent efficiency. She and Tania Dugatkin assembled all the figures. Stephanie Canada converted scribble into text, something only she can do because she reads my handwriting better than I do. In addition I wish to acknowledge the best editor by far, of the four I have worked with at Academic Press, Jasna Markovac. With her velvet hand and superb judgment and very cool efficiency she really kept this project going.

Finally, and of course most essentially, there are all the scientists whose terrific discoveries form the body of this work, and who gave me permission to reproduce their figures. Many people provided access to their yet unpublished work; my colleagues out there have been wonderfully collegial. I apologize to all of those whose work I should have cited but didn't, either out of ignorance or in a desire to use the minimum number of citations (as is, there are over 500 references). To this end I have tried wherever possible to cite recent reviews and other comprehensive secondary sources, or the most current experimental studies, rather than all the primary and original sources only. The text was written between Fall 1999 and mid-Summer 2000, and most of the references are to studies published between 1995 and 2000. But the worst crimes I committed, still unknown to me as I write this, are the mistakes I have surely made. It is almost impossible to be sufficiently expert in all the areas relevant to my subject, given its breadth: how gene regulation works in respect to the development and evolution of animals. I can only say what anyone in this position would, that I did my best.

A note on nomenclature: A large number of genes and gene products are referred to in this book, from many different organisms and many different fields. Each biological system has grown up with its own nomenclature. To make things as simple as possible I have referred to every gene in lower case italics (with the exception that if the organism's name is included as a two letter abbreviation where the generic letter is capitalized, so is the first letter of the gene name, e.g., *SpHox8*), and every gene product is given in roman with an initial capital letter. This means that some dearly loved local conventions have gone by the wayside. Some historically important distinctions, such as whether a gene was first identified by a dominant or recessive mutation are lost (*Ubx* is *ubx*, and *hox* genes, and Gata factors, are referred to as such). This usage can be viewed in several ways: as laziness on my part; as insufficient respect for tradition; or as a representation of the fact that regulatory genomic biology is one world; however the genes were originally discovered and named; from whichever animal they derived; and whatever they do.

Eric H. Davidson
August, 2000

CHAPTER 1

Regulatory Hardwiring: A Brief Overview of the Genomic Control Apparatus and its Causal Role in Development and Evolution

Some problems direct the intellectual explorer toward the heart of a matter, and in 20th C. bioscience one such problem has been the diversity of form in animal life. By the end of the 1950s it was clear that the causal differences between the body plans of a fish and fly, or a sea urchin and a mouse, are somehow encoded in their DNA genomes. But this fact alone does not provide the mechanisms by which various animal forms arise in development, or arose in evolution, though it does constrain the possibilities. Now we know a good bit about how the genome actually works in development, and the same question continues to lead us forward: in what sequences of the genome do in fact reside the causal differences responsible for morphological diversity, and how exactly do they function?

A large part of the answer lies in the gene control circuitry encoded in the DNA, its structure, and its functional organization. The regulatory interactions mandated by this circuitry determine whether each gene is expressed in every cell, throughout developmental space and time and if so, at what amplitude. In physical terms the control circuitry encoded in the DNA is comprised of *cis*-regulatory elements, i.e., the regions in the vicinity of each gene which contain the specific sequence motifs at which those regulatory proteins which affect its expression bind; plus the set of genes which encode these specific regulatory proteins (i.e., transcription factors). Sometimes these two parts of the immediate gene regulatory system are referred to as the *cis*- and *trans*-regulatory apparatus, respectively. Of course the *trans*-regulatory apparatus can be considered much more broadly. If one relinquishes the constraint of considering only those *trans*-regulatory molecules which directly interact with DNA, by recognizing and binding at *cis*-regulatory target site sequences, then large components of both nuclear and cytoplasmic cellular biochemistry might also be included. Among these would be all those signaling pathways, adaptor proteins, cofactors, and other entities that affect the activity of transcription factors. But it seems clear that most of this cellular machinery is in general ubiquitous or in any case relatively nonspecific; that it is always utilized for many diverse regulatory tasks in each organism; and that by far the most important genomic determinants of animal diversity are the regulatory elements which encode the genetic program for development. In the *cis*-regulatory target sites in the DNA of each species are to be found the major heritable sequence differences which underlie its form, from the most general to the most minute and particular of its morphological aspects. Because *cis*-regulatory elements consist of genomic DNA sequence, they are to be thought of as hardwired; they are the same in every cell of the animal, and their organization is a heritable species character. Among the most important of these elements are those which control regulatory genes (i.e., genes encoding transcription factors). These genes lie at the nexus of large regulatory networks, consisting of all of their target genes, and of the regulatory genes that encode the proteins controlling their own activity. How the hardwired control systems of the genome work; how their function underlies developmental processes; and how they provide an explanation for evolutionary change in animal body plans are the subjects of this work.

THE REGULATORY APPARATUS ENCODED IN THE DNA

The Genes and Gene Regulatory Components of Animal Genomes

Animal species vary enormously from one another in the amount of DNA per haploid genome, even within a given clade, or phylogenetic branch of related species. Examples are the greater than tenfold differences in genome size seen amongst insects, fish, and amphibians. This was already known by the end of the 1960s, as a result of measurements carried out on dozens of species (see Britten and Davidson, 1971, for review). On the other hand, measurements of the amount of genetic information read out into the mRNA populations of organisms of diverse genome size soon showed that large differences in genome size are not at all reflected quantitatively in mRNA population complexity (the classical meaning of the term "complexity," as used here, is RNA sequence diversity in nucleotides; i.e., the total sequence length if single molecules of each of the different mRNA species represented in the population considered were laid end-to-end). The best example is a set of RNA complexity measurements carried out on eggs of various species of animal of very different genome sizes (Hough-Evans *et al.*, 1980; Davidson, 1986). A summary of these results is reproduced in Fig. 1.1. It is reasonable to imagine that all eggs have similar tasks to perform, and indeed as Fig. 1.1 shows, all of the values fall within $35 \pm 10 \times 10^6$ nucleotides, except for *Drosophila*, in which the maternal RNA complexity was reported as about twofold lower. But these are small differences, compared to the > 100-fold range in genome size of the animals included in the comparison. Note particularly the two amphibians, *Xenopus* (point 8 in Fig.1.1) and *Triturus* (point 9 in Fig. 1.1), the genome sizes of which differ about tenfold. The equally complex maternal RNAs

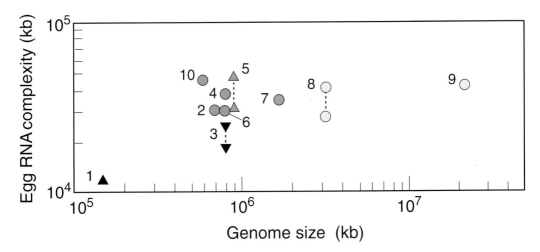

FIGURE 1.1   Complexity of egg RNA plotted against genome size for different species.   Data are from RNA excess hybridization against single-copy DNA or cDNA hybridization kinetics, and are expressed in terms of length of single-copy DNA sequence represented in the egg RNA. Echinoderms are shown in red: 2, *Arbacia*; 4, *Strongylocentrotus*; 6, *Lytechinus*; 7, *Tripneustes* (all sea urchins); 10, *Pisaster* (a starfish). Dipteran insects are in black: 1, *Drosophila*; 3, *Musca* (the housefly). Amphibians are in blue: 8, *Xenopus*; 9, *Triturus* (a newt). 5, *Urechis* (a nonsegmented worm) is in green. Points connected by dashed lines indicate two independent measurements. [Adapted from Davidson (1986) "Gene Activity in Early Development." Academic Press, Orlando, FL, where original references to both genome size and RNA complexity measurements can be found.]

of these two creatures are generated during the lampbrush chromosome stages of oogenesis, in which the transcription complexes are spectacularly displayed in the thousands of lateral chromosomal loops. Transcriptional processes in the large lampbrush chromosome loops of *Triturus* and the much smaller ones of *Xenopus* were thoroughly studied by cytological and biochemical methods, particularly by O. Miller and J. Gall and their colleagues (see Davidson, 1986, for review). As a result, we understand something of how such differently sized genomes generate cytoplasmic RNAs of such similar complexity. A very short summary is that both the noncoding transcribed sequence (i.e., intergenic and intronic sequences) and the amount of nontranscribed sequence is far greater in *Triturus* lampbrush chromosomes than in the equivalent *Xenopus* chromosomes. Thus, as one might expect, in the two species about the same amount of genetic information indeed ends up being used for the equivalent jobs of making an egg that will develop into an amphibian embryo.

Global estimates of gene content per genome have now become available for several species of different genome size as a result of genome sequencing, and these are summarized in Fig. 1.2 (reviewed by Rubin *et al.*, 2000; see legend for detailed references). In this book we are to be concerned almost exclusively with bilaterally organized animals, sometimes referred to as the "triploblasts," here as the "bilaterians;" this excludes protozoans, sponges, cnidarians, and ctenophores. Only the *Caenorhabditis elegans*, *Drosophila*, and human values can be taken literally, since the others are all based on small samples of genomic sequence. However, the general import is not very dependent on the exact accuracy of the estimates. Even though the data in Fig. 1.2 pertain to only a few species, we can see from them the size of the basic "package" of protein coding genes needed in the genome of a bilaterian animal. Figure 1.2 shows that the *C. elegans* genome contains about 18,400 genes, and the invertebrate chordate *Ciona* and the arthropod *Drosophila*, apparently a slightly smaller number. Estimates for *Strongylocentrotus purpuratus*, a sea urchin, are about twice the value for *Drosophila*, which is estimated at 13,600 genes (Rubin *et al.*, 2000). Two vertebrates, i.e., ourselves and the puffer fish (*Fugu rubripes*, the genome size of which lies at the lower limits known for vertebrates) probably have 70,000 genes. It is widely believed that early in the evolution of the vertebrates at least two whole genome duplications occurred, mainly on the basis of the larger number of copies of genes of given gene families in vertebrate genomes. That is, vertebrate genomes include more genes of these families than do lower deuterostomes, i.e., ascidians, echinoderms, and cephalochordates (amphioxus). The clearest argument for this is afforded by the *hox* gene complexes: amniotes have four *hox* gene complexes located on four different chromosomes (Ruddle *et al.*, 1994; McGinnis and Krumlauf, 1992), while the invertebrate chordate amphioxus has one (Garcia-Fernandez and Holland, 1994), as does *S. purpuratus* (Martinez *et al.*, 1999). If we divide the estimate of 80,000 human genes (Rubin *et al.*, 2000) by four, we end up with a number slightly larger than the "basic package" value, that is, the 2×10^4 genes implied by the first three entries in Fig. 1.2.

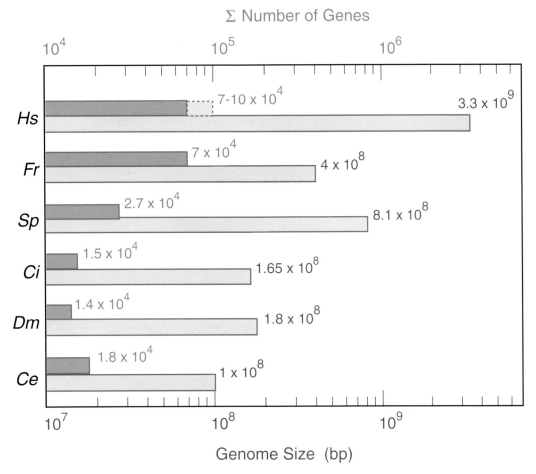

FIGURE 1.2 Gene numbers and genome size. Data include only estimates of total gene number based on genome sequencing. The estimates for *Fugu*, *Strongylocentrotus*, and *Ciona* are extrapolations from minor fractions of the genome sequenced; data for others are from almost complete genomic sequences. *Hs, Homo sapiens*; *Fr, Fugu rubripes* (puffer fish); *Sp, Strongylocentrotus purpuratus*; *Ci, Ciona intestinalis* (ascidian); *Dm, Drosophila melanogaster*; *Ce, Caenorhabditis elegans*. The number of genes indicated by red bars (top scale) are from the following sources: *Hs*, Collins, 1995; Rubin *et al.*, 2000; *Fr*, Brenner *et al.*, 1993; *Sp*, gene number calculation of Cameron *et al.*, 2000 (based on *S. purpuratus* Genome Project data), and genome size from Hinegardner, 1974; *Ci*, Simmen *et al.*, 1998; *Dm*, Rubin *et al.*, 2000; *Ce*, *C. elegans* Sequencing Consortium, 1998. Dashed lines indicate larger estimates also consistent with current data. Genome sizes, indicated by blue bars (bottom scale), are given in base pairs (blue numerals), shown within the figure. Gene number estimates are also given (red numerals).

Suppose we concern ourselves specifically with the two classes of gene which it is now clear are most directly engaged in setting up the spatial domains of gene expression which underlie all aspects of the developmental process. These are genes encoding transcription factors and genes encoding intercellular signaling ligands, receptors, and some downstream components of signaling pathways. Regulatory and signaling pathway genes together function in bilaterian development to transduce spatial intercellular signaling events into spatial changes in transcription factor presence or activity. How much do bilaterians of very different body plan and very different phylogenetic affiliation differ from one another in their repertoires of genes encoding these key classes of protein? An enormous amount of recent data indicates that the answer is, basically very little. Such is the aggregate outcome of thousands of laborious studies in which orthologous genes (i.e., genes belonging to the same immediate family, descendant from the same common ancestor gene) have been recovered from different organisms. Generalizing from what we know from the huge mammalian database; from *Drosophila*, in which developmental roles of many transcription factors and signaling pathway components were first revealed by mutational phenotypes; from the almost complete *C. elegans* (*C. elegans* Sequencing Consortium, 1998) and *Drosophila* (Adams *et al.*, 2000; Rubin *et al.*, 2000) genome projects; and from studies on specific genes in other vertebrates, invertebrate chordates, sea urchins, and a few other creatures, the following quite remarkable statement can be made: Except for a few clade-specific losses, the genetic repertoire of every bilaterian is likely to include genes encoding every known major family of transcription factors, and components of every known signaling pathway. Results from the *C. elegans* genome sequence are particularly revealing, since only when the complete sequence is available can we determine what is not present in a genome (see review of Ruvkun and Hobert, 1998). For example, the *C. elegans* genome is lacking a gene encoding the Hedgehog (Hh) intercellular signaling ligand, which is essential in both *Drosophila* and chordate development, and the *hh* gene is undoubtedly an example of loss of an otherwise panbilaterian signaling component in the evolutionary line leading to *C. elegans*. But a gene encoding the downstream transcription factor which Hh signaling affects, that is, the gene encoding the Cubitus Interruptus or Gli transcriptional regulator, is present in the *C. elegans* genome. So are genes encoding proteins similar to the Patched transmembrane receptor, which is involved in Hh signaling in other organisms (Ruvkun and Hobert, 1998).

Comparison of the known repertoires of signaling and regulatory genes permits as well a second, equally significant generalization (Rubin *et al.*, 2000; Ruvkun and Hobert, 1998). In each bilaterian clade, though all the regulatory and signaling gene families are present, these gene families have diversified differently; e.g., the different bilaterian genomes may have different numbers of genes encoding transcription factors belonging to the various subfamilies of homeodomain regulators, ETS regulators, T-box regulators, nuclear receptors or, winged helix regulators, and different numbers of Dpp or TGFβ ligands.

Furthermore, the duplication and diversification of these gene families and subfamilies are always accompanied by—or driven by—diversification of their functional roles in development. This is a major process in bilaterian evolution.

So, we can exclude the proposition that given bilaterian body plans and morphological structures differ from others because each has its own specific classes of gene regulatory protein and its own set of signaling pathways. Instead the opposite is true. A common repertoire of types of transcriptional regulator and of signaling pathways constitutes the shared bilaterian heritage, the utilization of which underlies all forms of bilaterian development. This argument can be broadened to include many other classes of gene, e.g., genes encoding the cytoskeletal proteins or enzymes that carry out metabolic functions; and most importantly, genes encoding the properties of many panbilaterian differentiated cell types, such as are muscle cells, neurons, photoreceptors, secretory cells, and so forth.

The working parts of the genome are the genes and their *cis*-regulatory elements. Since the key classes of gene are shared amongst bilaterians, we come down to the regulatory apparatus. The internal architecture of *cis*-regulatory elements determines how each gene will function, as we explore in the following; and the architecture of the networks in which they are interconnected determine the deployment of sets of genes in developmental space and time. It was possible to deduce that genomic regulatory architecture constitutes the structural, genetic basis for the morphological features of animals 30 years ago (see e.g., Britten and Davidson, 1969, 1971); now we know it for a certainty.

Overview of Regulatory Architecture

cis-Regulatory elements can be thought of as information processing units "wired" into the regulatory network so that they receive multiple inputs, in the form of the multiple transcription factors which bind within them. The inputs vary depending on when it is in development and in what cell a given gene is located, and the output is the "instruction" given to the basal transcription apparatus which determines whether the gene is to be silent, or active at a specified rate. If the gene encodes a transcription factor the output leads to all the other *cis*-regulatory elements within which there are functional target sites for that transcription factor. We speak of these as "downstream" linkages, while the inputs to a regulatory gene are the termini, in its own *cis*-regulatory system, of its "upstream" linkages. The internal architecture of a *cis*-regulatory element is that which enables it to "process" the various inputs it receives, resolving these inputs into a single output. Perhaps this seems an unnecessarily baroque mode of description. As everyone knows one can usually find a piece of DNA, often described as an "enhancer," which will generate a certain pattern of developmental expression when introduced into a recipient egg or animal: so why is it important to worry about its internal functional architecture and its information processing activities? There are two direct answers to this question. The first is that only by verifying the functional meaning of the specific transcription factor target sites within a *cis*-regulatory element do we understand what the genomic DNA sequence of the element means, and understanding the functional meaning of the genomic DNA sequence might well be considered the most important problem in bioscience. The second is that the information processing function of the *cis*-regulatory element constitutes the link between the diverse circumstances presented in each cell, and the response capacities hardwired into the genomic regulatory sequence.

The fundamental requirement for *cis*-regulatory information processing is easy to see *a priori*. Consider the problem faced by a given gene in a given cell at a given moment in development. Somehow this gene has to "know" when it is in the developmental process; what cells are adjacent and what they are doing or saying; what is the lineage or developmental status of its own cell. It must "know" whether the cell is in cycle, if that affects the need for transcripts of our gene; and also what regulatory events (mediated by other genes) have occurred earlier which would causally affect its own activity. The set of transcription factors that bind within a *cis*-regulatory module can be considered not only as biochemical effectors of function, but also as incident bearers to the gene of these kinds of relevant biological information. Note that "transcription factor" is used here as a neutral term, denoting any protein which displays a high specificity for a particular *cis*-regulatory DNA sequence, and which performs some function that affects transcriptional output. Transcription factors execute a variety of functions, e.g., repression, activation, transduction of external signaling, or architectural alteration of *cis*-regulatory complexes, and they mediate diverse *cis*-regulatory logic functions. For example, a commonly observed *cis*-regulatory format is one in which two different transcription factors responding to two different inputs, perhaps an intercellular signal and a lineage marker, must both be bound in order for there to be any output ("and" logic). For a transcription factor to affect the output of a given *cis*-regulatory element its target site must of course be included in the sequence of that element, and this is the genetic component of the system. But the factor also must be presented in the cell nucleus at a concentration that promotes occupancy of these target sites a significant fraction of the time, and it must be presented in an active form. Transcription factor concentrations and activities depend on circumstances. A factor may be synthesized only in certain spatial domains of the organism, and it may be active only after a signal transduction pathway has modified either it or a bound cofactor. It is in this direct, mechanistic sense that transcription factors convey circumstantial information to *cis*-regulatory elements.

Bilaterian organisms are complex, and have many parts, and they execute many developmental processes. Genes at all levels or positions in the regulatory operating system, for instance genes encoding signal pathway components, are typically utilized many times over during the life cycle, at different stages, and in different cells. A general feature of *cis*-regulatory architecture in bilaterians is that the diverse phases of expression of given genes are frequently mediated by diverse *cis*-regulatory elements, here referred to as regulatory "modules," strung out in the DNA flanking the gene or in its introns. A cartoon illustrating some principles of modular *cis*-regulatory organization is shown in Fig. 1.3. Use of the term "module" to denote individual *cis*-regulatory elements has some advantages compared to "enhancer," since the functions of these elements may be more complex and integrative than implied by the verb "to enhance," and indeed some modules function in some cells to repress rather than stimulate transcription. Nonetheless in the following both terms are to be found, "enhancer" usually in accord with its use in a cited study.

Every *cis*-regulatory module may be considered a control device which is called into play by those transcription factors for which it contains target sites, when and where these are present and active; and the module is otherwise silent. Each module is an information processing regulatory device in the sense just discussed. A survey of some well-characterized *cis*-regulatory modules active in developmental processes (Arnone and Davidson, 1997) yielded the conclusion that four to eight different transcription factors typically service each module, as symbolized by the vertical input arrows in Fig. 1.3. Sites for given factors are often multiple within a module. To a first approximation this can be regarded as a means for increasing the probability that that factor will at any one moment be bound within the module. Modular regulatory organization provides the organism with the services of a given gene in multiple developmental contexts, but as also implied in Fig. 1.3 the "price" is an additional layer of complexity: some *cis*-"trafficking" controls must operate so that the module relevant in any given context is the one which communicates its output to the basal transcription apparatus. This undoubtedly requires control of DNA looping, mediated by direct protein:protein interactions involving the transcription factors bound within the active module, perhaps via cofactors bound in turn to them. The most proximal region of the whole *cis*-regulatory system may have a special importance for the function of the whole system. This is suggested by several examples in which the proximal *cis*-regulatory region is required for activity, in addition to one or another of the upstream regulatory modules, but how general is this feature remains to be seen. As will be seen in later chapters, modularity in *cis*-regulatory systems is essential to their developmental function, and is also a key to understanding the evolution of developmental processes. For now we leave this discussion with the simple but essential point that modular *cis*-regulatory structure constitutes a discretely organized, DNA map, which represents in physical terms the different phases of gene expression that are to be installed throughout the life cycle, for every gene.

The morphological structures of bilaterians are of course never the product of single genes or single regulatory systems. Direct RNA complexity measurements (reviewed by Davidson, 1986), and now an increasingly enormous bank of EST data obtained from cDNA libraries of given tissues, confirm the *a priori* assumption that many hundreds and often thousands of genes must be expressed in order to create any given tissue, body part, or multicellular structure. These genes are controlled in developmental space and time by large developmental gene

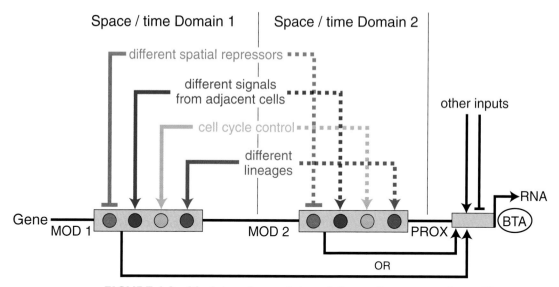

FIGURE 1.3 Modular *cis*-regulatory information processing.   The cartoon shows several kilobase (kb) upstream of a gene operating under the control of two *cis*-regulatory modules (MOD1 and MOD2), each of which is operative in a particular spatial domain at a particular time in development. Each module receives multiple parallel inputs (arrows). The diagram shows examples of the kinds of inputs each module might receive and of course is not meant to imply that every module utilizes these same particular inputs. The specific inputs of each kind will be different for MOD1 and MOD2, symbolized by the solid vs. dashed colored arrows. The inputs are of two types, positive and negative. The red barred inputs denote different spatial repressors which are utilized in each module to set boundaries of expression in the spatial domain where that module functions, i.e., these inputs repress the gene across the relevant boundaries. The blue activators are downstream of different intracellular signaling pathways; the tan activators turn on the gene when cells are in cycle; the different green activators are present in cells of the respective lineages that constitute the fields in which the gene will be active, but only when all inputs are present. These inputs can be thought of as bringing the indicated kinds of situational biological information to the gene. Each module acts by communication of its output to the proximal *cis*-regulatory module (PROX), which may receive other inputs; and may further process (e.g., amplify) the output of MOD1 or MOD2. PROX then communicates directly with the basal transcription apparatus (BTA). Note that the major spatial information processing occurs in the developmentally regulated, upstream *cis*-regulatory modules, and that the BTA simply responds to the various alternative outputs of the two upstream regulatory modules, as transmitted to it by PROX.

regulatory networks, as mentioned at the outset. Hence, returning to the question of what aspects of the genomic regulatory system are responsible for diversity in bilaterian morphologies, the most accurate and comprehensive answer is diversity in the architecture of developmental gene regulatory networks. "Network architecture" is a term that, in brief, denotes the organization of regulatory linkages which connect *cis*-regulatory elements by means of *cis-trans* interactions. The character of developmental regulatory network architecture is a problem of major importance, though the real nature of these networks is just beginning to emerge from experimental data. Suffice it to say that in principle genomic changes which alter network architecture have the power to create new developmental processes, because they can affect the activities of large sets of genes. Examples might include the insertion or appearance of new spatial *cis*-regulatory modules in the vicinity of a gene controlling a battery of other genes, thus causing these genes to be expressed in different spatial domains of an embryo; or *cis*-regulatory changes in target sites which result in the addition of genes to preexistent gene batteries; or target site changes that bring network subelements under control of different signaling functions. Network architecture is ultimately specified by the identity of the target sites within all the participating *cis*-regulatory elements. It follows that analysis of genomic *cis*-regulatory systems and their linkages holds the key to understanding how genomes encode the properties of organisms.

GENE REGULATORY FUNCTIONS IN DEVELOPMENT

The Regulatory Demands of Development

Development is the execution of the genetic program for construction of a given species of organism. The program itself is our concern, and those aspects of its execution that regulate its own readout from the DNA. The most cursory consideration of the developmental process produces the realization that the program must have remarkable capacities, for development imposes extreme regulatory demands. Of the thousands of genes indicated in the diagram of Fig. 1.2, most are utilized at some time during development, and all must be controlled accurately in space and time. But this is really the least of the problem: the essence of development is the progressive increase in complexity which it invariably entails. In informational terms the increase in developmental complexity is measured by the generation of new populations of cells, each of which reads out a defined genetic subprogram; and each of which arises in a particular spatial domain of the embryo. All the while these populations are being instructed to expand to given extents, by cell growth (and sometimes to contract, by cell death). Thus the spatial components of morphology are laid out.

Metaphors often have undesirable lives of their own, but a useful one here is to consider the regulatory demands of building a large and complex edifice, the way this is done by modern construction firms. All of the structural characters of the edifice, from its overall form to minute aspects that determine its local function-alities such as placement of wiring and windows, must be specified in the architect's blueprints. The blueprints determine the activities of the construction crews from beginning to end. At first glance the blueprints for a complex building might seem to provide a good metaphoric image for the developmental regulatory program that is encoded in the DNA. For example, just as in considering organis-mal diversity, it can be said that all the specificity is in the blueprints: a railway station and a cathedral can be built of the same stone, and what makes the difference in form is the architectural plan. But there is an interesting problem with this metaphor, in the way the regulatory program is used in development, compared to how the blueprint is used in construction. In development it is as if the wall, once erected, must then turn around and talk to the ceiling in order to place the windows in the right positions, and the ceiling must use the joint with the wall to decide where its wires will go, etc. Thus much of the genomic regulatory program for development is devoted to the progressive organization of spatial interactions between fields of cells of differing character. The morpho-logical outcome is the completed organism, and within a narrow anatomical range, for each species this outcome is the same, almost 100% of the time.

With the exception of unusual cases such as the syncytial early embryo of *Drosophila*, specification of cell fate in early development usually involves signal-ing between cells, and the same is true of adult body plan formation. Here developmental "specification" is defined as the process by which cells acquire the identities or fates that they and their progeny will adopt. Specification results in differential expression of genes, the readout of particular genetic subprograms. For specification to occur, genes have to make decisions, depending on the inputs they receive. This brings us back to the information processing capacities of bilaterian *cis*-regulatory systems. As illustrated in the next chapter, genes operat-ing at every level of the developmental process have to carry out information processing functions in the course of developmental specification. The point cannot be overemphasized: were it not for the ability of *cis*-regulatory systems to integrate spatial signaling inputs together with multiple temporal, and other inputs, bilaterian development could never occur. This is because development depends on creating new spatial and temporal domains of gene expression from preexisting information.

Until terminal form is achieved in each morphological element of a developing organism the genetic subprograms expressed in its progenitor cells are transient; they are way stations along the pathway to the final state of spatial organization. Transience implies temporal control, and in development temporal regulation is often exquisitely precise. There are many examples of *cis*-regulatory systems that operate for only a few hours of the whole life cycle, since the functions mediated by the genes they control are not useful at any other time. For instance, the establishment of transcriptional dorso/ventral (D/V) asymmetry in the frog embryo must occur within the crucial hour or so when the embryo genomes are activated at the end of the blastula stage. Genes encoding particular homeodomain transcriptional regulators, *viz, siamois, twin,* and *goosecoid,* are then transiently activated, specifically on the dorsal side of the embryo. The cell signaling and other events regulated by these factors are required even before gastrulation commences (for review, see Harland and Gerhart, 1997; Moon and Kimelman, 1998). Similarly, specialized gene products of individual differentiated cell types are often required only at specific stages, and only in those cells. A striking example is provided by the "glue proteins," which are produced by the salivary glands of *Drosophila* larvae just prior to the pupal stage (Beckendorf and Kafatos, 1976). These adhesive proteins, encoded by a small family of coordinately expressed genes, are necessary to fix the larva to a dry surface. There, immobile and enclosed within the pupal case, it will undergo metamorphosis. The glue proteins are of use only for this specific purpose, and are synthesized only during the last several hours of the 4-day larval period, out of the whole life cycle. Needless to say, the nature of the specific inputs which are utilized to effect temporal regulation in transiently expressed genes is determined explicitly by the target sites that are hardwired into the respective *cis*-regulatory systems of these genes.

Pattern Formation

"Pattern formation" is a term of many uses, because spatial patterns appear at every level of biological organization from subcellular to organismal. In this book it is applied specifically to the developmental process by which spatial domains of unspecified cells are assigned "regulatory states," thus ultimately creating fields of cells that will give rise to the diverse parts of a structure, an organ, an organism. The early events of pattern formation establish the basic elements of the body plan, e.g., the metameric segmentation which arises early in the embryogenesis of a fly; or the anterior/posterior (A/P) axis, and left-right asymmetry in a mouse. Later pattern formation events define the spatial organization of the main parts of the body plan, and still later ones define more detailed and smaller elements, e.g., the arrangement of the sensory bristles in the fly or of the limb digits in the mouse. Pattern formation processes are progressive in that as development proceeds, spatial elements are specified at increasingly fine resolution. Regardless of the scale on which it occurs, or at what stage of development, pattern formation requires the partitioning of existing domains of cells into novel subdomains. Until terminal differentiation processes are called into play, these developmental spatial patterns are therefore all transient. Their only significance is to enable the next stages of spatial subdivision to occur.

At the heart of pattern formation mechanisms is the process of regional specification. New subdivisions of an existing cellular domain are created by the expression of one or more particular transcription factors within a bounded region. This establishes a new regulatory state in this region, distinguishing it from the remaining cells, and setting up the program for subsequent developmental events. To view the process of pattern formation in this way is to orient our quest for causality in the pathways by which morphogenesis is organized; we now have an algorithm for tracking the genetic program for pattern formation to its lairs in the genome. The track leads to the *cis*-regulatory elements controlling spatial expression of genes encoding those transcription factors which execute regional specification processes. It is these *cis*-regulatory systems that perform the crucial function of integrating the signaling, lineage, or other spatial inputs which define the regions to undergo regional specification. In turn, the transcription factors which they control provide specific downstream inputs into the transcriptional activities of other genes, at the next level of the regulatory network. These are often genes encoding other spatially expressed transcription factors, as well as genes encoding elements of intercellular signaling systems; hence the progressive nature of the pattern formation process.

A very clear example of the way in which a pattern formation process is encoded in *cis*-regulatory DNA sequence is afforded by the process of metamerization early in the embryonic development of *Drosophila*. Metameric organization is a basic feature of the dipteran body plan, and the process by which it is created begins even before cellularization of the blastoderm. Among the genes which initiate this pattern formation process are the "pair-rule" genes which begin to be expressed in seven repeating "stripes" (alternating on and off) along the A/P axis of the embryo. These stripes define two-segment units from which the full complement of body segments is ultimately generated (see for review Rivera-Pomar and Jäckle, 1996). The primary pair-rule genes encode transcription factors. Each stripe of pair-rule gene expression constitutes a distinct regulatory subdivision, which is positioned specifically along the A/P axis within a much larger cellular domain at the cellular blastoderm-stage, i.e., the future "trunk" region of the body plan. The location of each stripe of expression of course depends on the *cis*-regulatory systems controlling these genes. In some well-studied cases the necessary control sequences for given stripes are contained within discrete *cis*-regulatory modules (Small *et al.*, 1991; 1992), and a demonstration of this is shown in Fig. 1.4.

Modules controlling expression in different stripes contain sites for different collections of transcription factors, and together they define all the regions of A/P cellular space in which the respective pair-rule genes are expressed. Each of the individual stripe modules reads and integrates the *trans*-regulatory inputs with which it is confronted, in each cell. As we discuss in detail in the following chapter, each of the modules contains target sites for positively acting transcription factors which cause it to be expressed; and for negatively acting transcription factors which set the anterior and posterior boundaries within which expression is allowed. That is, a given stripe module directs expression of the gene it controls when two external conditions are met: the transcriptional activators for which it includes binding sites are present and active, and the transcriptional repressors for which it includes binding sites are not present at significant levels. This is a paradigmatic case, since in the stripe modules of the pair-rule genes we are brought face to face with *cis*-regulatory elements of which the sole function is to organize spatial patterns of transcription factor expression.

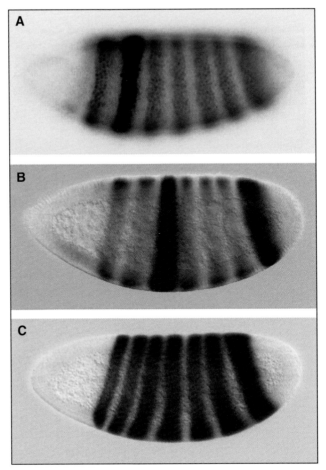

FIGURE 1.4   Accurate expression of eve stripe 2 and eve stripes 3+7 mediated by individual *cis*-regulatory elements. The *evenskipped* (*eve*) gene generates a metameric pattern consisting of seven stripes arrayed along the A/P axis, which appears in the precellular embryo. The brown stain displays the location of the seven *eve* protein stripes, as revealed by anti-*Eve* antibodies. (A) Transgenic embryo carrying a minimal 480 base pairs (bp) stripe 2 *cis*-regulatory construct with the *lacz* gene as a reporter. The mRNA generated by the *lacz* reporter is visualized by *in situ* hybridization (purple stain), and the construct can be seen to recreate the endogenous stripe 2 expression domain perfectly. [(A) From Small *et al.* (1992) *EMBO J.* **11**, 4047–4057 and by permission of the European Molecular Biology Organization.] (B) Transgenic embryo carrying a 500 bp stripe 3 + 7 *cis*-regulatory construct, *lacz* mRNA detected as in (A). The expression domain again corresponds exactly with the endogenous *eve* 3 + 7 stripes. (C) RNA *in situ* hybridization using a probe that detects *eve* RNA reveals the same seven stripes as displayed by immunostaining in (A) and (B). [(B, C) From Small *et al.* (1996) *Dev. Biol.* **175**, 314–324.]

All morphological features of adult bilaterian body plans are created by means of pattern formation process. The *cis*-regulatory systems that mediate regional specification are thereby the keys to understanding how the genome encodes development of the body plan. These systems also provide the most fundamental and powerful approach to understanding the evolution of bilaterian forms.

Terminal Differentiation

With the completion of development, populations of permanently specialized cells have been put in place throughout the organism. These express sets of genes encoding the definitive attributes of each component of the body plan. Terminally differentiated cells include familiar cell types: muscle cells, many kinds of hematopoietic cells, nerve cells, skeletogenic cells, gut cells, liver cells, light receptor cells, gland cells, and so forth. All metazoans produce some differentiated cell types, but only the bilaterians generate the three-dimensional morphological arrays of terminally differentiated cell types that we term "organs," and only in the adult body forms of bilaterians do layers or tubes of differentiated mesodermal cell types constitute major components of all the body parts. Under normal conditions, in bilaterians terminal differentiation is indeed just that: additional functions will never be expressed by these cells, and if they retain replication capacity they will only produce more of their own sort. There are some examples of highly differentiated cells which in the course of development de-differentiate and then re-differentiate in a second direction, but these cases are unusual ("trans-differentiation"; see review in Davidson, 1986). Terminally differentiated cells rely on diverse stabilization mechanisms to lock down their transcriptional patterns (i.e., what embryologists have always called "commitment"). These include stabilizing feedback circuits in the gene networks that regulate differentiation; changes in the state of chromatin structural domains that include the regulatory systems of the relevant genes; local DNA methylation or demethylation; decrease in mRNA turnover rates; in some cell types even loss or total inactivation of the whole nuclear transcriptional apparatus once the cell has been sufficiently loaded with the necessary specific mRNAs. But lockdown mechanisms are not our first concern; our focus is rather on the genomic regulatory apparatus that institutes given states of differentiation. Terminal differentiation can usually be defined in terms of a precise set of differentially expressed protein coding genes, such as the hemoglobins and enzymes of the vertebrate red blood cell, or the cell surface receptors and signaling molecules of immune effector cells. Our question is what features of genomic regulatory organization account for the coordinate expression of such dedicated sets of genes, in the same cells at about the same time.

A useful concept here (and in other contexts as well) is that of the "gene battery," defined as a set of functionally linked genes expressed in concert. The term "gene battery" was originally Morgan's (1934), and was later adapted by Britten and Davidson (1969) to denote sets of genes expressed together for the specific reason that their *cis*-regulatory systems respond to common *trans*-regulatory inputs. A good bit is now known of the *cis*-regulatory organization underlying expression of gene batteries in terminally differentiating cells, particularly in vertebrates (Arnone and Davidson, 1997). As a heuristic example, relevant *cis*-regulatory elements from four amniote muscle genes are diagrammed in Fig. 1.5. We see that several types of DNA-binding transcription factor are utilized by these regulatory systems, *viz*, a muscle specific bHLH factor of the MyoD family (MDF);

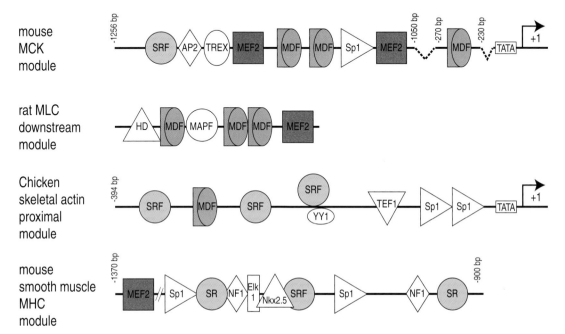

FIGURE 1.5 *cis*-**Regulatory elements of some muscle-specific genes.** These genes can be considered as members of muscle gene batteries: MCK, muscle creatine kinase; MLC, myosin light chain; MHC, myosin heavy chain. They share target sites for several specific transcriptional regulators, shown in color: MEF2, red; SRF, tan; MyoD family (MDF) bHLH factor, blue. In addition each gene has other specific target sites, shown as open symbols. Data for mouse MCK are from Donoviel *et al.* (1996), Fabre-Suver and Hauschka (1996), and Grayson *et al.* (1998); for MLC, from Rosenthal *et al.* (1992) and Rao *et al.* (1996); for chicken skeletal α-actin, from MacLellan *et al.* (1994); for mouse smooth muscle MHC, from Zilberman *et al.*, (1998). In the latter gene the target site sequences are highly conserved amongst rat, mouse, and rabbit smooth muscle MHC genes, and thus are liable to play significant regulatory roles; MyoD family factors are not expressed in smooth muscle (Olson, 1990) and its target sites are consequently absent from this regulatory element. [Adapted from Arnone and Davidson (1997) *Development* **124**, 1851–1864 and The Company of Biologists, Ltd.]

SRF (the "serum response factor"); and the muscle-specific MEF2 factor. Thus the basic principle of the gene battery: genes are indeed linked into given batteries by the presence of *cis*-regulatory target sites for common sets of transcriptional regulators. This principle is of large consequence, for it tells us where to look for the structural genomic features that assign given genes to given gene batteries, in both development and evolution.

Since *cis*-regulatory function devolves from *cis*-regulatory structure, one might expect there to exist structural features characteristic of elements that control terminal differentiation processes, and indeed there are. These elements need not act as regulatory pioneers, laying out new territories as do modules that execute pattern formation processes. Instead they are activated by transcription factors that have already been placed in the appropriate cells by prior specification processes. Single factors can sometimes provide the trigger needed to activate terminal differentiation genes; sometimes several different positive factors act synergistically, as in the examples shown in Fig. 1.5. When the spatial boundaries within which terminal functions are to be expressed have been set upstream, boundary-forming repressors are not required, and so some *cis*-regulatory elements that function in this situation are lacking any target sites for spatial repressors. Terminal *cis*-regulatory elements are often compact, and if a gene is to run in several different cell types, they may contain within the same module sites for diverse activators that are present in these cell types. Furthermore, terminal differentiation processes often mediate expression at very high levels, for the products of the genes they control are in some cases needed in relatively enormous quantities. The classic examples are the globin genes in mammalian erythrocytes. At certain stages these genes support > 90% of all protein synthesis in these cells (Hunt, 1974). Very high transcript levels imply that particular *cis*-regulatory features mediate maximal rates of initiation in the basal transcription apparatus, which of course must be instructed quantitatively as well as qualitatively (for transcription initiation rates in development see Davidson, 1986).

The special features of *cis*-regulatory elements that control terminal differentiation processes remind one of the general point that the kind of role played by a given element in the regulatory network is reflected in the nature of its complement of target site sequences. Thus we return to the conclusion that the regulatory logic of the system is directly encoded in the DNA (Arnone and Davidson, 1997).

GENOMIC REGULATORY SEQUENCE AND THE EVOLUTION OF MORPHOLOGICAL FEATURES

An explanation of evolutionary change in animal form in mechanistic terms rests on a classic syllogism. This goes as follows: If the morphological features of an animal are the product of developmental process, and if the developmental process in each animal is controlled by its genomic regulatory program, then the evolution of animal forms is the consequence of evolution of these genomic regulatory programs. This is scarcely a new thought (Britten and Davidson, 1971), but what is new is that now we can do something to study it experimentally. The fields of developmental molecular biology and evolution increasingly overlap, creating an exciting new domain, in fact a new field of bioscience. This is at present in process of redefining itself as it absorbs influences from new initiatives in other disciplines, including phylogenetics, paleontology, comparative genomics, and the newly resuscitated arts of invertebrate anatomy, amongst others. Most would concur with this description of present events. We shall take a step further, however, and imagine that what we are now seeing is the initial phase of a major intellectual realignment, in which the study of the mechanisms by which animal body plans evolve will ultimately be regarded as a branch of regulatory genomics, a central domain of bioscience. If we knew in functional terms the components of specific genomic control elements that result in different morphological outcomes in two animals of common ancestry, we would see exactly what are the essential causal differences in the DNA of these animals. So, by learning how these programs work, the objectives of both evolutionary and developmental inquiry are attained. In fact, these problems are so intertwined as to become indistinguishable, and the solutions can be seen to lie in the genomes of extant bilaterians.

Regulatory Evolution, and Evolution in General

Darwinian evolution has a long and important history, first because it drove the development of the whole concept of evolution per se, and proved that evolution can, did, and does happen; and second because over the last century it has generated a vast calculus for analyzing adaptive and selective processes. The objective has been to be able to treat the outcomes of confrontations between populations of organisms possessing given adaptive properties and their ecological environments, and to define the rules according to which heritable variations spread through populations. But classical Darwinian evolution could not have provided an explanation, in a mechanistically relevant way, of how the diverse forms of animal life actually arose during evolution, because it matured before molecular biology provided explanations of the developmental process. To be very brief, the evolutionary theory that grew up before the advent of regulatory molecular biology dealt with the problem of the origin of novel organismal structures in two ways. The first has been to treat the mechanisms generating novel morphological structures as a black box. New forms were considered to arise "because" the environment changed. But while changes in Precambrian or Ordovician weather, continental shifts, or temperature may have contributed crucial selective forces, they do not generate heads or appendicular forms; only genes do that. The second mode of classical argument was that organismal evolution is the product of minute changes in genes and gene products, which occur as point mutations and which accumulate little by little, providing the opportunity for selection and ultimately reproductive isolation. The major forms this argument has taken have focused on stepwise, adaptive changes in protein sequence, but this is probably largely irrelevant to the evolution of any salient features of animal morphology (see, e.g., Miklos, 1993). In hindsight we can see that since the families of genes encoding the key regulators of development, *viz*, transcription factors and signaling pathways, are all panbilaterian, as well as are representatives of most other gene families, the differences between animals displaying different body plans are not likely to be explained in general by differences in these proteins. Many remarkable examples of diverse usage of similar genes and gene families in the development and evolution of diverse organisms are to be found later in this book.

There is another version of this argument on which the jury remains out, however, and this concerns gradual mutational changes in *cis*-regulatory sequence. It is not at all clear as yet what is the relative importance of stepwise mutational change in *cis*-regulatory sequence relative to other kinds of alteration in DNA sequence, such as transpositional insertions of regulatory modules or of genes in the vicinity of these modules; sequence deletions; local genomic rearrangements; replication of genes or their *cis*-regulatory target sites; gene conversion, etc. (for general discussions see Dover, 1982, 1987; Britten, 1997; and the following text).

Interpretation of evolutionary change, it seems apparent, is going to take the form that changes in animal morphology, whether great or small, are generated largely by alterations in developmental regulatory sequence. There is no point here in distinguishing between changes in *cis*- as opposed to *trans*-regulatory elements, since a change in the expression of the *trans*-regulator of a given set of genes is simply a change in the *cis*-regulatory module controlling the relevant phase of its own expression. When clades of animal become extinct it is specific developmental gene regulatory networks that have disappeared from the earth; or alternatively, it is the genomic regulatory networks of evolutionarily successful clades that have become dominant.

Bilaterian Phylogeny

It is impossible to stress too much the importance of phylogeny for understanding developmental regulatory comparisons between different animals. If comparative regulatory genomics holds the key to evolution of morphologies, it does so only if the evolutionary relationships of the respective animals can be perceived correctly. These relationships must be known in order to deduce the polarity of changes; to determine whether apparent similarities are the consequence of conservation or of convergence; and, in general for any precise conclusions regarding evolutionary regulatory history.

In Fig. 1.6 is shown a current consensus phylogeny of all bilaterian phyla. This is essentially a ribosomal DNA (rDNA) phylogeny (see legend for references to the multiple studies supporting each individual region of the topology). The bilaterians are divided into three great clades. On the top of Fig. 1.6 (blue) are the ecdysozoans, i.e., animals which molt, including arthropods, nematodes, and many minor phyla. In the center (green) are the lophotrochozoans, i.e., animals displaying spiral embryonic cleavage, such as annelids, molluscs, and flatworms, plus a large number of clades distinguished by tentaculate head structures (lophophores) including brachiopods, entoprocts, and many others. On the bottom (red) are the deuterostomes, i.e., echinoderms, hemichordates, and chordates (including ourselves). The arrangement shown in Fig. 1.6 is supported by much evidence in addition to the sophisticated analyses of rDNA sequences on which the tripartite tree is based (Aguinaldo *et al.*, 1997; Adoutte *et al.*, 2000). First, and perhaps most importantly, it is largely consistent with appropriately performed cladistic analyses of morphological characters (Peterson and Eernisse, 2000). Very strong and completely independent support for the assignment of key species within the tripartite arrangement of animal phyla shown in Fig. 1.6 has subsequently emerged from phylogenetic analyses of *hox* gene sequences (de Rosa *et al.*, 1999; Finnerty and Martindale, 1998; Balavoine, 1997). Figure 1.6 is conservative, in that controversial branch placements have been avoided, so that many of the minor phyla are portrayed as equally related to one another (i.e., their relationships are portrayed as coterminal). Some obvious sister groups based on morphology are included in the diagram, such as annelids and echiurans; molluscs and sipunculids; nematodes and nematomorphs; and the panarthropods (i.e., arthropods proper, onycophorans, and tardigrades).

For what follows the most important points to be derived from Fig. 1.6 are its implications with respect to common ancestry. It is clear that the bilaterians as a whole are monophyletic: all bilaterians are more closely related to one another than any is to any other kind of animal. As Fig. 1.6 shows explicitly, they all derive from a common ancestor. This basic conclusion is of course also implied by the fundamental similarity in bilaterian gene repertoires, particularly those gene classes specifically involved in controlling development of the body plan, as discussed above. There is a deep cleavage to the nearest outgroup shown, the cnidarians (i.e., jellyfish, hydra-like creatures, sea anemones, and corals). This is clearly evident in rDNA sequence comparisons, and the differences between cnidarian and bilaterian morphological organization have long been appreciated (see e.g., Hyman, 1940). For example, cnidarians lack the mesodermal layers that are fundamental to the construction of all bilaterian body plans, and they also lack any organ-level morphological and functional structures. Nor do they possess a *hox* gene cluster that includes representatives of all the major groups of *hox* genes found in bilaterians, possessing only linked "anterior" and "posterior" type *hox* genes (Finnerty and Martindale, 1998, 1999; de Rosa *et al.*, 1999).

Each of the three great bilaterian clades (ecdysozoans, lophotrochozoans, deuterostomes) is defined by a unique set of shared characters. This turns out to be extremely important in thinking about the evolution of body plans, because it provides a framework for interpreting similarities and differences in use of regulatory mechanisms during development of various bilaterians. For example, until

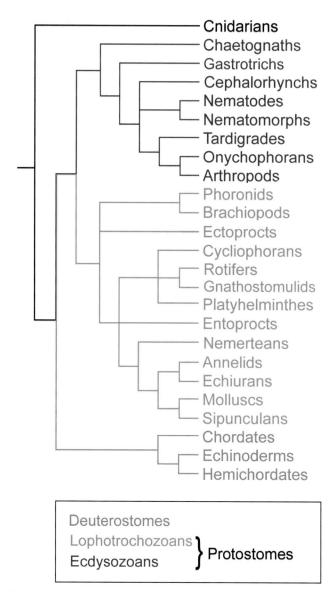

FIGURE 1.6   Phylogeny of the bilaterian phyla.   This tree is primarily derived from 18S rDNA and is supported by *hox* gene phylogeny and morphological cladistics. The bilaterians (shown in color) are divided into three major groups: the deutero-stomes (red); (for deuterostome phylogeny, Wada and Satoh, 1994; Turbeville *et al.*, 1994; Eernisse, 1997; Cameron *et al.*, 2000); the lophotrochozoans (green); (Halanych *et al.*, 1995); and the ecdysozoans (blue); (Aguinaldo *et al.*, 1997; Giribet and Ribera, 1998). The lophotrochozoans and ecdysozoans are each monophyletic, and together

recently it was assumed that annelids and arthropods are evolutionary sister groups, and that therefore their mechanisms of segmentation must be very similar; but now we see that annelids are lophotrochozoans and arthropods are ecdysozoans, and both of these large clades include many unsegmented animals. Not surprisingly, when looked at more carefully at a gene regulatory level, segmentation in annelids and arthropods looks a lot less similar than it used to (Shankland and Seaver, 2000). To cite another example, it is apparent from Fig. 1.6 that the deuterostome chordates cannot be considered a sister group of the ecdysozoan arthropods derived from an arthropod/deuterostome common ancestor, contra some earlier speculations (e.g., Arendt and Nübler-Jung, 1994). The phylogenetic positions of arthropods and chordates impinges directly on deductions regarding the evolutionary origins of genomic regulatory programs for many important processes, including development of head structures, CNS, heart, segmentation, etc.

Someday we are going to be able to write down the regulatory network architecture for key features of development that are characteristic of animals representing diverse regions of the phylogenetic diagram. This will take us far beyond the distribution of shared sequence motifs such as now provide the most reliable basis for these diagrams. While phylogenetic sequence analysis provides the quantitative indices of evolutionary relatedness in rDNA sequences that were used to construct the tree in Fig. 1.6, rDNA sequence differences in themselves of course play no causal evolutionary role. Differences in the regulatory network architecture are what cause phylogenetic diversification, and therein resides the "deep structure" that underlies diagrams such as that in Fig. 1.6.

------

they constitute the traditional protostomes (Giribet and Ribera, 1998). The echinoderm/hemichordate sister grouping is also supported by mitochondrial analyses (Castresana et al., 1998). The relationships shown within the ecdysozoan clade are derived primarily from Schmidt-Rhaesa et al. (1998). Inclusion of the chaetognaths into this clade is from Halanych (1996) and Eernisse (1997); the basal positions of these animals was proposed because although they possess a cuticle with chitin, they lack specific morphological details found in the cuticles of gastrotrichs and the other ecdysozoans. Within the lophotrochozoans the division into apical and basal groups is primarily from Halanych et al. (1995) and Aguinaldo et al. (1997). See Cohen et al. (1998) for the [phoronid + brachiopod] sister grouping. The [annelid + echiuran] and [mollusc + sipunculan] sister groupings are based on unique embryological characters found in each clade (see Ruppert and Barnes, 1994). The rotifer clade (which includes [acanthocephalans]; Garey et al., 1996) + cycliophorans is supported by 18S rDNA (Winnepenninckx et al., 1998); the [rotifer + gnathostomulid] clade is based on several unique morphological characters (e.g., Rieger and Tyler, 1995). Other relationships are left undefined, so that the various phyla are presented as equally related to a common ancestor. [From K. J. Peterson and E. H. Davidson, unpublished.]

CHAPTER 2

Inside the *cis*-Regulatory Module: Control Logic, and How Regulatory Environment is Transduced into Spatial Patterns of Gene Expression

Curiously, the great majority of developmental *cis*-regulatory inquiries stop at the borders of the regulatory module, or more accurately, at restriction sites which lie somewhere outside these borders. Often we read "...we recovered a fragment of DNA which confers accurate spatial expression on a reporter construct," but seldom do we learn how this fragment of DNA works, or exactly what lies within it. Knowledge of structure/function relations inside developmental *cis*-regulatory modules remains thin, a remarkable fact considering the fundamental importance of such knowledge for genomics, for evolution, and above all, for understanding how development is programmed. Partly this is due to focus elsewhere, and partly to practical experimental difficulties. *cis*-Regulatory elements that control physiological responses have been studied in elegant detail in cultured mammalian cells, e.g., the interferon gene response element (Kim and Maniatis, 1997) or the interleukin-2 gene response element (Garrity *et al.*, 1994). So have some regulatory modules that control expression of genes in terminal differentiation processes, e.g., the globin genes (Evans *et al.*, 1990), and other examples discussed below. But the regulatory basis of developmental specification, or of pattern formation, cannot be determined in cultured cell lines. For this, gene transfer into eggs or embryonic anlagen is required. To proceed beyond mere identification of a regulatory module, gene transfer must be sufficiently easy and the system sufficiently well behaved so that many experimentally built mutational variants can easily be assessed. Most of our knowledge of *cis*-regulatory systems that function in the context of developmental specification at present derives from *Drosophila* and from sea urchins, where these conditions are met. From these experimental systems are drawn the majority of the examples considered in this chapter. A useful method of gene transfer has recently been designed for ascidian embryos (Corbo *et al.*, 1997), and a number of murine *cis*-regulatory systems are also being characterized. Knowledge of these will now grow rapidly, but this essay is written a bit too soon to profit therefrom. Though gene transfer is possible in *Xenopus*, *Caenorhabditis elegans*, zebrafish, and chick eggs or embryos, these systems have so far not contributed greatly to our understanding of *cis*-regulatory mechanisms.

OPERATING PRINCIPLES FOR *CIS*-REGULATORY SYSTEMS THAT MEDIATE DEVELOPMENTAL SPECIFICATION EVENTS

Specification processes are those by which initially pluripotent embryonic cells choose among diverse fates, and are allocated to a given spatial territory or anlage. To understand the role of *cis*-regulatory elements in specification it is necessary to think of them as devices that make choices: they produce alternative outputs depending on the sets of inputs with which they are confronted in different cells. As discussed in Chapter 1, in each cell the *cis*-regulatory inputs consist of the concentrations and the activities of the nuclear transcription factors for which the element includes target sites. Regional differences in transcription factor presentation that are set up in different spatial domains of a pluripotential field of cells usually depend on intercellular signaling, or very early in development, on the location of the nucleus with respect to the spatial coordinates of the egg.

The primary principle of the mechanisms by which specification functions are executed is that specification depends on *cis*-regulatory transformation of input patterns into spatial domains of differential gene expression. This is the essential job performed by *cis*-regulatory elements at the initiation of each phase of development.

A second principle is that *cis*-regulatory elements that function in specification always consist of assemblages of diverse target sites, because they always require multiple inputs. No such module consists of sites for a single "master regulator," a fantasy of earlier days. This is because *cis*-regulatory elements that make specification choices work by binding sets of factors, the presence of which individually may depend on signaling events, cell cycle activity, temporal state, lineage, or spatial position (Fig. 1.3). *cis*-Regulatory elements that execute specification functions work by interpreting multiple inputs of these and other kinds, to produce a single output in each nucleus.

A third principle is that in specification processes *cis*-regulatory output is novel with respect to any one of the incident inputs, and is more precise in space and time than these inputs. It is because of this feature that we use terms such as "information processing" in respect to *cis*-regulatory specification events.

A fourth principle is that if a high specificity binding interaction between *cis*-regulatory DNA and nuclear proteins from the relevant cells can be detected *in vitro* by a physical measurement, then this interaction will have some function *in vivo*. If there are multiple sites for a given factor it may (or may not) be difficult to distinguish the roles of individual sites, but that species of interaction will in some manner be necessary for the overall function of the element. Thus the qualitative complexity of the internal functional elements of the *cis*-regulatory module is given by the complexity of the assemblage of target sites within the module. That every specific type of interaction within a regulatory system that can be detected *in vitro* is fundamentally significant was shown explicitly for two sea urchin *cis*-regulatory systems, both discussed below. These are the *cyIIIa* gene, which encodes a cytoskeletal actin, and the *endo16* gene, which encodes a secreted embryonic gut protein. In both cases all sites of interaction were mapped for which the affinity of the regulatory protein for the specific site, as opposed to its affinity for a double-stranded synthetic polydeoxynucleotide, exceeded $5–10 \times 10^3$ (for *cyIIIa*, Calzone *et al.*, 1988; Thézé *et al.*, 1990; for *endo16*, Yuh *et al.*, 1994). Detailed gene transfer studies subsequently revealed that each of the nine species of interaction in the *cyIIIa* regulatory system, and each of the ten species of interaction in the region of the *endo16* system so far examined contributes to the regulatory performance of these systems. A philosopher (or an evolutionist) might with some justification remark that it had better be so. It is very unlikely that highly specific site clusters, which are of improbable random occurrence within a few hundred base pairs of DNA, would have no function. In fact it is so, and there are some practical consequences. The most important is that at our present level of knowledge it is difficult to identify all the subfunctions executed within a *cis*-regulatory element, except by first mapping its target sites.

A fifth principle is that *cis*-regulatory elements which execute specification functions usually utilize negative as well as positive inputs, i.e., interactions with transcriptional repressors as well as with activators. The repressors set spatial boundaries of expression, while the activators are spatially more widespread (and are thus used for multiple genes and multiple purposes). Like much else that we have learned by probing within the *cis*-regulatory module, this aspect of DNA design logic has been more easily appreciated in hindsight, and was not initially anticipated.

SPATIAL REPRESSION IN *CIS*-REGULATORY SPECIFICATION

Several sorts of *cis*-regulatory design for spatial repression have been uncovered. In some systems whole regulatory modules are devoted to repression of activity in given domains of the embryo, their only detected function, and we meet some examples of this further on. In others, sites for spatial repressors and for activators are interspersed within a single module, which carries out all the control functions required to specify a given phase of developmental gene expression. These are particularly illuminating examples: they illustrate each of the operating principles just noted for *cis*-regulatory specification elements. In their internal functional organization we can see directly how a clustered set of regulatory DNA target sites leads to the institution of a bounded spatial pattern of gene expression.

Two Very Different Examples of Similar Import

If the same mechanism is found in a *Drosophila* and in a sea urchin developmental regulatory system, there is a pretty fair case for its generality. The arthropods and the deuterostomes lie distant from one another in the bilaterian cladogram (Fig. 1.6), and the strategies they use for early development are very different. Yet at the *cis*-regulatory level the operating principles in these two systems are indistinguishable. Within the syncytial *Drosophila* embryo, inputs for the initial tiers of specification functions depend on internuclear diffusion of transcription factors. At the anterior and posterior poles of the egg, respectively, the maternal *bicoid* and *caudal* mRNAs are localized, and the proteins they encode diffuse inward from the two poles, forming opposing concentration gradients. Early zygotic mRNAs encoding transcription factors are generated in relatively broad bands of nuclei that are spaced along the anterior/posterior (A/P) axis ("gap genes;" for reviews, Lawrence, 1992; Rivera-Pomar and Jäckle, 1996). The sea urchin embryo, in contrast, is cellular from the start. The inputs for zygotic specification functions are again maternal and zygotic transcription factors, but their activities and/or presentation in the nuclei depend crucially on early intercellular signaling, as well as on localized maternal activation systems (for review, Davidson *et al.*, 1998). But for present concerns these distinctions do not matter: they lie upstream of the *cis*-regulatory systems that must differentially interpret the regulatory states that they encounter in the various spatial domains of the embryo. This they do by very similar means, as the following examples show.

The output of the *evenskipped* (*eve*) stripe 2 *cis*-regulatory element was illustrated in Fig. 1.4. It produces a thin circumferential stripe of gene expression a few nuclei wide located at an exact position along the A/P axis. Figure 2.1A displays the minimal hardwiring required to generate *eve* stripe 2, i.e., the sequence of transcription factor target sites necessary and sufficient to cause accurate expression of the reporter gene. Because it is well understood, due to an incisive series of studies by M. Levine and colleagues (Small *et al.*, 1991, 1992; Arnosti *et al.*, 1996), this regulatory element provides a classic example of a spatial specification system which sets its boundaries by repression. The stripe 2 module activates transcription in response to the positive Bicoid (Bcd) regulator, and a second zygotically transcribed positive regulator, Hunchback (Hb), is also required (the *hb* gene is activated by Bcd as well). The Bcd protein is distributed in a broad concentration cline that decreases from anterior to posterior, and the Hb protein is present in about the anterior 40% of the early embryo (Fig. 2.1B). As Fig. 2.1A shows, there are multiple Bcd target sites in the stripe 2 element, some of which bind the factor more tightly than others. However, the multiple Bcd activators binding within the element work synergistically with one another, and with the Hb factor; these sites all contribute, but some are more important than others. Their synergism is required to obtain strong activation of the element in spite of low endogenous activator concentrations (Arnosti *et al.*, 1996).

The distribution of the activators to which the *eve* stripe 2 element responds would suffice for expression extending from far anterior of *eve* stripe 2 to far posterior of it, i.e., throughout almost the whole anterior half of the embryo. The actual boundaries of *eve* stripe 2 depend entirely on repression, mediated by the Krüppel (Kr) and Giant (Gt) target sites shown in Fig. 2.1A. The posterior boundary is produced as a response of the *cis*-regulatory element to the Kr repressor (Small *et al.*, 1992; Gray and Levine, 1996). If the Kr expression domain is caused to expand in an anterior direction, so that it overlaps rather than abuts *eve* stripe 2, it wipes out all expression of the *eve* stripe 2 element (Fig. 2.1C and D). Thus the location of the Kr repressor in the syncytial embryo constitutes an essential spatial input to the system. The anterior boundary is produced by interactions with the Gt repressor, working together with another as yet unknown corepressor which requires Gt binding for function (Small *et al.*, 1992; Arnosti *et al.*, 1996; Wu *et al.*, 1998). The Gt repressor is normally present in the embryo in a band extending up to the anterior border of *eve* stripe 2, as illustrated in Fig. 2.1E. Figure 2.1F shows that *eve* stripe 2 expression expands in the anterior direction when the Gt sites of the element shown in Fig. 2.1A are mutated. These and many other experiments demonstrate the crucial role of spatial repression in placing *eve* stripe 2 in the appropriate position. Institution of *eve* stripe 2 is an excellent example of a *cis*-regulatory specification function, in that expression of this transcription factor in the seven initial stripes is a key initial step in the embryonic pattern formation process that leads to the metamerization of the body plan. The role of the *eve cis*-regulatory element controlling stripe 2 expression is to "read" and integrate the broad patterns of its activators and repressors into a sharp

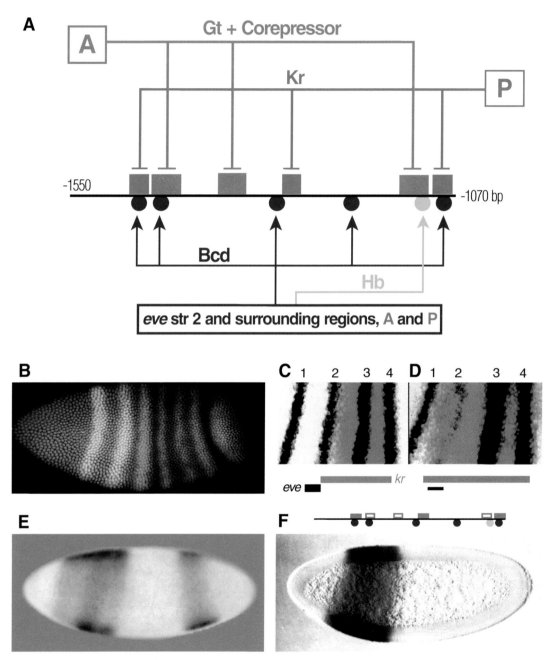

FIGURE 2.1 *cis-Regulatory control of evenskipped* **stripe 2 in** *Drosophila*. (A) Diagram of "minimal stripe element" (−1070 to −1550 of the upstream sequence). When introduced into the germ line this DNA fragment suffices to produce accurate

design: it creates a novel spatial regulatory domain, which did not previously exist, in a field of pluripotential nuclei (cf. Fig. 2.1C–F).

Spatial repression plays an essentially similar role in regulation of the *cyIIIa* gene of *Strongylocentrotus purpuratus*, as shown in Fig. 2.2. *cyIIIa* encodes a terminal product of the regulatory apparatus, *viz* a cytoskeletal actin, one of five in this genome. The *cyIIIa* gene is regulated in a way that distinguishes it from any other of the actin genes: it is expressed exclusively in the aboral ectoderm and its precursor cells from mid-cleavage on. The aboral ectoderm ultimately forms the single cell-thick squamous epithelial wall of the late embryo and larva, which is bounded by the neurogenic ciliated band and the oral ectoderm surrounding the mouth. The regulatory problem for the *cyIIIa* gene is to ensure its expression only in aboral ectoderm, and to prevent expression in oral ectoderm progenitors, in skeletogenic cells, and in all endomesodermal cells (see Fig. 2.2B, C). This problem is compounded by the constraint that oral/aboral (O/A) polarity cannot be set up in the sea urchin egg prior to fertilization. Any point on the equatorial circumference of the egg may serve as either oral or aboral pole, and in typical sea

---

expression of a *lacz* reporter in the exact location of endogenous stripe 2 (see Fig. 1.4). Target sites for transcription factors are indicated. Repressors, *viz*, Krüppel (Kr) and Giant (Gt), are shown as solid rectangles above the line representing the DNA. Activators, *viz* Bicoid (Bcd) and Hunchback (Hb), are shown as solid circles below the line. A, anterior; P, posterior. [(A) From Small *et al.* (1992) *EMBO J.* **11**, 4047–4057 and by permission of the European Molecular Biology Organization.] (B) Immunofluorescence display of the location of the Hb activator and the anterior Bcd domain, with respect to the stripes of Eve protein, 14th cleavage. Hb is stained in dark green; Bcd in blue fading to magenta more posteriorly; and Eve in orange-red; the overlap of Eve and Hb appears yellow. The anterior domain of Hb includes Eve stripes 1, 2, and 3. [(B) From J. Reinitz, personal communication.] (C, D) *eve* mRNA (black) and *kr* mRNA (red) as displayed by *in situ* hybridization. In (C) and (D) the numerals indicate the *eve* stripes. (C) The Kr domain directly abuts the posterior boundary of *eve* stripe 2 in a normal embryo. (D) Forced expression of *kr* in the region overlapping *eve* stripe 2 (by placing the *kr* gene under control of the stripe 2 *cis*-regulatory element) almost completely represses *eve* stripe 2. [(C, D) From Wu *et al.* (1998) *Development* **125**, 3765–3774 and The Company of Biologists, Ltd.] (E) Gt protein revealed by immunostaining (brown), together with a *lacz* mRNA stripe generated by the *cis*-regulatory element shown in (A), as revealed by *in situ* hybridization (blue-green). The Gt domain abuts *eve* stripe 2 at its anterior border. (F) Effect on expression of mutating all three of the Gt target sites. *cis*-Regulatory element of (A) at top, same color coding of target sites. Open boxes indicate mutated sites. Lacz mRNA is now transcribed in an anteriorly expanded region, since repression mediated by Gt (and its unknown cofactor; Wu *et al.*, 1998) can no longer occur. Anterior is to the left in (B–F). [(E, F) From Small *et al.* (1992) *EMBO J.* **11**, 4047–4057 and by permission of the European Molecular Biology Organization.]

FIGURE 2.2   Spatial repression in the regulation of the *Strongylocentrotus purpuratus cyIIIa* gene. (A) Target sites and spatial inputs in the *cyIIIa cis*-regulatory system (Kirchhamer and Davidson, 1996; Coffman *et al.*, 1997). Sites for a DNA looping protein (SpGcf; Zeller *et al.*, 1995) are omitted. The proteins which bind the sites shown are: R, SpRunt1; M, SpMyb1; Z, SpZ12-1; P, SpP3a2; CT, SpCtf1; T, SpTef1; ▷, SpOct1; and the open downward triangles represent an unknown positive regulator. Repressors are shown in red; activators in dark blue. The spatial domains where these regulators are functional are indicated (open boxes). (B) Normal distribution of endogenous *cyIIIa* mRNA, displayed by *in situ* hybridization in a late gastrula-stage embryo. *CyIIIa* transcripts are confined to the aboral ectoderm; oral ectoderm is indicated as the region between arrowheads in panels (B–D, F). (C–F) Whole mount *in situ* hybridization display of a CAT mRNA reporter, transcribed under control of various *cyIIIa cis*-regulatory elements. [(A, B) From Bogarad *et al.* (1998) *Proc. Natl. Acad. Sci. USA* **95**, 14827–14832; copyright National Academy of Sciences, USA.] (C) Expression of complete wild-type (control) *cyIIIa•CAT* transgene containing the regulatory system shown in (A), plus some distal sites for SpGcf1 which have a mild enhancing effect on the level of activity. CAT mRNA is confined to clones of aboral ectoderm cells. When DNA is injected into eggs it is usually incorporated into single blastomeres early in cleavage, resulting in mosaic expression patterns. Incorporation is stable, and completely random with respect to cell lineage (Livant *et al.*, 1991). The total expression pattern is obtained from observations on a population of injected embryos. (D) Effect on expression of mutating one of the P3a2 sites in the otherwise complete construct. Expression now occurs in oral ectoderm as well as in aboral ectoderm at the normal level; about 30% of embryos (i.e., the fraction in which DNA will be incorporated in clones generating oral ectoderm on a random basis) display only ectopic oral ectoderm expression, as in this particular embryo. [(C, D) From Kirchhamer *et al.* (1996) *Development* **122**, 333–348 and The Company of Biologists, Ltd.] (E) Expression in skeletogenic mesenchyme, as well as in a patch of ectoderm, in an embryo bearing the *cyIIIa•CAT* vector in which the two Z sites were mutated. The embryo is viewed from the bottom (i.e., blastopore end). [(E) from Wang *et al.* (1995) *Development* **121**, 1111–1122 and The Company of Biologists, Ltd.] (F) Ectopic expression in oral ectoderm and endoderm, plus normal expression in patch of aboral ectoderm, caused by a mutation of the M site which prevents binding of the Myb repressor [(F) From Coffman *et al.* (1997) *Development* **124**, 4717–4727 and The Company of Biologists, Ltd.] (G) Immunocytological display of Myb protein, in a late pluteus-stage embryo. The bilaterally organized oral ectoderm forms the upper wall of the embryo, and the aboral ectoderm constitutes the remainder of the external wall of the embryo. The Myb repressor is confined to oral ectoderm and gut (the circular midgut is clearly visible in this image), plus in some mesenchyme cells (not clearly visible here). [(G) From J. A. Coffman and E. H. Davidson, unpublished data.]

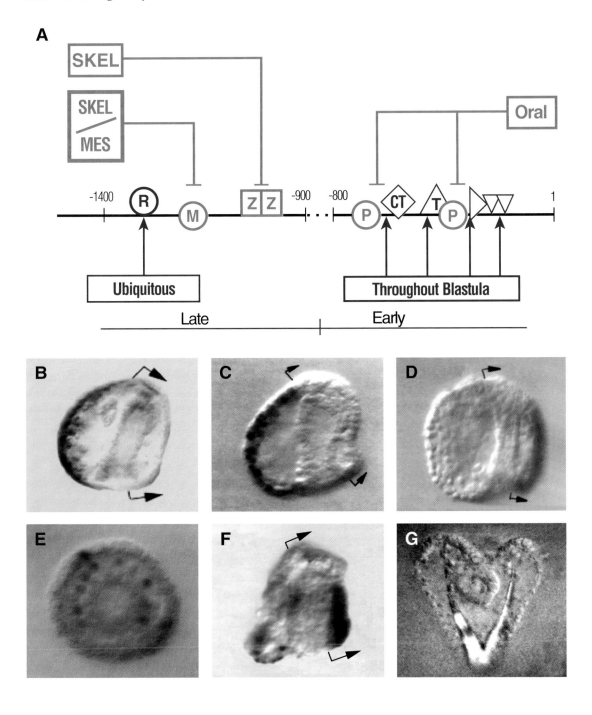

urchin species O/A polarity is not established until after cleavage has begun (Cameron *et al.*, 1989; Davidson *et al.*, 1998). The *cyIIIa cis*-regulatory system is relatively well understood (Kirchhamer and Davidson, 1996; Coffman *et al.*, 1997). Figure 2.2A illustrates the two *cis*-regulatory modules that control *cyIIIa* expression in the embryo. The proximal module initiates function early on, during the initial processes of ectoderm specification, and a more distal module drives expression in the blastula-gastrula stage as the aboral ectoderm differentiates into squamous epithelium. The endogenous pattern of *cyIIIa* expression, as it appears in the late gastrula, is shown in Fig. 2.2B.

Four different positive regulators work together to activate the early module, all of maternal origin, while expression of the late module depends on a largely zygotic Runt family transcription factor, which is transcribed actively following the blastula stage (Coffman *et al.*, 1996). But the activity of none of these positive regulators is confined to the aboral ectoderm cell lineages. Although the spatial details are not known for some, they all appear to be functional in multiple embryonic territories, including both oral and aboral ectoderm. It will no longer surprise (though it did on discovery, Hough-Evans *et al.*, 1990) that expression is confined to the aboral ectoderm by repressive *cis*-regulatory interactions. Each of the two modules includes target sites for its own spatial repressors. The two sites labeled "P" in the early module of Fig. 2.2A bind a repressor called "P3a2." If either site is deleted, the reporter is found to be expressed ectopically in the oral ectoderm [Kirchhamer and Davidson, 1996; in this case a gene encoding chloramphenicol acetyltransferase (CAT) is the reporter]. Results from such an experiment are illustrated in Fig. 2.2D and its control, Fig. 2.2C. Exactly the same effect is observed if instead of removing the target sites for P3a2, the P3a2 factor is inactivated *in vivo*. This has been accomplished by introduction of a vector encoding a single chain antibody against the P3a2 DNA-binding site (Bogarad *et al.*, 1998). But there are other repressive interactions required as well. When the late module becomes dominant, i.e., when the Runt factor is prevalent, two different repressors come into play: a 12–Zn finger repressor binding at the site labeled "Z" in Fig. 2.2A prevents ectopic expression in skeletogenic cells (Wang *et al.*, 1995; Fig. 2.2E); and a Myb-class repressor binding at site "M" prevents expression in oral ectoderm and also gut (Coffman *et al.*, 1997; Fig. 2.2F). By late in embryogenesis the Myb repressor is localized, as shown rather dramatically in Fig. 2.2G, to those tissues where it functions as a *cyIIIa* repressor, *viz*, oral ectoderm, gut, and mesenchyme cells. In keeping with our theme of intramodular repression functions, it is an important experimental result that if the whole of the late module is deleted, repressor sites and all, no ectopic expression is observed, only a low level of spatially correct expression. Therefore the late module repressors are just needed to control expression driven by the late module activator (i.e., the Runt factor). In other words, ectopic expression is seen only if the Runt site is present, and the Myb or Zn finger repressor sites are also mutated or absent from the construct.

The role of the P3a2 repressor is additionally interesting because its asymmetric activation appears to play a role in the initial process by which O/A (oral/aboral) polarity is set up early in cleavage. The activity of P3a2 depends on ambient redox state (Coffman and Davidson, 2000). As has long been known, the sea urchin egg displays an early O/A asymmetry in redox potential (Czihak, 1963; Coffman and Davidson, 2000) possibly caused by cytoskeletal rearrangements that concentrate mitochondria on one side of the egg following fertilization. If artificial means of altering the redox gradient are imposed, O/A specification is correspondingly constrained. The redox gradient may explain how the maternal P3a2 factor acts as a repressor on the oral side of the early embryo, wherever the oral pole comes to lie. It is the job of the early *cis*-regulatory module of *cyIIIa* to interpret the spatial distribution of regulatory activity; and thus to transduce the polarized cline of P3a2 activity into an asymmetric and ultimately lineage-specific pattern of zygotic gene activity.

cis-Regulatory Design for Autonomous Modular Function

In the two *cis*-regulatory modules we have so far discussed, the repressors can be thought of as acting by canceling the output of the positive regulators that bind within the same module, when and where these repressors are present. These modules are independent devices with self-contained positive and negative spatial control functions. The further examples summarized in Figs. 2.3 and 2.4 illustrate their autonomy, and provide some idea of the constraints on this kind of design.

The functional organization of the neuroectodermal control module of the *Drosophila* gene *rhomboid* (*rho*) is described in Fig. 2.3. This gene encodes a cell surface component of a signaling pathway, which begins to be expressed during specification of the neuroectodermal territory from which the ventral CNS will derive (about one fourth of the cells will become neuroblasts; the remainder ectoderm). The embryo is at this stage still syncytial, and the neuroectodermal regions consist of two longitudinal stripes of nuclei in which *rho* is expressed, overlying the ventral mesodermal domain on either side. The relevant *cis*-regulatory module of the *rho* gene contains binding sites for several activators. Of these the most important is Dorsal (Dl), a Rel domain regulator which is the key early transcription factor in dorsoventral (D/V) patterning. As discussed below in more detail, by late cleavage Dl is present in the syncytial nuclei in a graded concentration cline from ventral to dorsal. Other positive regulators which function synergistically with Dl in the *rho* neuroectoderm *cis*-regulatory element are Twist, a bHLH activator, and some additional bHLH factors (Ip *et al.*, 1992a; Jiang and Levine, 1993). Figure 2.3A shows that this *rho cis*-regulatory element generates the two lateral bands of expression that constitute the neuroectodermal territories if incorporated in a *lacz* fusion construct (Gray *et al.*, 1994). Exclusion of expression from the ventral mesodermal domain is due to the Snail (Sna) repressor, which binds at the indicated sites within the *rho cis*-regulatory module. The *sna* gene is itself activated in the ventralmost regions of the embryo of the future mesodermal domain by high levels of nuclear Dl protein (Ip *et al.*, 1992b; Jiang and Levine, 1993). If the Sna target sites in the *rho* regulatory element are mutated, *lacz*

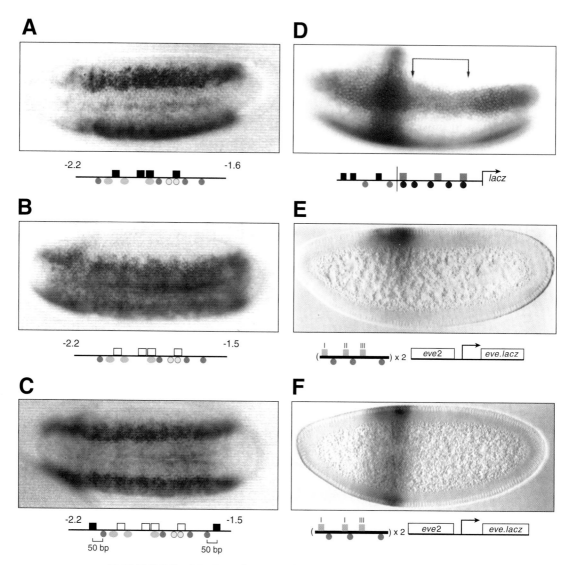

FIGURE 2.3 Modes of *cis*-regulatory expression in *Drosophila*. (A–D) Short-range repression; (E, F), long-range repression. (A) *rhomboid* (*rho*) gene neuroectodermal module, and its expression revealed by a *lacz* transgene, displayed by *in situ* hybridization; ventral view of an early cellularization-stage embryo. The construct is expressed in two broad lateral stripes directly abutting the ventral mesodermal domain, in which it is silent. Factors that bind the target sites shown are Dorsal (red); a bHLH factor (green); Twist (yellow); (all activators are shown below the line representing the DNA); and Snail (black boxes), a repressor. (B) The Snail sites are responsible for repression in the mesoderm, since ectopic expression occurs there if

expression spreads throughout the ventral mesoderm (Fig. 2.3B), a dramatic effect. The experiment reproduced in Fig. 2.3C provides insight into the structural organization of *cis*-regulatory elements of this sort. Here two Sna target sites have been restored to the mutated construct in Fig. 2.3B, but they are positioned 50 bp outside all the positive regulatory sites on either end of the sequence. Nonetheless, ventral repression is restored, and in fact, a single Sna site suffices. However, if the Sna sites are removed to distances of 150 and 120 bp, the spatial repression that they exert is greatly weakened (Gray *et al.*, 1994). The Sna sites do not have to be in any particular position, e.g., overlapping with, or immediately adjacent to, the sites for positive regulators, but they have to be within about 100 bp of at least some of those sites: there is considerable design flexibility, but also a crucial design constraint in the sequence organization of this regulatory module.

The term "autonomy" here means that the package of target sites within a module suffice for it to execute its function, irrespective of its regulatory environs. *cis*-Regulatory autonomy is demonstrated precisely in the experiment shown in Fig. 2.3D (Gray and Levine, 1996). The *eve* stripe 2 module and the *rho* neuroectoderm module are here combined in a single *lacz* expression vector: they both operate, independently, producing a summed pattern of *lacz* mRNA.

---

these sites are mutated (open black boxes). (C) Snail still functions to repress expression in the mesodermal domain when its target sites are moved to positions 50 bp proximal and distal to the nearest activation sites; the natural Snail target sites mutated as in (B). [(A–C) Adapted from Gray *et al.* (1994) *Genes Dev.* **8**, 1829–1838.] (D) Autonomy of short-range repression. Two different modules, each including short-range repressors, are physically linked in a single *lacz* expression construct. This contains a version of the *rho* element of (A), distal, and the *eve* stripe 2 element, proximal (see Fig. 2.1 for structure; blue, Kr repressor; black, Bcd activator). The two modules function independently of one another, generating both the A/P *rho* pattern and the D/V *eve* stripe 2, so as to produce a crossed pattern. The Sna repressor does not prevent *eve* stripe 2 expression in the mesoderm, and the Kr repressor does not prevent *rho* expression, though it has a mild weakening effect (bracket). [(D) Adapted from Gray and Levine (1996) *Genes Dev.* **10**, 700–710.] (E) Expression of a vector containing two copies of the ventral long-range repression element of the *zen* gene, distal to the same *eve* stripe 2 module (*eve2*) as above. "*eve/lacz*" denotes the promoter/*lacz* reporter fusion. The *zen* element contains sites for Dorsal (red circles) and three sites (I, II, III) where other factors bind (colored boxes). Though the *zen* elements are positioned hundreds of bp upstream of the BTA, and outside of the *eve* stripe 2 element, *lacz* mRNA is seen only in the dorsal region of the *eve* stripe 2, and expression is silenced elsewhere. (F) Expression of *eve* stripe 2 is restored throughout if the second of the three sites in the *zen* repression element that are shown as colored boxes (I, II, III) is changed to be the same as the first; this is the site at which Deadringer and Cut corepressors interact (Valentine *et al.*, 1998). [(E, F) Adapted from Cai *et al.* (1996) *Proc. Natl. Acad. Sci. USA* **93**, 9309–9314, copyright National Academy of Sciences, USA.]

The experiment proves that the repression functions that set the anterior and posterior boundaries of *eve* stripe 2 (i.e., those mediated by Kr and Gt regulators), and the repressor setting the ventral boundaries of the *rho* stripes as well (i.e., Sna), are dedicated to the control of their own respective modules. There is no gap in the *eve* stripe where Sna, the *rho* repressor, is active, and only a very slight attenuation in the *rho* stripes where the Kr repressor is active. With respect to the disposition of their *cis*-regulatory target sites, all of these repressors work over a short range, in that their effect is confined to the modules within which they bind.

A similar example from the sea urchin is shown in Fig. 2.4. Here again two different *cis*-regulatory modules are combined. One of these, which includes the only basal transcription apparatus in the construct, is from the *sm50* skeletal matrix protein gene, and this regulatory module confers expression of the CAT reporter exclusively to skeletogenic mesenchyme cells (Fig. 2.4A). The second is from the *endo16* gene, which at the stage shown functions only in the gut (Fig. 2.4B). The workings of the *endo16* system are discussed below, and for now the only important point is that it includes target sites for repressors required to prevent expression of *endo16* in skeletogenic mesenchyme cells. But again this repression function is pointed at its own *cis*-regulatory system, and it does not interfere at all with skeletogenic expression driven by the *sm50* module. When combined, each module operates as if the other were not there: the output is just the sum of the two expression patterns (Fig. 2.4C).

Other *cis*-regulatory repressors function in a different way, silencing all transcriptional activity mediated by the basal transcription apparatus (BTA) which

FIGURE 2.4   Autonomy of repression in a synthetic *Strongylocentrotus cis-regulatory vector.* (A) Expression of a vector in which a CAT gene serves as reporter, driven by a *cis*-regulatory module from the skeletogenic gene *sm50*. CAT mRNA is visualized in a late gastrula-stage embryo viewed from the side by *in situ* hybridization. Expression is confined to skeletogenic mesenchyme cells (white arrowheads). The *sm50* module contains sites for positive mesenchyme specific factors but no negatively acting elements (Makabe *et al.*, 1995). [(A) From Lee *et al.* (1999) *Dev. Growth Differ.* **41**, 303–312.] (B) Expression of *endo16 cis*-regulatory system in gut (white arrowheads). This system specifically represses *endo16* expression in skeletogenic mesenchyme cells, as well as in ectoderm (see text). [(B) From Yuh *et al.* (1994) *Mech. Dev.* **47**, 165–186; copyright Elsevier Science.] (C) Synthetic fusion of regulatory systems shown in (A) and (B). The *endo16* system is distal and the *sm50* module including the BTA are proximal. Expression is additive, and there is no repression of the transcriptional activity in skeletogenic mesenchyme which is mediated by the *sm50* module. Black arrowheads indicate staining in skeletogenic cells; white arrowheads indicate expression in gut cells. Similarly skeletogenic expression driven by the *sm50* module remains unaffected when an *endo16* DNA fragment containing only the skeletogenic repressor sites is directly linked to it. [(C) Adapted from Kirchhamer *et al.* (1996) *Proc. Natl. Acad. Sci. USA* **93**, 13849–13854; copyright National Academy of Sciences, USA.]

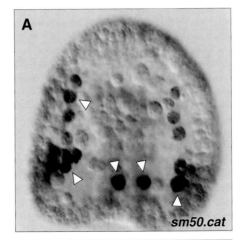

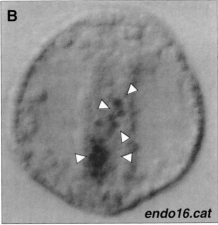

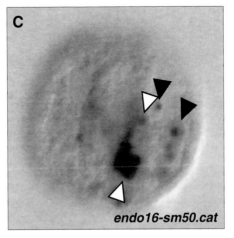

they service, whether the positive inputs originate in their own or another *cis*-regulatory module. These are termed "long-range" as opposed to "short-range" repression systems. This is illustrated in Figs. 2.3E, F (Cai *et al.*, 1996), which demonstrate long-range repression mediated by a *cis*-regulatory element from the *Drosophila zerknüllt* (*zen*) gene. The *zen* gene encodes a homeodomain regulator that carries out an early pattern formation function required for specification of the embryonic dorsal ectoderm. This gene is activated by ubiquitously present transcription factors, and its expression is confined to the dorsalmost region by a "ventral repression" module that contains sites both for Dl and for two other factors that together with Dl constitute a repressor complex (*viz* the Cut [Ct] and Dead Ringer [Dri] factors; Jiang *et al.*, 1993; Valentine *et al.*, 1998). Figures 2.3E, F show that the *zen* ventral repressor module prevents ventral and lateral expression of an *eve* stripe 2 construct to which it is linked, and that this effect depends on the Ct/Dri binding sites (Valentine *et al.*, 1998). Repression of *eve* stripe 2 by the *zen* element occurs even if the latter is placed 5 kb away, downstream of the BTA, indeed justifying the term "long-range." The *zen* repressor element will silence multiple regulatory elements linked in the same construct (Cai *et al.*, 1996; Gray and Levine, 1996). This is an example of a regulatory module that acts by shutting down the whole BTA, which contrasts sharply with the short-range mechanisms utilized in the systems illustrated in Figs. 2.3A–D and 2.4. Silencer modules are found in many genes, and are a major feature of modular *cis*-regulatory control apparatus.

Not unexpectedly, the difference between these two classes of repressor function depends on the identity of the cofactors that mediate their activities. This in turn depends on the identity of the transcription factors to which the cofactors bind. Hence the mode of repression is ultimately determined by the *cis*-regulatory target sites that specify which repressor will be bound, and thereby which kind of repression function is to be executed. Several short-range repressors bind a particular corepressor, the "C-terminal binding protein" (CtBP). In mammalian cells CtBP inhibits the transcriptional activation function of the viral E1A protein (Schaeper *et al.*, 1995), and to do this CtBP must bind to the sequence PXDLSXK in the E1A protein. The PXDLSXK sequence is also found in two of the *Drosophila* short-range repressors we have already met, *viz*, the Sna and Kr proteins (Zhang and Levine, 1999). This sequence is present as well and is required both for CtBP binding and for repressor function in the Knirps (Kni) protein of *Drosophila* (Nibu *et al.*, 1998). Kni is another short-range repressor which controls spatial expression of various target genes in the embryonic abdomen, e.g., the Kr gene (Hoch *et al.*, 1992). These three transcription factors are otherwise very different in structure: Sna and Kr are unrelated Zn finger proteins, and Kni is a factor of the nuclear receptor family. The distinct *cis*-regulatory target sites at which these three different factors (and a large number of others) bind, all provide regulatory access to a common corepressor. This may act by interfering with common coactivators, which similarly mediate the activation functions of diverse transcription factors (see Torchia *et al.*, 1998; Mannervik *et al.*, 1999 for reviews)

In the same way, diverse long-range repressors share common cofactors and include common sequence motifs that specify interaction with these cofactors. One example of these is the *Drosophila* protein Groucho (Gro), and its relatives in other creatures. This protein binds to the sequence WRPW, which for instance is present in the Hairy repressor, a regulator of the spatial expression of other (pair-rule) genes required for metamerization (Paroush *et al.*, 1994; Jimenez *et al.*, 1997; Zhang and Levine, 1999), as well as of various genes required for nervous system development at later stages (Van Doren *et al.*, 1994; Ohsako *et al.*, 1994), and of the *sex lethal* gene in females (Fisher and Caudy, 1998). Gro services many bHLH repressors which contain WRPW motifs, e.g., the *Enhancer of split Complex* repressors (Fisher *et al.*, 1996), and also other repressors of entirely different structure. Among these are Runt, which has instead a WRPY binding site; Huck-ebein, which utilizes an FRPW site; and also Engrailed, the *zen* ventral repressor complex described above, and the Wnt pathway transcription factor Lef1, all of which bind Gro by other short sequences (Goldstein *et al.*, 1999; Cavallo *et al.*, 1998; Roose *et al.*, 1998; Dubnicoff *et al.*, 1997; Jimenez *et al.*, 1997). There is a protein related to Gro in yeast, Tup1, and detailed studies indicate that the mode of repression of such proteins is indeed direct interference with polymerase function in the BTA (Herschbach *et al.*, 1994; Kuchin and Carlson, 1998). Groucho is one of a zoo of different species of corepressor, and a variety of coactivators exist as well (reviewed by Torchia *et al.*, 1998; Lemon and Freedman, 1999; Knoepfler and Eisenman, 1999).

Our focus here is on the DNA regulatory sequences that encode developmental functions by specifying the binding of transcription factors, rather than on the biochemical processes which these factors then entrain off the DNA. Corepressors and coactivators are often ubiquitous. Like TAFs, TATA binding factors, and the many polymerase subunits, they are to be regarded as essential elements of the transcription complex, without which nothing we are considering here would work at all. But we must not confuse the issue of how these effector systems work with the problem of understanding the genetic control apparatus that at each gene specifically determines the deployment of these ubiquitous or nearly ubiquitous effector systems. *cis*-Regulatory elements that mediate the activities of intra-modular or dedicated short-range repressors, and those that execute long-range silencing functions represent two different classes of regulatory design, two kinds of spatial control device in the regulatory tool kit for development. There are many more, as we survey below. But first it is important to appreciate the great diversity of developmental contexts in which *cis*-regulatory elements that utilize spatial repression operate.

The Generality of Repression

A glance at Table 2.1 yields the main points. First, and most importantly, it is clear that repressors are found interacting within the control elements of genes that carry out a great variety of functions, in both embryonic and postembryonic

TABLE 2.1 Examples of Spatial Regulatory Functions in Development that Depend on Interactions with Transcriptional Repressors

Type of protein encoded by gene	Gene	Organism	Spatial regulatory function of module	Repression required to prevent expression in:	Repressor	Ref.
Transcription factor						
Pair rule factor	evenskipped	Drosophila	Stripe 2	Outside of stripe boundaries	Gt, Kr,	1
			Stripes 3 + 7	Outside of stripe boundaries	Hb, Kni	2
			Stripes 4 + 6	Outside of stripe boundaries	Hb, Kni	3
			Stripe 5	Outside of stripe boundaries	Kr, Gt	3
	ftz	Drosophila	7 stripes	Interstripes	Slp1	4
	paired	Drosophila	14 stripes	Center cells of initial 7 stripes	Odd	5
	hairy	Drosophila	Stripe 6	Outside stripe boundaries	Kr, Tll, Hb	6
			Stripe 7	Outside stripe boundaries	Kni, Tll	7
Gap gene factor	knirps	Drosophila	P expression domain	Most of embryo	Hb	8
	tailless	Drosophila	Early A and P domains	Terminal region	Elf1	9
			Specific band regions	Specific neuroblasts	—	10
Domain-specific regulators	spalt	Drosophila	Early A and P bands	Outside of boundaries	Kr, Hkb	11
	engrailed	Drosophila	Even parasegments	Odd parasegments	Odd	12
	cubitus interruptus	Drosophila	A compartments of parasegments	P compartments	En	13
			A compartments of imaginal discs	P compartments	En	13
	tinman	Drosophila	Dorsal mesoderm	Dorsal ectoderm, amnioserosa	—	14
			6 early mesodermal stripes	Interstripe domains	Eve	15
	proboscepedia	Drosophila	Head	Hemocytes	Btd	15
			Labial discs	Antennal and thoracic discs	—	16

	Gene	Organism			Factor	
	ultrabithorax	Drosophila	Central band in blastoderm	Outside A and P boundaries	Hb	17
	achaete-scute	Drosophila	Peripheral sensory organs	Ectopic bristles	H	18
	hoxb1	Mouse	r4	r3 and r5	RA response system	19
	hoxb3	Mouse	r5	r6	—	20
	nkx2.5	Mouse	Cardiac domains	Various ectopic locations	—	21
	brachyury	Ciona	Notochord	Tail muscle cells	Sna	22
	decapentaplegic	Drosophila	Dorsal region of embryo	Ventral domains	NTF1	23
Signaling ligand						
	fgf4	Mouse	Midgut, parasegment 7 P AER of limb bud	A and P midgut	Abd-A	24
				A AER	—	25
	rhomboid	Drosophila	Neuroectoderm	Mesoderm	Sna	26
Signal receptor apparatus						
Cell type-specific proteins						
	cardiac α myosin HC	Rat	Cardiac myocytes	Nonmuscle cells	Ets factor	27
	cyIIIa cytoskeletal actin	Strongylocentrotus	Aboral ectoderm	Oral ectoderm, skeletogenic cells	P3a2, Z12-1, SpMyb1	28
	endo16	Strongylocentrotus	Vegetal plate, midgut	Ectoderm, skeletogenic cells	CREB factor, others	29

Note. 1—See Fig. 2.1; 2—Small et al., 1996; 3—Fujioka et al., 1999; 4—Topol et al., 1991; Yu et al., 1999; 5—Gutjahr et al., 1994; 6—Langeland et al., 1994; Häder et al., 1998; 7—La Rosée et al., 1997; 8—Rivera-Pomar et al., 1995; 9—Liaw et al., 1995; 10—Rudolph et al., 1997; 11—Kühnlein et al., 1997; 12—Florence et al., 1997; 13—Schwartz et al., 1995; 14—Xu et al., 1998; 15—Yin et al., 1997; 16—Kapoun and Kaufman, 1995; 17—Qian et al., 1991; Zhang et al., 1991; 18—Van Doren et al., 1994; 19—Studer et al., 1994; 20—Manzanares et al., 1997; 21—Reecy et al., 1999; 22—Fujiwara et al., 1998; 23—Huang et al., 1993, 1995; 24—Manak et al., 1995; 25—Fraidenraich et al., 1998; 26—See Fig. 2.3; 27—Gupta et al., 1998; 28—See Fig. 2.2; and 29—Yuh and Davidson, 1996; unpublished data.

Abbreviations for factors, in order of appearance (where more extensive names exist): Gt, Giant; Kr, Krüppel; Hb, Hunchback; Kni, Knirps; Slp1, Sloppy-paired; Odd, Odd-skipped; Tll, Tailless; Hkb, Huckebein; En, Engrailed; Eve, Evenskipped; Btd, Buttonhead; H, Hairy; Sna, Snail; NTF1, Elf1 or Grainyhead product; Abd-A, Abdominal A.

Other Abbreviations: A, anterior; P, posterior; D/V, dorsal/ventral; AER, apical ectodermal ridge; HC, heavy chain; r, rhombomere.

development. Table 2.1 includes genes that encode transcription factors, signaling ligands, cytoskeletal proteins, cell surface receptors, etc. Second, although the majority of cases so far known are from *Drosophila*, the same strategies are utilized in mammals, ascidians, and sea urchins. Spatial repression is a pan-bilaterian device. For inclusion in this table, the requirement imposed is that a specific *cis*-regulatory element conferring the listed expression pattern has been identified in a gene transfer experiment. That is, ectopic expression must have been observed when the target sites for the repressor were mutated or if they were absent from the construct, or the repressor is absent or inactive. Were the requirements further relaxed, a huge swath of additional cases could be added, where there is less defined evidence of negative interactions. But as is, Table 2.1 speaks for itself. If a gene operates in a developmental context in which cell fate or cell type decisions are being made, i.e., if its expression or nonexpression mediates such decisions, it is liable to rely on repression as well as on activation to set its transcriptional boundaries. The generality is that both are usually required to execute developmental specification.

DOWNSTREAM OF SPECIFICATION

Once a state of specification has been installed, there remains the job of making something out of it. Cohorts of genes must be activated within the newly created spatial domains. Particularly as the system approaches terminal differentiation, the regulatory task is to respond at the appropriate rate and at the right time to multiple regional activators. Most of the mammalian *cis*-regulatory elements which have been characterized in order to understand how cell type-specific genes work display the features of "postspecification" modules. Indeed, the majority of *cis*-regulatory elements in the genome are likely to be of this type.

Though postspecification *cis*-regulatory elements do not have to perform complex boundary-setting functions, they still need to process information and integrate various inputs. For example, they often function combinatorially, i.e., they run only when several factors are present at once. Combinatorial response means integration of diverse inputs, as for example when factors to which these elements respond are the products of different upstream developmental processes. In addition, such regulatory elements often need to direct a kinetic transcriptional program. A rising rate, a peak rate, and a declining rate commonly succeed one another in differentiation processes. Postspecification *cis*-regulatory elements always depend on diverse interactions, at multiple target sites, just as do elements that program specification functions. This is so even though particular cell type-specific regulators can often be identified that seem so important for activation that it is easy to ignore all else.

Some Examples

The batteries of genes that are expressed specifically in muscle, in pituitary, and in eye lens of amniotes suffice to illustrate the main points. In each case these batteries are called into play downstream of pattern formation processes by which are defined the fields of cells giving rise to the respective morphological domains of expression. In each of these cases the specification processes are typical: they involve progressive signaling events that establish bounded transcriptional domains, and they require both repressive and positive regulatory interactions (some accessible reviews are as follows: for muscle, Molkentin and Olson, 1996; Yun and Wold, 1996; Borycki and Emerson, 1997; Firulli and Olson, 1997; Schwartz and Olson, 1999; for pituitary, Treier *et al.*, 1998; Dasen and Rosenfeld, 1999; Ericson *et al.*, 1998; for lens, Cvekl and Piatigorsky, 1996). *cis*-Regulatory elements of many differentiation genes that function downstream of the specification process have been characterized. Several muscle-specific regulatory elements were illustrated in Fig. 1.5, and many more are known from genes encoding skeletal, smooth muscle, and cardiac muscle contractile proteins and enzymes (Arnone and Davidson, 1997; Zilberman *et al.*, 1998; Kovacs and Zimmer, 1998). Similarly, *cis*-regulatory systems controlling expression of genes encoding the endocrine products of pituitary cell types have been much studied (Dasen *et al.*, 1999; Andersen and Rosenfeld, 1994; Simmons *et al.*, 1990; Treier and Rosenfeld, 1996; Dasan and Rosenfeld, 1999), as have genes encoding the crystallin proteins of eye lens (Cvekl and Piatigorsky, 1996).

One striking generalization that emerges is that in each case several transcription factors synergistically and combinatorially establish the specification of gene expression; no one factor does it all. That is, the gene batteries are defined by shared sets of regulators, just as in the example already considered in Fig. 1.5. There we saw that several muscle gene regulatory elements share sites for Mef2, myogenic bHLH factors, Srf, and other factors. In pituitary genes the shared regulators are the homeodomain factor Pit1, and either Gata2, or factors of the nuclear receptor family. For lens crystallins they are Pax6, Sox family factors, and a number of others.

A second general aspect is that detailed cell type specificity within muscle, pituitary and lens is encoded by further combinatorial interactions. For example, the distribution of target sites for Pit1 and Gata2 factors distinguish thyrotrope, somatotrope, and lactotrope genes (Dasen *et al.*, 1999; this case is discussed at more length in Chapter 4). Similarly, *cis*-regulatory elements from lactotrope genes utilize Pit1 synergism with estrogen hormone receptor, and somatotrope genes utilize Pit1 synergism with thyroid hormone receptor (Andersen and Rosenfeld, 1994; Dasen and Rosenfeld, 1999). In specific cell types of the lens, combinations of Pax6 with different other factors promote the expression of different crystallin genes (Cvekl and Piatigorsky, 1996). The developmental origin of the tissue is reflected in the presence of *cis*-regulatory sites for the widely shared factors (e.g., Pit1 and Pax6). Superimposed on this is the finer scale target site code that determines cell type specificity by the requirement for additional interactions with factors of more confined distribution.

Other programming devices are used by downstream *cis*-regulatory systems as well. For example, a bHLH repressor serves as a temporal controller in the muscle creatine kinase gene. This factor, "MyoR," binds to the same site as later occupied by the MyoD activator, but instead functions as a potent transcriptional silencer. It is evidently used to delay skeletal muscle differentiation during the primary phase of myogenesis (Lu *et al.*, 1999). Another common feature of *cis*-regulatory modules that control expression of battery-specific transcription factors is autoregulation, i.e., control of a gene encoding a transcription factor by its own product. Autoregulatory functions can of course only be positioned downstream of specification. An excellent example is at hand in the *pit1* gene. The *cis*-regulatory module that initially activates this gene early in pituitary development is separable from that required for maintenance of expression later on. The latter is driven by its own gene product, together with additional factors (Rhodes *et al.*, 1993; DiMattia *et al.*, 1997). Examples of special regulatory features like these could easily be multiplied, and each of the regulatory systems mentioned here is fascinating in detail. Just as every individual of a species is different in detail, so is every *cis*-regulatory element, even those belonging to the same gene battery. One thing can be said in general about the class of *cis*-regulatory element that we denote "postspecification": they may operate as slaves of the regulatory states erected in prior processes of specification, but they are surely not dumb slaves. They too refine and process the regulatory information they confront. It is they that provide the sharp, finely delineated patterns of cell type-specific gene expression which underlie subtle differences in cell type within a tissue, and which are among the terminal outputs of the whole genomic regulatory system.

At the Beginning of Embryogenesis

"Downstream of specification" need not be taken to mean "later in development." Specification, whenever it occurs, generates a specific spatial array of nuclear transcription regulators, and embryogenesis begins as well as ends with specification functions. At the outset the regulators are of maternal provenance (Davidson, 1968, 1986). In order to produce regionally distinct regulatory states, these regulators must be either localized or activated in the domains of the egg or very early embryo (see Davidson, 1986, 1989, 1990 for review of this phenomenon throughout the bilaterians). The job of the *cis*-regulatory systems that "read" and directly respond to the initial patterns of maternal spatial regulators is to transduce these patterns into the first tiers of zygotic gene expression.

A *Strongylocentrotus cis*-regulatory element that responds directly to the localized nuclear prevalence of maternal activators is shown in Fig. 2.5. The gene is called *SpAn*, and it encodes a metalloprotease expressed only in the animal tiers of blastomeres, as shown by *in situ* hybridization (Reynolds *et al.*, 1992). Expression begins as early as the eight-cell stage. A diagram of the single *cis*-regulatory

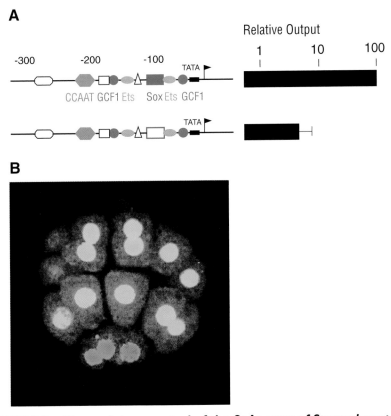

FIGURE 2.5 *Cis*-regulatory control of the *SpAn* gene of *Strongylocentrotus*. *SpAn* encodes a metalloprotease, the gene for which is transcribed in cleavage and blastula stages. Expression is confined to nonvegetal blastomeres, i.e., it is not expressed in the cells that give rise to skeletogenic mesenchyme and sometimes the adjacent portions of the overlying prospective endomesoderm. After blastula stage the gene is silenced. (A) *Cis*-regulatory module, and effect of mutation in SoxB1 site. The wild-type *cis*-regulatory element necessary and sufficient to recreate the normal spatial system is shown at top, with transcription factors indicated (data from Kozlowski *et al.*, 1996; Kenny *et al.*, 1999). The relative activity of the expression vector was measured by assay of CAT enzyme, as shown at right on a log scale. When the SoxB1 site was mutated (open green box, below), only 6% of the wild-type level of expression is retained. [(A) Adapted from Kozlowski *et al.* (1996) *Dev. Biol.* **176**, 95–107.] (B) An explanation for the initial spatial pattern of SpAn expression: distribution of SoxB1 protein in embryo nuclei at 5th cleavage. A confocal microscope image is shown, indicating location of an α-SoxB1 antibody (green) and of nuclear DNA (orange). The transcription factor is confined to nonvegetal cell nuclei (yellow) [(B) From Kenny *et al.* (1999) *Development* **126**, 5473–5483 and The Company of Biologists, Ltd.]

module which controls *SpAn* expression is given in Fig. 2.5A (Kozlowski *et al.*, 1996). Four different transcriptional activators contribute to expression, and no negative regulatory interactions can be detected in this system. The SoxB1 factor is the key regulator, and if its target sites are obliterated the activity of a *SpAn* expression construct is nearly abolished (Fig. 2.5A). Furthermore, if by incorporation of the Engrailed repressor domain, the SoxB1 factor is converted into a repressor and this form is translated in the egg the endogenous *SpAn* gene is silenced (Kenny *et al.*, 1999). Figure 2.5B dramatically displays the localization of the SoxB1 factor at the 32-cell stage. It is present in all of the blastomere nuclei, but only at low levels in the nuclei of the skeletogenic blastomeres at the vegetal pole of the embryo. After this, zygotic expression of the *soxB1* gene further strengthens the nonvegetal pattern of nuclear SoxB1 factor. Additional evidence indicates that another of the regulators that animate the *SpAn* gene is also localized to nonvegetal nuclei. The zygotic genes that encode these factors either auto-and/or cross-regulate, so as to lock in the initial nonvegetal pattern shown for the maternal SoxB1 factors in Fig. 2.5B (reviewed by Angerer and Angerer, 2000).

The *cis*-regulatory elements that interpret the gradient of the Bicoid (Bcd) regulator in early *Drosophila* embryos provide another example. As noted previously, the *bcd* gene is expressed during oogenesis and its maternal mRNA is localized at the anterior end of the egg, from which there is generated an anterior-to-posterior gradient of Bcd protein (Driever and Nüsslein-Volhard, 1988). The *cis*-regulatory element of the *knirps* (*kni*) gene is organized to permit it to respond to the low levels of Bcd protein in the future abdominal region of the egg, toward the posterior pole (Rivera-Pomar *et al.*, 1995). The particular *cis*-regulatory feature that is responsible is the presence of six closely apposed Bcd target sites, organized in face-to-face pairs so as to promote cooperative binding of Bcd. If this arrangement is subtly altered by inversion of two of the sites, expression constructs fail to express in the posterior regions of the egg (Burz *et al.*, 1998).

These are examples of *cis*-regulatory elements which perform the tasks of initiating zygotic control of development. Such elements are built to respond to the initial distributions of maternal factors in the blastomere nuclei.

Polyfunctional Downstream Modules

Some genes encode proteins that are required in several different kinds of cells, the products of several different specification pathways. The *cyIIa* cytoskeletal actin gene of *Strongylocentrotus* is an interesting example in which diverse phases of expression depend on a single *cis*-regulatory module. In the embryo this gene is expressed transiently in the invaginating archenteron and then in migrating mesenchyme cells of several kinds, skeletogenic and otherwise (Arnone *et al.*, 1998). Finally it is expressed stably in the midgut and hindgut endoderm of the late embryo and larva. All of the phases of this relatively complex pattern of expression occur after the specification and initial differentiation of the cell types that apparently require its gene product. A clue to the use of this particular gene might be that the transient aspects of its expression all occur at the point when the expressing cells become motile, while its terminal phase is simply endoderm-specific. The whole embryonic pattern of *cyIIa* expression is controlled by a single 450 bp *cis*-regulatory module. This contains target sites for at least eight different factors, most of these sites occurring multiply (Arnone *et al.*, 1998; Martin *et al.*, 2000). One of these factors specifies expression in hindgut and midgut. Another directs expression to all migrating mesenchyme cell types. This includes both skeletogenic and other intra-blastocoelar mesenchymal cell types, despite the entirely separate lineage origins of the skeletogenic cells (Cameron *et al.*, 1987). Most strikingly, just as for the *Span* gene, there are no spatial repressors in the *cyIIa* regulatory system. The positive regulators of the *cyIIa* gene derive from several different prior states of specification which are of distinct embryological origin, and it uses its machinery to respond to these diverse regulators: it is a polyfunctional control device.

The contrast with the regulatory organization of *cyIIIa*, its sister cytoskeletal actin gene, is revealing. *cyIIIa* utilizes two different *cis*-regulatory modules, each of which requires spatial repressors to properly confine its locus of expression (Fig. 2.2). Yet its pattern of expression is much simpler than that of the *cyIIa* gene, in that under normal circumstances it is confined to a single polyclonal domain and cell type, the aboral ectoderm. But the job the *cyIIIa* regulatory apparatus has to do is more complicated: it participates in a specification process, while the *cyIIa* system simply obeys orders, so to speak. The total complexity of each of the *cyIIIa* modules and of the single *cyIIa* embryonic module is more or less equivalent, in terms of number of diverse factors each employs. Perhaps there is a natural limit on this, something less than eight factors per module (Arnone and Davidson, 1997). In any case the comparison serves to illustrate a hopeful thought. This is that someday we will be able to determine the kind of role a *cis*-regulatory element performs in the overall network by inspection of the nature and internal organization of its target sites.

THE "POWER" OF THE *CIS*-REGULATORY MODULE

Their internal designs endow *cis*-regulatory elements with the power to generate unique spatial interpretations from what is often rather sloppily graded input information, or from overlap of individually crude regulatory domains. The developmental process depends on this property of *cis*-regulatory systems, and thus on their exact DNA sequence features. Understanding development obviously requires appreciation of these structure/function relationships. But that is not all: these structure/function relationships also define the evolutionary problem of how given *cis*-regulatory elements originated in the genomes of taxa to which they are specific.

Diverse *cis*-Regulatory Outputs from a Simple Input

The declining ventral-to-dorsal gradient of nuclearized Dl factor in *Drosophila* embryos can be regarded as a simple input, in that it appears to a first approximation to have a diffusion-limited form, and it has a unique, localized source. This gradient results from the activation of the Toll receptor along the ventral surface of the egg, initially triggered by the regional proteolytic activation of an extracellular ligand (Stein and Nüsslein-Vollard, 1992). Diffusion of the Toll pathway signaling components ultimately results in the phosphorylation and then degradation of a protein to which the maternal Dl factor is bound in the cytoplasm. It is thereby released for transit into the nuclei. In this way a graded ventral-to-dorsal concentration cline of nuclear Dl protein is produced (Anderson *et al.*, 1985; Steward and

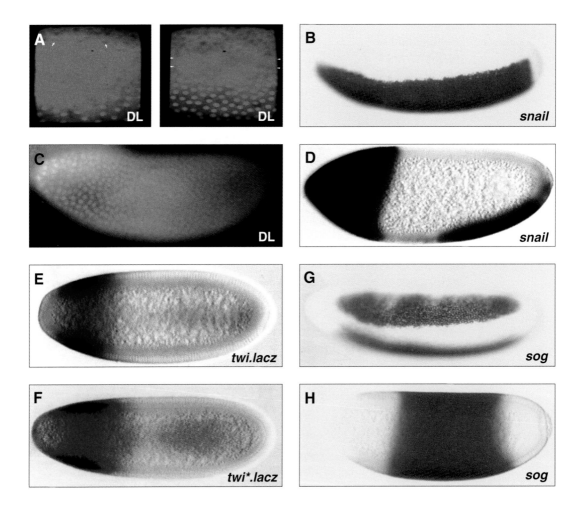

Govind, 1993; Steward, 1989; Rusch and Levine, 1996). This is revealed by immunological staining in Fig. 2.6A (Rushlow *et al.*, 1989).

We have already met two *cis*-regulatory elements that respond in distinct ways to this simple Dl input, *viz* the *zen* repressor element, and the *rho* neuroectoderm element (Fig. 2.3). But these are only two of a series of diverse *cis*-regulatory designs, all of which enable their genes to "read" the Dl gradient, and each of which produces a different output (Jiang and Levine, 1993; reviewed by Rusch and Levine, 1996; Huang *et al.*, 1997). The genes that encode the transcriptional regulators Twist and Snail (Sna) are expressed only in the ventral mesodermal domain: *twist* because its *cis*-regulatory element includes a series of low-affinity binding sites for the Dl activator so that they are occupied only at high Dl concentrations; and the *sna* gene because its *cis*-regulatory element includes both low-affinity Dl target sites and also sites for the positively acting Twist factor. Endogenous *sna* expression in a normal egg is illustrated in Fig. 2.6B. The

---

FIGURE 2.6 Diverse *cis*-regulatory responses to normal and reoriented gradients of Dorsal in the *Drosophila* embryo. (A) The endogenous ventral-to-dorsal gradient of nuclear Dorsal (Dl) transcription factor. The factor is displayed in syncytial embryos by fluorescence immunocytology. The left panel is 12th cycle; nuclearization has just begun. The right panel is 13th cycle. Higher nuclear Dl concentrations are ventral. Nuclei lacking Dl appear dark, and solid red indicates cytoplasmic Dl. [(A) Adapted from Rushlow *et al.* (1989) *Cell* **59**, 1165–1177.] (B) Endogenous *snail* expression in ventral mesodermal domain displayed by *in situ* hybridization in a normal embryo. *Snail* expression pattern depends on a *cis*-regulatory element that contains sites for Dl and Twist activators (Ip *et al.*, 1992a). (C) Reoriented anterior-to-posterior gradient of nuclearized Dl, displayed in a late syncytial-stage embryo by yellow-green immunofluorescence (cf. A). The view is dorsolateral, and the highest Dl concentration is anterior (the endogenous Dl gradient is invisible). In the anterior third of the egg the Dl factor is almost entirely localized to the nuclei; it is partly nuclear and partly cytoplasmic in the middle region; and it is largely cytoplasmic at the posterior end. (D) Expression of *snail* gene visualized by *in situ* hybridization, in egg with an anterior-to-posterior Dl gradient superimposed on the endogenous gradient (cf. B). (E) Anterior expression of a *lacz* construct driven by *cis*-regulatory elements containing two low affinity sites for Dl from the *twist* gene, visualized by hybridization of *lacz* mRNA. (F) Anterior expression of a derivative *twist lacz* construct driven by two copies of the element in (E), but altered so that the Dl sites bind the factor with higher affinity (*twi**); the anterior cap of expression now extends further in the posterior direction. (G) Expression of endogenous *short gastrulation* (*sog*) gene in a normal embryo, displayed by *in situ* hybridization. The gene is expressed in a broad pair of lateral stripes overlying the mesodermal domain. (H) Expression of *sog* in an embryo bearing a reoriented Dl gradient, as in (C). In this egg the endogenous gradient has been prevented from forming by mutational interference with the endogenous Toll signaling process. [(B–H) From Huang *et al.* (1997) *Genes Dev.* **11**, 1963–1973.]

expression of the *rho* gene higher up on the embryo (i.e., more dorsally), where nuclear Dl concentrations are lower, is due not only to higher affinity Dl sites, but also to the inclusion of sites for additional bHLH factors that synergize with D1 (*viz,* Scute and Daughterless; see Fig. 2.3A). The *cis*-regulatory elements of the *twist, sna,* and *rho* genes include the specific apparatus that is necessary and sufficient for their diverse interpretations of the Dl gradient.

The autonomous responses of the *cis*-regulatory elements of these and other genes have been demonstrated, very dramatically, by redirecting the Dl gradient with respect to the egg axes (Huang *et al.,* 1997). A gene encoding a constitutively active Toll receptor was introduced under the control of a maternally active *cis*-regulatory element, and the encoded mRNA was also equipped with the 3' trailer sequence of the bcd mRNA, which mediates mRNA localization to the anterior pole. The result is an anterior-to-posterior gradient of nuclear Dl. This is seen by immunofluorescence in Fig. 2.6C (compare Fig. 2.6A). *cis*-Regulatory systems responding to Dl which normally generate stripes of expression parallel to the A/P axis now produce stripes parallel to the D/V axis. Expression of the endogenous *sna* gene in such an egg is shown in Fig. 2.6D. An anterior cap of *sna* expression appears, and indeed these cells are respecified as mesoderm (Huang *et al.,* 1997). Anterior expression is superimposed on the normal ventral *sna* mRNA stripe. However, note that a gap appears between the ectopic and the original *sna* expression domains, betraying the existence of a repressor which is apparently also under Dl control, and which delimits the distal boundary of *sna* expression (i.e., what would normally be the dorsal boundary). Figure 2.6E illustrates the behavior of a different expression construct bearing the low affinity Dl-binding sites from the *twist* gene, in an embryo in which the highest levels of nuclear Dl are at the anterior end: the result is an anterior cap of expression. But if the sequences of these Dl target sites are altered so that they bind Dl with higher affinity, expression of the construct spreads in the posterior direction (Fig. 2.6F). A further example is given in Figs. 2.6G and H. The normal lateral band of expression of the endogenous *short gastrulation* (*sog*) gene, is shown in (G); and when the Dl gradient is reoriented to form an anterior to posterior cline, a circumferential band of *sog* expression instead appears (H; in this case in an embryo lacking the endogenous D/V Dl gradient). These examples illustrate a profound principle of development: diversity in the structure of *cis*-regulatory elements is the cause of diversity in gene expression patterns that arise in response to much less diverse prior regulatory states.

Direct Integration of Noncoincident Spatial Inputs

There is a class of *cis*-regulatory elements that function only where two different transcription factors overlap, each representing a different prior regulatory pattern. In this way these elements create new domains of gene expression. Such elements are "wired" so that they are incapable of responding adequately to either input alone (at normal levels of input activity), but when both are present they stimulate meaningful transcription. In biological terms, these elements are used to integrate different developmental processes. For instance field "A" could arise as a stripe of gene expression along the A/P axis of an organism, while field "B" is independently set up orthogonal to field A, along the D/V axis. Where they intersect a gene encoding a transcription factor required for specification of a new population of cells is activated. Or field A might define the progenitors of a new structure while field B overlaps field A only in its anterior portion, thus defining this region of the structure. The spatial coordinates of many developmental processes in both insects and vertebrates appear to be set in this way, though only a few have been analyzed at the *cis*-regulatory level.

In *Xenopus* embryos a gene encoding a homeodomain transcription factor called Siamois is activated on the future dorsal side when transcription resumes at the midblastula stage. Siamois is required for specification of the Spemann organizer (Lemaire *et al.*, 1995; Carnac *et al.*, 1996; reviewed by Kimelman, 1999). Its transcription is confined to the vegetal dorsal quadrant by a combination of positive and negative interactions. The gene is repressed on the ventral side by a ubiquitous maternal Tcf factor (Tcf3) which acts by way of the Groucho corepressor. Activation requires two spatially confined positive inputs that function synergistically in the *siamois cis*-regulatory element where they overlap (as well as a third positive regulator that is ubiquitous). These are a vegetal input downstream of a TGFβ signaling system; and a dorsal input consisting of a β-catenin-Tcf3 complex. β-catenin interferes with Gro function and in complex with Tcf3 acts instead as an activator (Brannon *et al.*, 1997; reviewed by Kimelman, 1999). β-catenin activity is confined to the dorsal side by a mechanism which is triggered by the cortical rotation toward that side (reviewed by Moon and Kimelman, 1998). Immediately downstream of the *siamois* gene lie other regulatory genes, one of which is *goosecoid*. The *goosecoid* (*gsc*) gene is expressed in cells that will form the head organizer, and which themselves are progenitors of dorsal mesoderm (Harland and Gerhart, 1997). Its locus of expression is also defined by integration of two inputs within the relevant *gsc cis*-regulatory element: this element responds directly to a vegetal TGFβ family signal, and also to the dorsal input provided by Siamois or another similar factor, Twin (Watanabe *et al.*, 1995; Laurent *et al.*, 1997). Both the *siamois* and *goosecoid cis*-regulatory elements thus include mechanisms for integration of vegetal and dorsal signals (Moon and Kimelman, 1998; Kimelman, 1999). The same form of "and" logic continues to be utilized downstream. *Xnr3*, a gene encoding a TGFβ family ligand expressed in the Spemann organizer, is also controlled by such a system. In this case the inputs, both required, are a Tcf factor plus another regulator that is expressed zygotically only at gastrula stage (McKendry *et al.*, 1997). Thus we have a series of multiple input systems, each adding spatial, or spatial plus temporal, regulatory value to what was there before.

Integrative summing of diverse spatial regulatory inputs is utilized over and over in *Drosophila* development. Pair-rule and gap genes are regulated in this way in the syncytial embryo (Baumgartner and Noll, 1991; Häder *et al.*, 1998; Rivera-Pomar and Jäckle, 1996; Wu and Lengyel, 1998). Another example is the integration of Hedgehog and Wingless signaling inputs by a *dpp* enhancer active in wing imaginal discs (Hepker *et al.*, 1999). *cis*-Regulatory elements that are targets of *hox* gene cluster regulators often display this same kind of program feature. For example, the *teashirt* gene, which encodes a Zn-finger transcriptional regulator required for trunk specification, is expressed in trunk epidermis in response to Antennapedia plus an epidermis-specific factor; and in thoracic mesoderm in response to Ultrabithorax (Ubx) plus a mesoderm-specific factor (McCormick *et al.*, 1995). Similarly, the midgut regulatory element of the *dpp* gene responds positively to the combined inputs of Ubx plus a visceral mesoderm-specific factor (Capovilla *et al.*, 1994; Manak *et al.*, 1995).

A visually beautiful example of *cis*-regulatory integration of geometrically distinct input patterns is illustrated in Fig. 2.7 (Kim *et al.*, 1996, 1997a,b). The *vestigial* (*vg*) gene encodes a transcriptional coactivator which works together with the Scalloped transcription factor (Halder *et al.*, 1998). The *vg* gene is expressed and is required for growth in all imaginal disc cells of the developing wing pouch. But its control is more mosaic than meets the eye, in that its expression throughout the wing pouch depends on two separate regulatory modules. A "boundary enhancer" initially sets up expression along the D/V boundary, in response to Notch signaling, and along the A/P boundary as well. The cross-like pattern generated by a *lacz* construct controlled by this element is shown in Fig. 2.7A. Expression in the remainder of the wing pouch is controlled by the "quadrant enhancer." This element integrates two signals, *viz*, a Dpp signal emanating from the A/P boundary, and a second signal emanating from D/V boundary. A Mads box transcription factor which mediates the Dpp signal response binds directly within this element, and its interaction is required for function (Kim *et al.*, 1997a,b). The integrative response of the quadrant enhancer can be seen pictorially in Figs. 2.8B and C: note the initial pattern of expression, which begins just where the A/P and D/V boundaries intersect, and then spreads concentrically outward. The geometry of the expression pattern described by the quadrant enhancer follows from its activation in response to dual, orthogonal inputs.

cis-Regulatory elements that integrate inputs can be thought of as control units which execute logic functions. But in essence, that is what all *cis*-regulatory elements do: they are genomic sequences that specify logic operations.

A *cis*-Regulatory Logic Device

Portions of the *endo16 cis*-regulatory system of *Strongylocentrotus* are to date the most extensively explored of any, with respect to the functional meaning of each interaction that takes place within them. What emerges is almost astounding: a network of logic interactions programmed into the DNA sequence that amounts essentially to a hardwired biological computational device.

As will be recalled, *endo16* is expressed in the endoderm of the embryo. At first glance its regulation appears a rather typical tissue-specific process, and indeed, it probably is rather typical. But the number and variety of regulatory transactions required for this gene to achieve its expression profile is surprising. The spatial

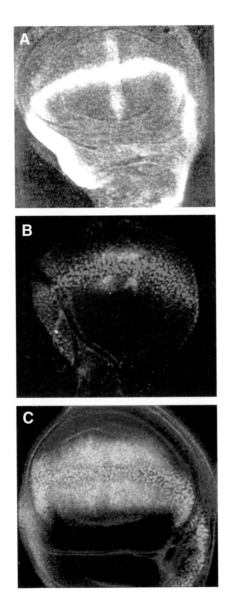

FIGURE 2.7   Integrative control of the *Drosophila vestigial* gene. (A) Expression of a *cis*-regulatory construct in the wing blade region of a third instar wing imaginal disc. In this construct a *lacz* reporter is controlled by the "boundary enhancer," a *cis*-regulatory module from the 2nd intron of the *vestigial* (*vg*) gene. The boundary enhancer initiates expression along the D/V boundary in response to Notch signaling. The expression pattern marks the D/V boundary (horizontal, ventral to top) and the A/P boundary (vertical, anterior to left). *lacz* expression is shown by

(Continues)

pattern of *endo16* activity is shown in Fig. 2.8A (Ransick *et al.*, 1993). It is not known whether the protein has earlier functions as well, but the gene is activated long before there is a midgut, or any gut at all, in mid-to-late cleavage, during or soon after the initial processes of endoderm specification (Godin *et al.*, 1996). Figure 2.8A1 shows the initial expression of *endo16* in the vegetal plate of a blastula-stage embryo, which is comprised of the cells that will give rise to the endodermal and mesodermal cell types of the embryo, except for skeletogenic cells. The skeletogenic progenitors (plus a few cells of the future coelomic sacs) appear as the nonexpressing patch within the ring of expressing endomesodermal cells in Fig. 2.8A2. We can see here that there are two boundaries that must be established: one within, between the endomesodermal cells and the skeletogenic cells; and one without, between the endomesodermal cells and the overlying ectoderm. It will not surprise that these boundaries depend on spatial repression (Yuh and Davidson, 1996). The subsequent panels of Fig. 2.8A show that *endo16* is later expressed throughout the archenteron, remaining silent in the skeletogenic mesenchyme and the surrounding ectoderm. Expression is then extinguished in the foregut, and later in the hindgut, while intensifying in the midgut, its terminal locus of activity (Nocente-McGrath *et al.*, 1989; Ransick *et al.*, 1993).

Figure 2.8B is a protein-binding map of the *endo16* cis-regulatory system, which is included in a 2.3 kb fragment of DNA that suffices to reproduce the normal pattern of expression when linked into an expression vector (Ransick *et al.*, 1993; Arnone *et al.*, 1997). Thirteen different proteins bind with high specificity within this region (Yuh *et al.*, 1994), and it has been divided experimentally into six functional regions (G–A). The role played by each region is indicated briefly below the map (Yuh and Davidson, 1996; Yuh *et al.*, 1996). The boundary beyond which expression is precluded in the skeletogenic cells requires

---

immunofluorescence. [(A) From Kim *et al.* (1996) *Nature* **382**, 133–138; copyright Macmillan Magazines, Ltd.] (B) Expression of *lacz* construct controlled by the quadrant enhancer, early third instar. This enhancer integrates distinct inputs, one emanating from the A/P boundary of the wing blade portion of the imaginal disc, and the other from the D/V boundary. The A/P boundary input depends on Dpp, which is expressed along this boundary (see text), and is mediated by target sites for Mads transcription factors. *lacz* is shown in blue; endogenous *vg* gene expression in red. At this early stage *vg* expression is mainly due to the boundary element, and activity of the quadrant element, blue and purple, is just beginning, at the intersection of D/P and A/P boundaries. [(C) From Kim *et al.* (1996) *Nature* **382**, 133–138; copyright Macmillan Magazines, Ltd.] (D) Final pattern of quadrant enhancer expression, visualized as in (B). The Lacz product now fills all four quadrants of the wing blade (blue overlying red). The dual spatial source of these inputs is clearly implied by the central location of the early pattern shown in (A). The total pattern of *vg* expression in the wing blade region of the disc is the sum of the outputs of the quadrant and boundary enhancers. [(D) From Kim *et al.* (1997a) *Nature* **388**, 304–308; copyright Macmillan Magazines, Ltd.]

repressive interactions in the DC region; and for the boundary with the overlying ectoderm interactions in both regions E and F are required. The protein responsible for repression in the F region is a factor of a class which is usually associated with signaling systems, *viz*, a Creb factor, so the boundary with the ectoderm is at least in part likely to be established by a signaling interaction (as implied by other evidence as well; Davidson *et al.*, 1998). The *endo16* system includes three positive regulatory regions. These are the relatively unimportant Module G, which acts as a booster; Module B, which is solely responsible for the late rise in midgut-specific expression; and Module A. Module B is activated by an endoderm-specific regulator (UI), the same that causes *cyIIa* gene expression in the gut late in development (Arnone and Davidson, 1997). Module A acts as the central processing unit for the whole upstream system, G–B, and it is on structure/function relations within Module A and its intricate linkages to Module B that we now focus.

Module A carries out multiple functions. The basal promoter (Bp) of the *endo16* gene consists of the sites where the transcription apparatus assembles plus a few proximal target sites (Fig. 2.8B). The Bp is entirely promiscuous with respect to the inputs it will service, and it is almost inactive without upstream inputs (Yuh *et al.*, 1996; R.A. Cameron and E.H. Davidson, unpublished data). The only inputs the Bp receives from the whole *endo16* *cis*-regulatory system are normally channeled through Module A; their processed result is the output of Module A. The value of this output determines whether transcription of the gene will occur, and if so at what rate, in every cell throughout embryogenesis (Yuh *et al.*, 1996, 1998). Module A also serves as a terminus for the upstream modules that mediate spatial repression, *viz*, E, F, and DC, so that in its absence they do not work. The outputs of these repressor subsystems can also be considered inputs to Module A. As we have already seen, the skeletogenic cell repressors in region DC do not act as transcriptional silencers (or long-range repressors; Fig. 2.4), but are rather dedicated to control of the positive regulatory activity of Module A. The positive regulatory function of Module A is mediated by an Otx factor which binds within it (Yuh *et al.*, 1998; Li *et al.*, 1999), and the repressive controls on spatial expression that Module A mediates are necessary because Otx is active in many regions of the embryo (Mao *et al.*, 1996; Li *et al.*, 1997). However, later in development the DC, E, and F regions can all be discarded, as can Module G, and a construct consisting only of B and A runs accurately in the midgut at almost the same rate as does the whole system. By this stage the system utilizes only the gut-specific positive input of Module B. But this requires two additional functions of Modules A and B. The first is a switch which is sensitive to the level of input of the positive regulator of Module B, and when this becomes significant the switch turns off the input of the Otx regulator, ensuring that only the input of Module B will count. The second is an amplification function by which Module A steps up the amplitude of the Module B input. These various functions of Modules A and B could never have been enumerated except by assessment of the functional meaning of each of its target sites (Yuh *et al.*, 1998, 2000).

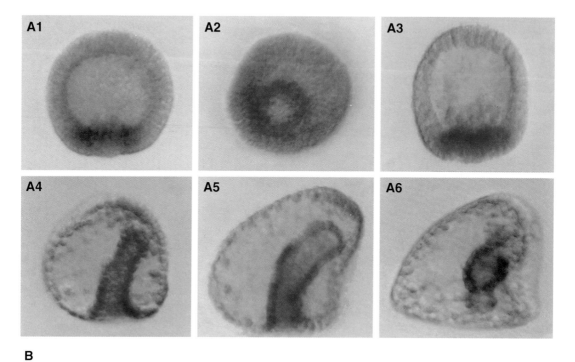

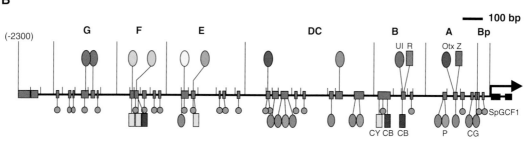

G   Positive booster

F ⎫
  ⎬ Repression in adjacent ectoderm
E ⎭

DC Repression in skeletogenic mesenchyme

B   Expression in midgut of late embryo

    Controls late rise in expression

    Activates switch resulting in exclusive use of its own input

A   Expression in vegetal plate in early embryo

    Sole communication to BTA for whole system

    Synergistic amplification of B input

    Transduction of FE, DC repression

FIGURE 2.8 (*Continued*)

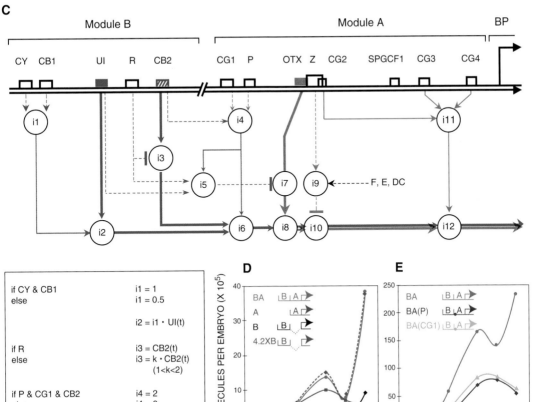

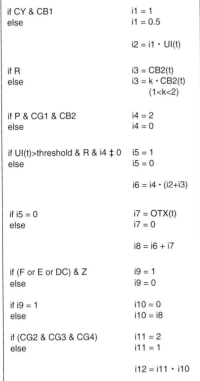

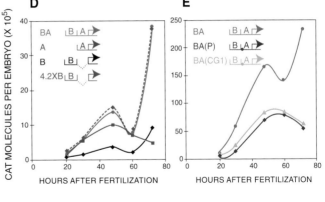

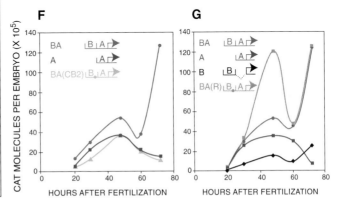

FIGURE 2.8 *cis*-**Regulatory logic in the *endo16* gene**. (A) *In situ* hybridization display of *endo16* expression pattern in *Strongylocentrotus* embryos (Ransick *et al.*, 1993). The gene is expressed in the vegetal plate (A1), but not the skeletogenic cells and small micromeres located at the vegetal pole in the early blastula (A2), nor is it active in the skeletogenic cells as they ingress (A3). After gastrulation it is expressed throughout the archenteron (A4). Expression is later shut off in the foregut and secondary mesenchyme (A5), and the hindgut (A6). The terminal pattern of expression is confined to midgut. [(A) From Ransick *et al.* (1993) *Mech. Dev.* **42**, 117–124; copyright Elsevier Science.] (B) Map of protein interactions in the *endo16* cis-regulatory system. The 2300 bp DNA sequence indicated as the horizontal line is necessary and sufficient to provide accurate expression of a reporter construct. Proteins that bind at unique locations are shown above the line, and proteins that bind at several locations are indicated below (Yuh *et al.*, 1994). Different colors indicate distinct proteins. "G–A" indicate the functional regions or modules discussed in text. Below the map the functions of each module are briefly indicated (Yuh and Davidson, 1996). [(B) Adapted from Yuh *et al.* (1994) *Mech. Dev.* **47**, 165–186; copyright Elsevier Science.] (C) Logic model for Modules B and A (Yuh *et al.*, 1998, 2000). The regulatory DNA of *endo16* is shown as a horizontal strip at the top of the diagram. The individual binding sites are indicated by labeled boxes. Module B and its effects are shown in blue; Module A and its effects are shown in red. Logic interactions (i) are indicated by numbered circles. Each represents a specific regulatory interaction modeled as a logic operation. Note the two types of regulatory input: time-varying interactions (colored boxes) which determine the temporal and also spatial pattern of *endo16* expression, and time invariant interactions (open boxes) which affect the level of expression and control intra-system output and input traffic. In the diagram interactions that can be modeled as Boolean are shown as dashed lines; those which are scalar as thin solid lines; those which are time-varying quantitative inputs as heavy solid lines. The individual logic interactions are defined in the set of statements below the diagram. Here statements of the form "If X," where X is the name of a target site, means that this site is present and occupied by the respective factor. If the site has been mutated (or if the factors were inactivated or eliminated) this is denoted by zero; or as the alternative ("else") to the site being present and occupied. The statements afford testable predictions of the output for any given mutation or alteration of the system. Briefly, the system works as follows: CB1 and CY1 interactions together (at i1) synergistically increase the output of the positive spatial and temporal regulator of Module B which binds at the U1 site. The output of the U1 subsystem is at (i2). An additional smaller time-varying positive input, which peaks at about 40 h, is generated by the interaction at CB2 (i3). An interaction at R is required for the BA intermodule input switch, which shuts off Otx input (i5, i7), but this switch operates only if there is input from U1, and the CB2 site is present and occupied (i5). Furthermore, the proteins binding at the adjacent R and CB2 sites apparently interact, in that if the R site is mutated CB2 input (at i5) is somewhat enhanced. In Module A the CG1 and P sites together with CB2 in Module B are all required for linkage of Module B to Module A (i4), and for synergistic amplification (by a factor of about 2) of the Module B input (at i6). If the switch mediated by R does not function (i.e., in an R mutation) the summed input of Modules A and B at i8 is observed. If CB2, CG1, or P are mutated Module B is unlinked from Module A. That is, i4 = 0, so i6 = 0 and in this case i8 is just the output of the Otx interaction (at i7). If

The sites of Modules A and B are shown explicitly in Fig. 2.8C. This is a logic model that shows all the inputs into Modules A and B, and indicates explicitly the operations carried out on them by the various elements of the system (i.e., by the proteins bound at each target site and their cofactors, whatever they may be). Each function was established by measurement of the output of the system after mutating one or more target sites, or substituting synthetic target sites for the natural sequence (Yuh *et al.*, 1998; 2000). Examples of such experiments are reproduced in Fig. 2.8D–G.

The intimate linkages between Modules B and A are shown explicitly in Fig. 2.8C. Here we see how the output of Module B runs through Module A to the BTA, and therein is multiplied severalfold. Interactions with three proteins that bind adjacently, CB2 of Module B, and CG1 and P of Module A (Fig. 2.8C) are required for the linkage and amplification functions. A fourth interaction at the site R of Module B mediates the switch which cuts off Otx input into Module A when UI

---

the gene is in a location where the repressive interactions in any of the upstream F, E, or DC modules are operating, the sytem is shut off via an interaction at the Z site of Module A (i9, i10). Finally, the CG2, CG3, and CG4 sites combine (at i11) to boost whatever output is present (at i12) by an additional factor of 2. The result at i12 is transmitted to the basal transcription apparatus. [(C) From Yuh *et al.* (2000). In preparation.] (D–G) Expression of CAT reporter as a function of time after fertilization, in batches of 100 eggs per time point. The eggs were injected with expression constructs consisting of Modules B and A linked to one another and to the basal promoter as in the normal gene (BA; red curves); with a construct driven by Module B alone (B; black curves); with a construct driven by Module A alone (A; blue curves); or with constructs in which specific target sites in Modules B or A were mutated, as indicated by parentheses. In the color-coded cartoons in each figure the dots indicate the location of the mutated target sites given in the parentheses. (D) Wild-type constructs. The dashed line shows the time function generated if the B curve is multiplied by the factor 4.2 at each point. (E) CAT activity profile in batches of eggs injected with BA, with BA in which the P site had been mutated [BA(P); violet]; or with BA in which the CG1 site had been mutated [BA(GC1); pink]. The latter two time curves are almost identical to that generated by wild-type A module in isolation from any other modules. [(D, E) From Yuh *et al.* (1998) *Science* **279**, 1896–1902; copyright American Association for the Advancement of Science.] (F) CAT activity profiles generated by BA control; and by BA in which the CB2 site was mutated [BA(CB2); orange]. The profile of expression generated by this construct can be seen to be almost identical to that produced by Module A alone. As predicted in C: if CB2 = 0, i4 and i6 are 0, and the only input is from Module A, i.e., at i8 = i12). (G) Derepression of Module A when the intermodule switch function is cancelled by mutation of the R site of Module B [BA(R); green]. The output is the sum (at i8) of the enhanced CB2 input (i3), Otx input (at i7), plus the amplified input from Module B, including (at i3) the enhanced CB2 input: i.e., [A + i4 · (i2 + i3)]; see logic table. [(F, G) from Yuh *et al.* (2000). In preparation.]

input attains a certain level (i.e., functions i5 and i7, of Fig. 2.8C). Of the target sites in Module B one, UI, conveys the main spatial and temporal activation input. The remainder are ancillary to this input (CB1, CY1) or perform linkage and internal control functions (R, CB2). Figure 2.8D demonstrates the augmentation by Module A of the input from Module B: we see that (after very early in the profile, not visible on this scale) the output of a BA construct is equal to about a fourfold linear amplification of the output generated by Module B alone. The Module A output is also comparatively low, but has a different time course, as its activity dies away late in development. Figure 2.8E demonstrates that the functional linkage of Module B to Module A, and the step-up in activity, depend specifically on interactions at both the CG1 and P sites of Module A. If either is mutated the output of BA is about equal to that of A alone and has the same kinetic form; although the physical linkage between B and A is undisturbed in these mutants, the functional linkage is gone. Figure 2.8F similarly shows that if the CB2 site is mutated, the B-A linkage is again cut. Therefore, exactly as predicted, the output is just that of Module A alone, even though the remainder of Module B is present, including the site where the UI domain binds. In Fig. 2.8G an experiment is shown demonstrating the effect of mutation of the R site of Module B: now the switch which normally cuts off Otx input is inactive and the output is higher than normal, i.e., it is the sum of the normal output of BA plus that of A (see legend). Other experiments (Yuh *et al.*, 1998; 2000) show that the "Otx" site mediates the only positive spatial regulatory input within Module A itself; that the "Z" site is required to mediate the repressive inputs to Module A; that the CG2, CG3, and CG4 sites are required for communication of the system output with the Bp.

Modules A and B of *endo16*, servicing a garden variety gene, so to speak, are unlikely to be exceptional. The apparatus represented in Fig. 2.8C serves notice of the nature, and the complexity, of the logic processing functions carried out by *cis*-regulatory systems in general; and of the significance of their internal organization. There is also a lesson to be drawn in respect to the way these systems have to be treated. Some of the inputs into Modules A and B are time-varying functions, the amplitude of which should, to a first approximation, depend on target site occupancy (see legend); examples are the kinetic curves shown in Figs. 2.8D–G. Though others of the inputs can be considered as either present or not, a wholly Boolean analysis would be mythological.

The variety of *cis*-regulatory modules that we have touched on in this chapter have one thing in common: they are all information processing devices. Many genes are controlled by multiple *cis*-regulatory modules, and hence they require the services of multiple genomic information processors. The genome must contain tens of thousands of these elements. We have seen by example in this chapter just how their internal organization allows them to determine events at every stage of the developmental process.

CHAPTER 3

Regulation of Direct Cell Type Specification in Early Development

Organized increase in biological complexity is what the developmental process achieves. Our metric of complexity is the diversity of the programs of gene expression that are installed and executed as the embryo develops. The phrase "program of gene expression" implies coordination of a set of individual *cis*-regulatory systems, linked to one another in such a way that the genes they control do some developmental act that we can name: for instance, they carry out a specification function, or they generate a differentiated cell type. The programs fit the jobs they do, and indeed it is the program architecture that defines what kind of job it is to be. This chapter is about gene regulation programs that control processes of direct cell type specification early in embryogenesis. In the Bilateria the most widespread, basic, and simple means of creating an embryo out of an egg depends on these direct specification mechanisms.

THE BASIC PACKAGE FOR BILATERIAN
EMBRYOGENESIS: TYPE 1 SPECIFICATION PROCESSES

Bilaterian body plans are amazingly diverse, and yet a shared property of most of the clades included in Fig. 1.6 is the common set of mechanisms that they use for early embryonic development. The aim of their developmental strategy is to divide the three dimensional cytoarchitecture of the egg into polyclonal lineages of differentiating cells, at the earliest possible time. The differentiated cells directly form the morphological elements of the embryo, which become manifest after gastrulation, such as its ectodermal wall; its gut, mouth, and anus; its mesodermal cell types. The similarities in the processes used to effect this strategy, across the Bilateria, were not recognized until recently, because people were focused instead on phylotypic differences in embryonic anatomy, and on special features of the diverse juvenile body plans which emerge. But the similarities are of mechanism, not outcome: the same microcircuits can be used to build all manner of device. The term "Type 1 embryogenesis" is used for that process of early development which consists of direct cell type specification during cleavage (Davidson, 1991).

It is easy to summarize the telltale characters that bespeak Type 1 embryonic specification processes. The tempo and nature of early embryonic gene expression is of primary importance, and its control is the subject of much of this chapter. In Type 1 embryonic processes the zygotic genomes are activated at once, and throughout cleavage the blastomere nuclei remain transcriptionally active. For example in the sea urchin, where detailed transcription rate measurements were made, activation occurs even in the zygote pronuclei, prior to fusion, and the level of transcription is higher (per nucleus) by mid-late cleavage than ever again (reviewed by Davidson, 1986; 1990). There is no such thing as the "midblastula transition" characteristic of many vertebrate embryos, in which there is near complete transcriptional quiescence until thousands of cells have been produced.

A second earmark of these processes is activation, even during cleavage stages, of genes that encode differentiation proteins, and these genes continue to be expressed in the completed morphological structures of the embryo. Two examples from the sea urchin embryo that we have already considered (in Chapter 2) are the *endo16* and the *cyIIIa* genes. As will be recalled, the *endo16* gene encodes a protein secreted into the lumen of the midgut in the late embryo and feeding larva, but the gene is activated back at 7th cleavage or so, when the future endomesoderm lineages are segregated from lineages of other fate. Likewise, the *cyIIIa* gene encodes a cytoskeletal protein of the mature aboral ectoderm, and this gene is also activated during cleavage. When these two genes are first turned on, there is no gut and there is no aboral ectoderm, and at this point a microscopist could perceive nothing but an indifferent ball of cells. One might say that the timing of activation of differentiation genes such as these is essentially "as soon as possible," i.e., right after the relevant lineage segregations have occurred.

In the following we encounter other examples of a similar nature, in other organisms. They are revealing because they betray the basic point of Type 1 embryonic process: direct cell type specification in the cleavage-stage embryo.

The use of cell lineage in the cleavage-stage specification process is another important feature. Embryos of different species display a range in the variance of the lineage from individual to individual of the species; that is, in the relation between cell lineage and cell fate. At one extreme is the almost complete invariance of ascidian and *Caenorhabditis elegans* embryonic lineages and at the other is the less rigid sea urchin embryo lineage. But even in the sea urchin embryo, while the later boundary regions that separate certain territorial domains of cell fate are variable in exact lineage, the lineages of some parts of all territories, and of all parts of some territories, is invariant (Davidson *et al.*, 1998). Invariant lineage means, in essence, that given sectors of the maternal cytoarchitecture are inherited by given blastomeres, since the cleavage planes occur in fixed positions with respect to the cytoarchitecture of the egg. Thus invariant lineage is a positional device: founder blastomeres are specified in given positions. Their progeny occupy the same relative positions, since in Type 1 embryos there is no cell migration or motility until well after the whole initial tier of embryonic specifications has been installed. The corollary is that this kind of mechanism works mainly for small embryos consisting of small numbers of cells, that is, a few hundred or less at gastrulation. In other words, lineage-space relations can be maintained for only a modest number of blastomere divisions. So most embryos that display Type 1 processes are smallish embryos with an invariant early cleavage and a more or less invariant later cleavage. By the time cleavage is complete, the embryo consists of a mosaic of largely specified lineage subelements.

Some very important groups (to us), particularly the insects and the vertebrates, depart in important ways from Type 1 strategies. Insects develop at least in part syncytially, and in their syncytial domains they of course cannot use intercellular signaling for early development. Their embryos contain thousands of cells at gastrulation, compared to the few hundred (or less) typical of Type 1 processes of embryogenesis. Blastomere lineage plays almost no role in early insect development, and differentiated cells appear only following the spatial definition of the fields within which they are to arise. Vertebrates also produce embryos with relatively large numbers of cells. While yet in an incompletely specified state, large populations of these cells characteristically migrate widely within the embryo before their specific differentiated fates are assigned by regional inductive processes. Thus in contrast to embryos developing by Type 1 processes, vertebrate embryos also generally lack invariant relations between early cell lineage and given states of postgastrular differentiation. A glance at the phylogenetic chart in Fig. 1.6 shows that both the insect and vertebrate strategies must be derived, since all their current sister groups and cousins are clades that utilize Type 1 processes, and thus so must have done their ancestors (Davidson, 1991; Davidson *et al.*, 1995).

In Type 1 embryos specification occurs within the fixed spatial matrix that the cleavage-stage blastomeres constitute, by a combination of autonomous and conditional processes. Specification is here a cell-by-cell operation. A blastomere learns who it is (and what will be the differentiated state to be executed by its progeny) from the maternal components it has inherited (autonomous specification), and/or by means of signals from immediately adjacent blastomeres (conditional specification, see Davidson, 1990). What usually happens is that autonomously specified blastomeres which arise at one or another pole of the egg axes emit signals to adjacent blastomeres, and thence to one another as well. There is no detectable embryonic requirement for the kind of processes by which adult bilaterian body plans are constructed (these are the subjects of the following chapter), and which are also used in both vertebrate and insect embryogenesis (reviewed by Davidson, 1990, 1991). These are the regional specification processes used for defining developmental spatial domains out of large populations of indifferent cells, a problem not faced in Type 1 embryogenesis. Developmental usage of the *hox* gene complex provides a nice example of this dichotomy. As everyone knows, the *hox* complex is involved in regional specification mechanisms over and over again during adult body plan development in all Bilateria. But the whole *hox* complex as such is not needed for completion of the embryo per se in the sea urchin *Strongylocentrotus purpuratus* (Arenas-Mena *et al.*, 1998); nor for development of embryonic structures in the polychaete annelid *Chaetopterus* (Peterson *et al.*, 2000a); nor to produce an essentially complete embryo in ·*C. elegans* (Salser and Kenyon, 1994; Brunschwig *et al.*, 1999; Van Auken *et al.*, 2000); nor for anterior/posterior (A/P) patterning of the embryonic gut in this organism (Schroeder and McGhee, 1998). Development of sea urchin, polychaete, and nematode embryos up to their respective larval stages occurs by Type 1 specification processes that do not require the level of regulatory mechanism of which the *hox* gene complex is an important and almost ubiquitous component.

The nature of the genomic regulatory processes required for Type 1 embryogenesis is illustrated in the following. Our examples come from sea urchin, ascidian, and *C. elegans* embryos.

REGULATORY MECHANISM IN TERRITORIAL SPECIFICATION OF THE SEA URCHIN EMBRYO

The Definitive Territories of the Embryo

The zygotic mechanisms that lead to the formation of the sea urchin embryo can be defined as a set of territorial specification processes, followed by the downstream events of differentiation. By late cleavage each territory is a unique domain of gene expression, and the cells of each are the unique progenitors of a specific structure of the embryo. The territories constitute a spatially simple map, as shown in Fig. 3.1A–D. Except for some interterritorial boundary regions in which individual cell fates are sorted out only later, there are no overlapping domains of cell fate because the territories are all specified during cleavage, before there is any cell migration. Initially five territories are set up, *viz*, the oral and aboral ectoderm territories; the skeletogenic mesenchyme territory; the endomesodermal or "*veg₂*" territory; and the "small micromere" territory, the cells of which in the late embryo become incorporated in the coelomic pouches. By the end of cleavage the endomesodermal territory has been further divided into a peripheral endodermal territory, which produces most of the gut or archenteron, and a central mesodermal territory, which produces several mesenchymal cell types and the remainder of the coelomic pouches. These territories and the structures to which each gives rise are indicated by color in the diagrams of Fig. 3.1.

The simplicity of the outcome is the key to the process: there are only about 12 differentiated cell types in the completed embryo or larva, and this consists mainly of single cell-thick structures. Though it is composed of a total of only a couple of thousand cells (1800 in *S. purpuratus*, the species after which Fig. 3.1 is drawn), this small organism is capable of feeding and of prolonged free-living existence in the wild. As shown in Fig. 3.1E–G, the completed embryo is equipped with a tripartite gut, consisting of a foregut which is connected to the mouth, a large midgut or stomach, and a hindgut terminated at the site of the original blastopore. The coelomic sacs protrude bilaterally from either side of the foregut (Fig. 3.1E, F). The blastocoel and its mesenchymal cell types are enclosed by the squamous aboral ectoderm, and at the top by the oral ectoderm, which surrounds the mouth. Within the blastocoel are skeletal rods that maintain the triangular shape of the larva and extend its cilia-bearing arms. At the intersection of the oral and aboral ectodermal territories is the ciliated band. This consists of some neurogenic precursors and of specialized cells that produce a dense belt of long cilia. A few neurons are also generated from elsewhere within the oral ectodermal territory.

For what follows a brief description of the main events leading up to the definition of the embryonic territories will be useful (some further details are in the legend to Fig. 3.1; see reviews of Davidson *et al.*, 1998; Davidson, 1986, 1989; Wilt, 1987; and for the initial, "preterritorial" stages, Angerer and Angerer, 2000). At the outset of development the egg is polarized in the animal/vegetal (A/V) axis. This polarization is reflected in the diverse fates of the blastomeres inheriting given sectors of the maternal cytoarchitecture located along this axis. Thus oral and aboral ectoderm (green and yellow in Fig. 3.1) are formed from lineages arising from the upper (i.e., by convention toward the animal pole) two thirds or so of the egg, and endodermal and mesodermal cell types from the lower portion. An oral/aboral (O/A) polarization, which occurs after fertilization, sets up the second axis of the egg very early in development. This polarization has already taken place by third cleavage, since by then there have arisen founder blastomeres whose progeny will contribute exclusively to either oral or aboral lineages (Cameron *et al.*, 1987; Henry *et al.*, 1992; in *S. purpuratus* O/A specification is already set in train by 1st cleavage, as shown by Cameron *et al.*, 1989). Much evidence indicates the presence of a complex assemblage of maternal transcription

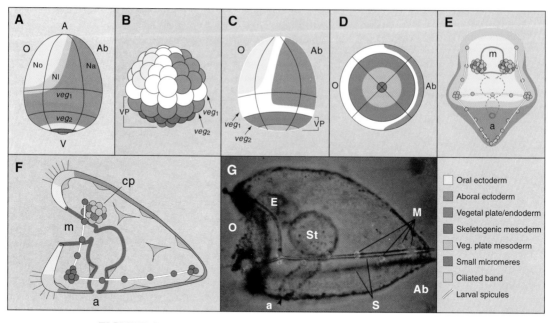

FIGURE 3.1   Territorial origins of major embryonic structures in the sea urchin.   (A) Fate map of the embryo, lateral view, projected on an external image of the blastula (~400-cell stage in *S. purpuratus*); see color key at lower right. The black lines indicate lineage compartments (Cameron *et al.*, 1997, 1993; Hörstadius, 1939; Logan and McClay, 1997; Ransick and Davidson, 1998; Ruffins and Ettensohn, 1993, 1996). No, Nl, and Na are 3rd cleavage blastomeres of the animal half (Cameron *et al.*, 1987); the *veg*₁ and *veg*₂ layers consist of the progeny of two 6th cleavage rings of eight cells each, which are sister cells; A, animal pole and V, vegetal pole; O, oral and Ab, aboral sides of the embryo. The skeletogenic (and barely visible small micromere) domains at the bottom consist of the progeny of the 4th cleavage micromeres. All cells are not finally allocated to the indicated territorial domains until the late blastula stage. The structures to which the territories give rise are shown in (E) and (F). (B–F) The diagrams show in color the already specified polyclonal domains at the respective stages. White areas in (B–D) denote border regions between territories the boundaries of which are not yet finally specified; i.e., progeny of different cells in the white regions may participate in both the adjacent programs of differentiation. (B) State of specification at 6th cleavage. The small micromere lineages (violet) have been set aside; the skeletogenic territory (red) is specified; and the *veg*₂ cell ring is specified as endomesoderm (for lineage relationships see text and references therein). At least the central polar clones of oral (yellow) and aboral (green) ectoderm are by this stage specified. *veg*₁ is not, and its progeny will be allocated, by subsequent lineage-independent specification processes, both to endoderm and ectoderm territories. The bracket indicates the lineage constituents from which the vegetal plate (VP) will form. (C) Mesenchyme blastula stage, about 500 cells. The skeletogenic cells have ingressed into

factors in the egg. But the subdivision of the egg into diverse territories cannot be attributed solely to regional sequestration of maternal factors, since blastomere recombination experiments show that with one exception, blastomeres from any region of the egg have the capacity to produce either endomesodermal or ecto- dermal cell types, depending on where they are placed and who their neighbors are (Hörstadius, 1973; reviewed by Davidson, 1989; Davidson *et al.*, 1998). The exception is the polyclonal skeletogenic domain which arises at the south pole of the egg (Fig. 3.1). These cells are autonomously specified and they are bound to execute their skeletogenic fate no matter where they are transplanted and even in culture, isolated from other cells. Specification of the remainder of the blastomeres involves both autonomous and conditional functions. An important mechanism of autonomous specification is regional nuclearization or regional activation of maternal transcription factors (recall, for example, the unequal nuclear distribu- tion of the maternal SoxB1 factor illustrated in Fig. 2.5). Conditional specification

---

the blastocoel and can no longer be included in this external view. The vegetal plate derived from veg_2 is now subdivided into a peripheral endodermal (blue) and a more central mesodermal (lavender) domain (Ruffins and Ettensohn, 1996). (D) Territorial domains of the vegetal plate viewed from the vegetal pole. The small micromere territory (violet) is at center. The secondary or vegetal plate mesodermal domain (lavender) will give rise to four mesodermal cell types: pigment cells, blastocoelar cells, muscle cells, and cells of the coelomic pouches (to which the small micromeres also contribute). Some veg_1 cells which have been recruited to the aboral ectoderm domain are also visible in (C). Most of the remaining veg_1 cells visible in (D) will subsequently be specified as endoderm. These will constitute the final population of future arch- enteron cells to invaginate at gastrulation, in the course of which they contribute mainly to the future hindgut and anus. (E, F) Final state of specification in early pluteus (about 1800 cells in *S. purpuratus*); (E) oral view; m, mouth; a, anus; (F) Lateral view. The ciliated band forms from cells at the oral/aboral interface, some of which are respecified from a prior state as aboral ectoderm (Cameron *et al.,* 1993). Blastocoelar mesoderm cells, pigment cells lying along the ectodermal wall, and a coelomic pouch are indicated. The aboral ectoderm consists of a single squamous epithelial cell type. The gut is tripartite, consisting of esophagus, midgut, and hindgut. Skeletogenic cells are in process of generating the skeletal rods. The transverse skeletal rods can be seen in (E), as can the bilateral arrangement of the coelomic pouches, and the neurogenic oral epithelium and ciliated band. [(A–F) From Davidson *et al.* (1998) *Development* **125**, 3269–3290 and The Company of Biologists, Ltd.] (G) Fluorescence photomicrograph of transgenic pluteus, lateral view. The embryo is expressing CAT protein under control of the *cis*-regulatory system of the *sm50* skeletal matrix protein gene (cf. Fig. 2.4), here visualized with an anti-CAT antibody (Zeller *et al.*, 1992). Only skeletogenic cells display fluorescence. Parts of embryo are indicated: abbreviations as above; and st, stomach; S, skeletal rod; M, skeletogenic mesenchyme cell; E, esophagus. [(G) From Zeller *et al.* (1992) *Dev. Biol.* **151**, 382–390.]

in the early sea urchin embryo operates by short-range interblastomere signaling, and some particular interactions of this kind are touched upon in the following.

Very briefly, the origins and the fates of the definitive territories are as follows. The autonomous skeletogenic territory (red in Fig. 3.1) consists of the progeny of four 5th cleavage founder cells, which are the daughter cells of the four micromeres separated from the macromeres by the unequally positioned 4th cleavage plane. At blastula stage the descendants of the skeletogenic micromeres ingress into the blastocoel: their progeny (32 in *S. purpuratus*, 64 in *Lytechinus variegatus*) form a syncytium that is arranged in patterned lines against the blastocoel wall, thereby determining the positions of the skeletal rods that they secrete (Fig. 3.1G; McClay *et al.*, 1992; Ettensohn, 1992; Davidson *et al.*, 1998). The small micromere territory is founded by the 5th cleavage sister cells of the skeletogenic micromeres. After dividing once more they are carried into the interior by gastrular invagination. Ultimately, in postembryonic development, their progeny contribute to the mesoderm of the adult body plan, which derives from the embryonic coelomic sacs.

Both skeletogenic and coelomic sac constituents are mesodermal cell types, and the additional mesodermal cells of the embryo arise from progenitors that lie immediately peripheral to the small micromeres and the skeletogenic cells in the center of the vegetal plate (see Fig. 3.1). The mesoderm of the embryo (and of the postembryonic larva as well) thus all derives from blastomeres near the vegetal pole of the egg. But while the small micromere and skeletogenic territories are constituted of lineage elements which are autonomously specified and which segregate precociously during cleavage, the remaining mesodermal domain (lavender in Fig. 3.1) is defined only later in cleavage, by processes that are at least partly conditional. Notch signaling is required within the prospective mesodermal cells, and their specification depends on a late cleavage signal from the skeletogenic micromeres (Sherwood and McClay, 1997, 1999; Sweet *et al.*, 1999; McClay *et al.*, 2000).

Going back a step, we note that the endomesodermal territory as a whole consists of an invariant cleavage-stage lineage element, i.e., the veg_2 polyclone, though the later endodermal and mesodermal domains to which it gives rise are not defined according to lineage. That is, in every embryo, the veg_2 territory begins as a ring of eight 6th cleavage blastomeres which are the lower granddaughters (with respect to the A/V axis) of the 4th cleavage macromeres. Specification of the veg_2 ring as endomesoderm is completed soon after these cells are born, and as we take up in more detail below, this event again depends on both autonomous and conditional functions.

The more peripheral cells of the veg_2 endomesoderm territory become endoderm (blue in Fig. 3.1). Additional endodermal components are derived from those veg_1 cells that happen to lie adjacent to the prospective veg_2 endoderm (Logan and McClay, 1997; Ransick and Davidson, 1998). Late in gastrulation, following invagination of veg_2 endoderm, these cells roll inward and constitute the anus and hindgut, sometimes contributing to more anterior regions of the gut as well. The complex origins of the archenteron endoderm account for the endodermal fate map shown in Fig. 3.1A.

There remain the oral and aboral ectoderm territories, which differ greatly in the diversity of the cell types their progeny produce. The squamous epithelial cell type (dark green in Fig. 3.1F) is the sole differentiated product of the aboral ectoderm (except for some contributions to the ciliated band at the margins; Cameron *et al.*, 1993), but oral fates are more complex. From the oral ectoderm territory arise the mouth and esophagus; the columnar cells of the "face" of the late embryo (yellow in Fig. 3.1E, F); and neurogenic precursors located in the region surrounding the mouth and the ciliated band.

This will set the stage, and to begin we consider a little more extensively one of the most unusual aspects of Type 1 embryonic process, the precocious expression of territorial differentiation genes.

Early Transcriptional Activation of Cell Type-specific Genes

Many cell type-specific genes are activated during cleavage in sea urchin embryos, and some examples are listed in Table 3.1. As the territories of the cleavage and early blastula stage embryo each produce only one or a few cell types of the late embryo, the patterns of expression of these genes (where known) resemble the colored domains of Fig. 3.1A–F. That is, such genes are activated following specification in the early polyclonal territories, and they continue to be transcribed, often at stepped-up rates, within the differentiated morphological structure to which each territory gives rise. Table 3.1 includes genes encoding both differentiation proteins and cell type-specific transcription factors, expression of which ends up as a permanent feature of the respective differentiation states, as assayed in the late embryo.

Of course additional differentiation genes are called into play within each territory/differentiated structure as development proceeds, particularly those activated in response to novel intercellular interactions that come about as cells become motile and morphogenesis takes place. For example, several genes encoding proteins required for skeletogenesis begin to be expressed only as the skeletogenic precursors prepare to ingress into the blastocoel, and thereafter their rates of expression depend strongly on local skeletogenic activity (see review of Guss and Ettensohn, 1997). Nor is the expression of genes that in the end are utilized only in a given structure initially always territorial; e.g., the arylsulfatase gene is initially transcribed all over the embryo, then in all ectoderm territories, and finally only in the aboral ectoderm (Akasaka *et al.*, 1990). So the examples in Table 3.1 are not necessarily illustrative of the way most cell type-specific genes expressed in the late embryo are mobilized. Rather, these examples tell us something about the specification mechanism itself: they imply that the network of regulatory interactions in the early embryo is evidently quite shallow. Genes such as *SpKrox1* and *PlHbx12* that are activated in early cleavage, as illustrated in Fig. 3.2A, B and Fig. 3.2C, D, respectively, are likely to be controlled directly by

TABLE 3.1 Territory- and Cell Type-specific Genes Activated During Cleavage or Early Blastula Stage in the Sea Urchin Embryo

Gene	Early territory	Final expression domain	Protein function
spec1, spec2f,[1] LpS1f[1-3]	Aboral ectoderm	Aboral ectoderm	Ca[+2] binding
cyIIIa[4]	Aboral ectoderm	Aboral ectoderm	Cytoskeletal structure
metallotbionein[5]	nd[12]	Aboral ectoderm	Heavy metal binding
sm50,[6] sm37[7]	Skeletogenic	Skeletogenic mesenchyme	Skeletal matrix components
msp130[8]	Skeletogenic	Skeletogenic mesenchyme	Cell surface glycoprotein
endo16[9]	Endomesoderm	Midgut	Secreted glycoprotein
krox1[10]	Endomesoderm; endoderm	Blastopore and hindgut	Transcription factor
fkb[11]	Endomesoderm; endoderm	Foregut, midgut, hindgut	Transcription factor

Except where noted, all examples are from *Strongylocentrotus purpuratus*. The following references refer to location of embryonic gene expression, rather than to characteristics of the protein product, regulatory apparatus, or other aspects:

[1] f, family (of genes).
[2] *Lp, Lytechinus pictus spec1.*
[3] For *spec* and *LpS* genes see Tomlinson and Klein (1990); Hardin *et al.* (1985); Lynn *et al.* (1983); Gan *et al.* (1990).
[4] For *cyIIIa* see Cox *et al.* (1984); Kirchhamer and Davidson (1996); Bogarad *et al.* (1998).
[5] Nemer *et al.* (1991).
[6] Killian and Wilt (1989); Benson *et al.* (1987); Sucov *et al.* (1988); see Fig. 3.1G.
[7] Lee *et al.* (1999).
[8] Guss and Ettensohn (1997).
[9] Nocente-McGrath *et al.* (1989); Ransick *et al.* (1993); Godin *et al.* (1996).
[10] Wang *et al.* (1996); unpublished data from author's laboratory (see Fig. 3.2A).
[11] Unpublished data from author's laboratory.
[12] Not determined, except that gene is activated in late cleavage (10–12 h in *S. purpuratus*).

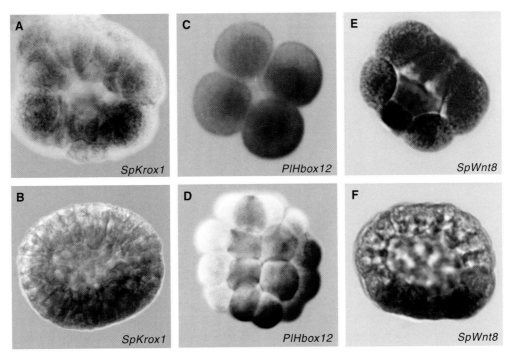

FIGURE 3.2 Transfer of developmental control to zygotic genomes in cleavage-stage sea urchin embryos. Gene expression is visualized by *in situ* hybridization. None of the genes included are represented significantly in the maternal mRNA. (A, B) *SpKrox1*, an endomesoderm and later endoderm-specific Zn finger transcription factor of *S. purpuratus* (Wang *et al.*, 1996; cf. Table 3.1). (A) *SpKrox1* expression in "macromeres" at 16-cell stage. The four macromeres give rise to the veg_2 and veg_1 lineages, and veg_2 in turn to endodermal and secondary mesenchymal (mesodermal) territories; see Fig. 3.1. [(A) From Wang *et al.* (1996) *Mech. Dev.* **60**, 185–195; copyright Elsevier Science.] (B) *SpKrox1* expression is soon confined to veg_2 progeny. A blastula-stage embryo is shown here. [(B) From C. Livi and E. Davidson, unpublished data.] (C, D) *PlHbox12*, a gene encoding a homeodomain transcription factor (Di Bernardo *et al.*, 1995). (C) *Paracentrotus lividus* embryo at 8-cell stage, animal pole view. This is among the earliest known zygotic gene expressions, which is confined to founder blastomeres for the oral ectoderm. [(C) From Oliveri (1996) Dottorato di Ricerca, Universita di Palermo, Italy.] (D) *PlHbox12* expression at 28-cell stage, animal pole view. The gene is expressed in animal pole blastomeres which are oral ectoderm progenitors (cf. Fig. 3.1); [(D) From M. Di Bernardo, R. Russo, P. Oliveri, R. Melfi, and G. Spinelli, unpublished data.] (E, F) *SpWnt8* (Wikramanayake *et al.*, 2000). (E) Expression of gene encoding Wnt8 ligand (*SpWnt8*) at 16-cell stage; expression is confined to the micromeres. (F) Expression of *SpWnt8* extends to veg_2 at the 60-cell stage and continues into early blastula, as shown here. [(E, F) From A. Wikramanayake, unpublished data.]

maternal transcription factors. The same is even true for some differentiation genes, such as *cyIIIa*, as noted above (Chapter 2). The inference is that there cannot be many layers of zygotic regulatory genes intervening between the maternal regulators and cell type-specific gene expression. The overall regulatory strategy here is to activate zygotic cell type-specific transcription factors, in order to control cell type-specific gene batteries, as directly as possible. In the remainder of this chapter we first consider the kinds of regulatory processes required to execute this strategy, and then take up examples which illustrate it in detail.

Initial Regulatory Processes

The process of development requires many layers of explanation. For the early Type 1 embryo, the number of layers is not so great as in the systems taken up in the following chapter, but the outcome is nonetheless miraculous: a few hours earlier the embryo was an egg, and now it has become a territorial mosaic of differently functioning genomes. The outer layer of explanation is the description of the territories: what they are; what maternal components and cellular interactions are required to set them up; what blastomeres they consist of; and so forth. This kind of information is rich and essential, but it is largely particular to each bilaterian clade, and it is external to genomic regulatory mechanisms. The innermost layer, where the buck stops at the DNA, was the subject of the last chapter. In between are a commonly used, canonical set of specific regulatory processes. These are elements of mechanism that are used all over the embryo to organize territories. They carry out certain distinct jobs, that need to be done in order for zygotic control of territorial functions to be established.

These basic elements of regulatory process can all be visualized directly in sea urchin embryos. The key mechanistic stepping stone, as illustrated in Fig. 3.2, is the early territorial activation of genes encoding zygotic transcription factors and signaling components. Thenceforth the embryo is a set of differentially functioning genomes organized in space, such that its lineage elements produce populations of cells that differentiate in the right places. In addition to transcription factors that mark endomesodermal (e.g., Fig. 3.2A, B) and oral ectodermal territories (e.g., Fig. 3.2C, D), Fig. 3.2E also shows lineage-specific activation of a zygotic gene, *SpWnt8*, that encodes a signaling ligand. At the 16-cell stage, this gene is expressed only in micromeres. Shortly after expression spreads to the veg_2 endomesodermal territory (Fig. 3.2F; Wikramanayake *et al.*, 2000).

As noted, the territorial specification processes in this embryo occur by both autonomous and conditional mechanisms. But at the very beginning only autonomous functions are available. By definition, these are regulatory events leading to regional gene expression that depend exclusively on the activity of maternal components (Davidson, 1990). In the sea urchin embryo, sequestration or localization of maternal components along the A/V axis clearly plays a part, since isolated animal pole blastomeres or half embryos produce only ectoderm, while isolated vegetal half embryos generate endomesoderm as well (Davidson *et al.*,

1998; Angerer and Angerer, 2000). In addition, localization either of transcription factors or of systems to activate them at the vegetal pole of the egg probably accounts for the autonomous specification of the micromere lineages. Right from the beginning things are different in the micromeres: their cytoplasm to nuclear volume ratios are several-fold lower than elsewhere in the embryo, and the complexity of their total nuclear RNA is significantly lower (Ernst *et al.*, 1980); unlike other blastomeres their nuclei lack maternal SoxB1 transcription factor (Kenny *et al.*, 1999; Fig. 2.5), but on the other hand they transiently accumulate maternal Otx transcription factor in their nuclei in advance of any other cells (Chuang *et al.*, 1996). Micromeres alone express the *SpWnt8* gene at the 16-cell stage (Wikramanayake and Klein, 2000; Fig. 3.2E). Furthermore, as noted above, micromeres are distinct in that they are already both specified and determined when they are born, since they execute their skeletogenic fate irrespective of where they are placed in the embryo, and their progeny normally do nothing else.

Many maternal transcription factors that are probably widely distributed in the egg at fertilization are evidently activated only in confined regions. Mechanisms could include the effect of differences in cytoplasm-to-nucleus ratio in different parts of the egg (cf. Fig. 2.5); cytoplasmic polarizations that ultimately affect the activity of enzymes which modify transcription factors; or differential translation of maternal mRNAs, as in the eggs of *Drosophila* (Gavis and Lehman, 1994) and *C. elegans* (Evans *et al.*, 1994). But the most general mechanism of this kind in Type 1 embryos is likely to be differential activation of maternal factors in consequence of interblastomere signaling. Conditional mechanisms of this kind must operate because of the plasticity of fates revealed in blastomere recombination experiments. These results in themselves require that given blastomeres possess maternal factors needed to promote more than a single state of specification, and show that the fate actually realized depends on signaling from whatever blastomeres are adjacent (Davidson, 1986, 1989, 1990; and see below). Signals affect cell fate only by affecting the activity of transcription factors, and hence transcriptional output, and in the beginning the relevant transcription factors are all maternal. The signals provide the spatial inputs that end in differentiating one blastomere from another.

Multiple Inputs for Endomesoderm Specification

Intertwined elements of conditional and autonomous regulatory mechanism drive the specification of veg_2 endomesoderm in sea urchin embryos, our paradigmatic sea urchin embryo example. veg_2 requires signals from the micromeres for its specification, and an experiment illustrating this by tracking expression of the *endo16* gene is shown in Fig. 3.3A (Ransick and Davidson, 1995; see Fig. 2.8A for normal course of *endo16* expression). *endo16* is here considered an early endomesodermal specification marker, since this gene is transcribed throughout the veg_2 domain and only there until gastrulation, when (under control of module B of its *cis*-regulatory system; cf. Chapter 2) it acquires an archenteron- and then midgut-specific function. If the micromeres are removed as soon as they appear or

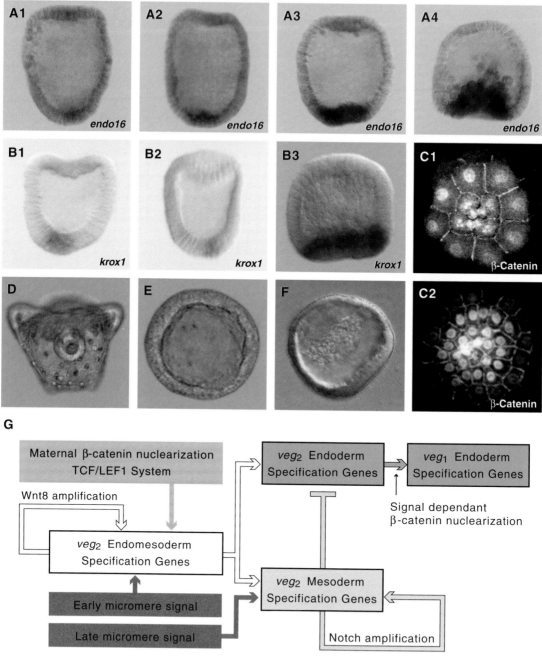

A1 *endo16*

A2 *endo16*

A3 *endo16*

A4 *endo16*

B1 *krox1*

B2 *krox1*

B3 *krox1*

C1 β-Catenin

C2 β-Catenin

D

E

F

G

Maternal β-catenin nuclearization
TCF/LEF1 System

Wnt8 amplification

veg2 Endomesoderm
Specification Genes

Early micromere signal

Late micromere signal

veg2 Endoderm
Specification Genes

veg1 Endoderm
Specification Genes

Signal dependant
β-catenin nuclearization

veg2 Mesoderm
Specification Genes

Notch amplification

4th - 6th cleavage   **7th - 9th cleavage**   **late blastula**

FIGURE 3.3   Conditional and autonomous specification of veg₂ endomesodermal domain in the sea urchin embryo.   (A) Expression of the *endo16* gene is conditional on signaling from micromeres in the 4th–6th cleavage interval. *endo16* activity is visualized by *in situ* hybridization in 24 h *S. purpuratus* blastulae, following microsurgical removal of all four micromeres at: (A1) 4th cleavage, (A2) 5th cleavage (about 45 min later), (A3) 6th cleavage, (A4) untreated control. [(A) From Ransick and Davidson (1995) *Development* **121**, 3215–3222 and The Company of Biologists, Ltd.] (B) Expression of *SpKrox1* also requires signaling from micromeres. Embryos were fixed for *in situ* hybridization at 24 h. (B1, B2) Two embryos from which micromeres were removed at 4th cleavage; (B3) control. [(B) From A. Ransick and E. Davidson, unpublished data.] (C) Autonomous accumulation of β-catenin in vegetal cell nuclei early in development, vegetal pole view. β-catenin is visualized immunocytologically (Logan *et al.*, 1999). (C1) *Lytechinus variegatus* embryo, vegetal pole view, at 32-cell stage; (C2) 60-cell stage. Nuclearized β-catenin is visible only in *veg₂* cells, and in large and small micromeres. (D, E) Endomesoderm specification requires β-catenin nuclearization. (D) Normal pluteus of *L. variegatus* (control embryo). (E) *L. variegatus* embryo of same age as (D), developing from egg injected with mRNA encoding the cytoplasmic domain of cadherin, which traps β-catenin in the cytoplasm and prevents its nuclearization. The embryo consists of a hollow epithelial sphere and contains neither archenteron nor any endomesodermal cell types. [(C–E) From Logan *et al.* (1999) *Development* **126**, 345–357 and The Company of Biologists, Ltd.] (F) Overexpression of GSK3 also prevents endomesodermal specification. An embryo of *Paracentrotus lividus* developing from an egg injected with mRNA encoding GSK3 is shown, at the same age at which controls are fully formed pluteus larvae. Overexpression of GSK3 also blocks β-catenin nuclearization. [(F) From Emily-Fenouil *et al.* (1998) *Development* **125**, 2489–2498 and The Company of Biologists, Ltd.] (G) Process diagram for *veg₂* endomesoderm specification. The boxes demarcated with black lines represent genes encoding transcription factors; no downstream genes are shown. Color coding for the various domains of the embryo are as in Fig. 3.1: *veg₂* and *veg₁* endoderm, blue; *veg₂* mesoderm, lavender; skeletogenic micromeres, red; the initial endomesodermal domain is shown in white. Evidence for the role of maternal β-catenin nuclearization and for the early micromere signal are reviewed in text. The evidence for the Wnt8 loop is that the gene encoding this ligand is expressed throughout *veg₂* early on (Fig. 3.2F), and that introduction of negatively acting forms of Wnt8 blocks endomesoderm specification (Wikramanayake *et al.*, 2000). The evidence for the late micromere signal and the role of signal mediated β-catenin nuclearization in the late blastular specification of *veg₁* endoderm is cited in text (McClay *et al.*, 2000; Sweet *et al.*, 1999; Davidson *et al.*, 1998). The late micromere signal is likely to be a Notch (N) ligand since mesoderm specification requires N signaling (Sherwood and McClay, 1997, 1999). The evidence for the N amplification loop is that N internalization occurs in the cells that are to become *veg₂* mesoderm after 7th cleavage; that if N signaling is blocked by introduction of a negatively acting form of this ligand then mesoderm but not endoderm specification is blocked; and conversely, that more mesoderm is

(Continues)

at the next (i.e., 5th) cleavage, *endo16* expression is severely reduced (Fig. 3.3A1–2). But by 6th cleavage it is already too late, and micromere removal now has little effect on subsequent *endo16* activity (Fig. 3.3A3; compare control Fig. 3.3A4). Conversely, if micromeres are transplanted to the animal pole of a host embryo at the 8- or 16-cell stage, ectopic *endo16* expression is induced to occur at the site of transplantation (Ransick and Davidson, 1993). This operation in fact results in complete respecification to endomesodermal fate of the adjacent animal pole blastomeres, which normally produce only ectoderm. A second vegetal plate appears at the top of the embryo and this presently gives rise to a complete tripartite archenteron that fuses with the endogenous one at the esophagus (Ransick and Davidson, 1993). These results cap off a long history of experimental embryology (Hörstadius, 1939; Khaner and Wilt, 1990) that had already indicated the conditional nature of veg_2 specification. Furthermore, the micromere signals directly affect transcription of the *SpKrox1* gene (Wang *et al.*, 1996), the zygotic endomesoderm regulator of Fig. 3.2A, B. This is illustrated in Fig. 3.3B1–3, in which we see that normal expression of *SpKrox1* requires direct contact with micromeres just as does normal expression of *endo16* (*endo16* is activated several cleavages later than is *SpKrox1* and is not a direct target of *SpKrox1*; these genes are components of interlocked but initially parallel pathways). The experiment of Fig. 3.3B shows that an early cleavage signaling event is directly required for initiation of a territorial regulatory state in the veg_2 blastomeres. Nor is this the only intercellular interaction required: the expression in veg_2 cells of the *SpWnt8* gene of Fig. 3.2F is also essential, since if the embryo is loaded with mRNA encoding a form of Wnt8 which acts negatively (i.e., it occupies Wnt receptors but fails to cause signaling) then endomesoderm specification is halted (Wikramanayake *et al.*, 2000).

But an autonomous component also contributes critically to endomesoderm specification. This is the nuclearization of maternal β-catenin, a cofactor for the Tcf/Lef1 family of transcriptional regulators. Maternal β-catenin is present in all the cells of the early embryo, where it is associated with membrane-bound cadherin. In midcleavage it appears in nuclei of all blastomeres that will contribute to the initial endomesodermal and mesodermal territories (Logan *et al.*, 1999). Thus at 5th cleavage, as illustrated in Fig. 3.3C1, an immunological stain reveals nuclearized β-catenin in nuclei of macromeres, skeletogenic micromeres and small micromeres (cf. Fig. 3.1); after 6th cleavage, when veg_1 and veg_2 blastomere tiers arise from the parental macromeres, it begins to fade from the veg_1 nuclei; and by 7th cleavage it is strictly localized to veg_2 endomesoderm and

---

formed with excess N signaling (Sherwood and McClay, 1997, 1999). The negative transcriptional interaction on endodermal genes within the mesodermal domain follows from the observation that LiCl and many other treatments enlarge the mesodermal domain at the expense of the more peripherally located endodermal domain (while the latter is also enlarged at the expense of the overlying veg_1 ectoderm; see text).

to the micromere mesodermal domains, as shown in Fig. 3.3C2. Nuclearization of β-catenin is a perfect marker of the segregation of lineages into the endomeso-dermal territories (blue + violet + lavender in Fig. 3.1), for elsewhere in the embryo it remains associated with the cell membranes. The early β-catenin nuclearization in these territories is transient, and disappears by the end of cleavage. Nuclear β-catenin reappears later in the blastula stage in a ring of veg_1 cells immediately external to the vegetal plate, i.e., in just those veg_1 cells that will shortly contribute to hindgut endoderm cf. fate map, Fig. 3.1A). The late nuclear-ization of β-catenin thus occurs in these cells at the time they are specified as endoderm, just prior to their invagination.

The early embryonic nuclearization of β-catenin is an autonomous process, according to two kinds of evidence. First, at 7th cleavage the same number of blastomeres display nuclear β-catenin as in controls even if the embryos are continuously dissociated, so that from the 2-cell stage onward the blastomeres divide in isolation from one another and sister cells are immediately separated. Second, removal of micromeres (as in the experiments of Fig. 3.3A, B) has no effect on the extent of β-catenin nuclearization in veg_2 cells (Logan et al., 1999).

The nuclearization of β-catenin is an essential process in endomesoderm specification. It can be prevented by overexpression of the cytoplasmic domain of cadherin, thereby trapping the β-catenin in the cytoplasm throughout the embryo. The consequence is total failure of endomesoderm specification, that is, absence of expression of endomesodermal markers (e.g., endo16) or of endo-mesodermal cell types (Logan et al., 1999; Wikramanayake et al., 1998). These embryos develop as hollow ciliated spheres, much as do the progeny of isolated animal-half blastomeres. This is a dramatic effect, illustrated in Fig. 3.3E (Fig. 3.3D shows a control embryo of the same species and same age). Furthermore, over-expression of GSK3, which inhibits β-catenin activation and nuclearization, pro-duces just the same phenotype (Emily-Fenouil et al., 1998; Fig. 3.3F). The system can be pushed back and forth: treatment of early embryos with LiCl, the relevant function of which here is to inhibit GSK3 (Klein and Melton, 1996), has the effect of expanding the endomesodermal domain so that it includes all of veg_1 and sometimes more (Hörstadius, 1973; Cameron and Davidson, 1997). LiCl teatment correspondingly expands the domain of nuclear β-catenin (Logan et al., 1999). Finally, expression of negatively active forms of Tcf/Lef1 also blocks endomeso-derm specification, while overexpression of the normal form expands it (Vonica et al., 2000; Huang et al., 2000). The function of β-catenin nuclearization in endo-mesoderm specification in fact depends entirely upon Tcf/Lef1.

As it affects endomesoderm specification the β-catenin system may be consid-ered an autonomous expression of maternal polarization along the A/V axis of the egg (Logan et al., 1999; Davidson et al., 1998; Angerer and Angerer, 2000), while the requirement for a 4th–6th cleavage micromere signal designates the exact lineages in which endomesodermal specification will occur, i.e., in the blasto-meres that abut the micromeres. So these are dual inputs, of different informa-tional consequence. Since only advanced echinoids such as our example

S. purpuratus make micromeres, the autonomous system is the basal one for echinoderms. Its remnants can still be seen in the residual *endo16* and *krox1* gene expression that occurs in the micromere-less embryos of Fig. 3.3A, B. The dual inputs that control *veg₂* specification in *S. purpuratus* must ultimately be integrated at the *cis*-regulatory level (cf. Chapter 2). This job will be performed by gene(s) that respond to both inputs. For example the expression of *SpKrox1* requires the micromere signal (Fig. 3.3B); and its domain of expression is expanded to include *veg₁* by LiCl treatment, but is obliterated if β-catenin nuclearization is prevented by excess cytoplasmic cadherin (A. Ransick, C. Livi, and E. Davidson, unpublished data). Whichever genes carry out this integrative function it can be concluded that the lineage-specific transcriptional output that defines *veg₂* and sets the endomesodermal fate of its progeny is controlled by a combination of spatial inputs. These are the maternal β-catenin nuclearization system and the zygotic 4th–5th cleavage mesomere signal.

Figure 3.3G shows an outline of the inputs into the stepwise endodermal specification system of the sea urchin embryo (see legend for further details). The diagram illustrates how a combination of maternal and zygotic inputs position endomesoderm fate in *veg₂* (4th–6th cleavage). Subsequently the late micromere signal (McClay *et al.*, 2000), plus an endomesodermal regulatory output, are responsible for *veg₂* mesoderm specification (7th–9th cleavage). The remaining *veg₂* cells become endoderm. Note that both the endomesodermal and the *veg₂* mesodermal domain seem to be equipped with signal-mediated state amplification devices; these constitute what could be termed a "community effect" (Gurdon *et al.*, 1993). That is, signaling amongst cells within these domains is required in order to maintain and perhaps amplify the respective states of specification.

So we have the flavor of the mechanism that quickly transforms the initial organization of the sea urchin egg into a specified mosaic of blastomeres. These mechanisms turn on regulatory genes in the nuclei of the early blastomeres even as the cleavage process generates the lineage elements that will constitute each territory. It is an interesting historical note that for generations the early sea urchin embryo was known as the quintessential "regulative" system in that (almost) any part could be made to recreate any other part if put in its place. But, as has been clear for a long time now, "regulative" character simply means that cell fate is conditional, neither more nor less; that is, that interblastomere signaling is important for specification (Davidson, 1986). Both conditional and autonomous mechanisms are required for endomesoderm specification, and as we see in the following this is typical.

REGULATORY MECHANISMS CONTROLLING SPECIFICATION IN ASCIDIAN EMBRYOS

The embryos of solitary ascidians also solve the problem of embryogenesis by operating Type 1 specification mechanisms. Like sea urchin embryos they carry out direct cell type specification during cleavage, within polyclonal lineage elements that are the territories from which specific larval structures arise after gastrulation. The initial tier of specifications have largely taken place by 5th–7th cleavage, depending on the lineage (for general review of experimental evidence regarding lineage and specification in these embryos see Satoh, 1994). Though they are both deuterostomes (cf. Fig. 1.6) ascidian embryos differ from those of sea urchins in that their cell lineage is more rigidly invariant further into embryonic development, but this is of no great importance for present considerations except that it provides an experimental leg up. A more significant difference is that ascidians are direct developers, in contrast to most sea urchin species, which develop indirectly. That is, the ascidian embryo gives rise directly to a small (and very simplified) version of an adult chordate body plan, rather than to a single cell-thick feeding larva, which bears no obvious morphological relation to the adult body plan (Peterson *et al.*, 1997, 2000a, b). The postgastrular outcome of embryogenesis in ascidians is therefore anatomically more complex than in sea urchins. It is complex in another sense as well: though the species considered here are motile, their larvae do not feed, but soon settle down and destroy some of their chordate-like structures. They then proceed to generate, in part from undifferentiated larval set-aside cells, a taxon-specific morphology that equips them for life as sessile filter feeders. These are postembryonic features, however, and our interest here is rather with how the initial specification functions of the embryo are controlled.

Territorial Specification

A territorial map of the ascidian embryo is shown in Fig. 3.4, based on a century of cell lineage analyses (reviewed by Satoh, 1994; Davidson, 1986). The 6th cleavage-stage embryo is portrayed in Fig. 3.4A, and an embryo that has partially completed 7th cleavage in Fig. 3.4B; by this point all but a few cells have been finally allocated to one or another of the definitive territories, in a predictable and invariant sequence. The immediate daughters of most (though not all) of these few cells are alternately assigned to one or another territory after the very next cleavage. The structures of the chordate-like larva to which each territory gives rise are indicated by color in Fig. 3.4C. The population of cells denoted "mesenchyme" (MCH) includes the mesodermal set-aside cells that serve as precursors of some adult (i.e., sessile phase) tissues (Satoh, 1994; Hirano and Nishida, 1997), and similarly the trunk lateral cells (TLC) are the precursors of adult hemocytes (Kawaminani and Nishida, 1997).

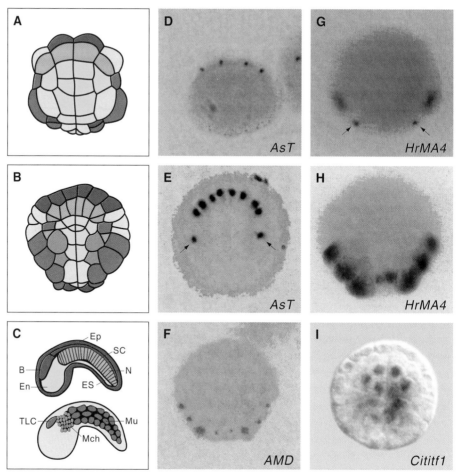

FIGURE 3.4 Specification and early territorial expression of cell type-specific genes in ascidian embryos. (A–C) Color-coded territorial map based on detailed knowledge of the embryonic cell lineage (Conklin, 1905; Ortolani, 1955; Nishida, 1987). Embryos are viewed from the vegetal pole, anterior up. The dorsal side, not visible, consists of cells fated to become epidermis, except for the anteriormost cells which will contribute to the cerebral vesicle. Gray regions represent yet incompletely specified cells, the progeny of which will contribute more than one cell type; colored regions have been uniquely specified at the stages shown so that all progeny will contribute to one cell type or structure. Specification of the brain, and completion of the anterior/posterior regionalization of the epidermis, occur later (Miya et al., 1996; Wada et al., 1999; review by Satoh, 1994). Territories are: purple, dorsal glial cell column ("spinal cord", SC); orange, notochord (N); blue, trunk lateral cells (TLC; these are precursors of the adult blood cells [Hirano and Nishida, 1997]); yellow, endoderm (En) and endodermal strand (ES); dark green, epidermis (Ep); light green, mesenchyme

The blastomeres which give rise to endodermal, epidermal, and tail muscle lineages are specified autonomously. Classical and modern evidence has led to this conclusion, which is mainly based on three kinds of result. First, the progeny of isolated blastomeres of these lineages are competent to generate appropriate structures and express cell type-specific markers in the absence of the remainder of the embryo. Second, cleavage arrest with cytochalasin at successive stages (the nuclei continue to divide but no new cell membranes are formed after the drug is added) shows essentially that so long as the nuclei are surrounded with given sectors of maternal cytoplasm they will express the cell type-specific marker genes that would normally be expressed by the blastomere lineages which inherit these same sectors (for review, Davidson 1986, 1990; Satoh, 1994). Third, and most incontrovertible, is the demonstration that fusion of given, localized sectors of egg cytoplasm with certain early blastomeres or nucleated egg fragments is sufficient to endow these with the capacity, in culture, to express muscle, endoderm, or epithelial markers, which they will otherwise not do (Nishida, 1992, 1993, 1994; Marikawa et al., 1994; reviewed by Nishida, 1997).

The remainder of the specification states indicated in the maps of Fig. 3.4A, B are conditional, depending on signaling interactions between specific blastomeres. These interactions take place mainly at 32–64 cell stage, though of course additional interactions follow on at later stages, e.g., within the nervous system. Among the prominent conditional specification events in which 5th–6th cleavage signaling interactions have been identified are the specification of notochord

(Mch); red, tail muscle (Mu); dark blue, brain (B). (A) 64-cell stage; (B) 110-cell stage; (C) Locations of the morphological structures to which these cell types give rise, seen in a midsaggital, and a more lateral saggital view of the completed larva. [(A–C) From Satoh et al. (1996b) Dev. Growth Differ. **38**, 325–340.] (D, E) Expression of AsT (brachyury) gene in notochord lineage founder cells. At 64-cell stage this gene is expressed in the A7.3 and A7.7 pairs of blastomeres (D), and at 76-cell stage (E) in the eight progeny of the A7.3 and A7.7 blastomere pairs plus the B8.6 pair, arrows (i.e., the notochord lineages shown in orange in B). [(D, E) From Yasuo and Satoh (1994) Dev. Growth Differ. **36**, 9–18.] (F) Expression of AMD (myoD) gene at 76-cell stage, in nuclei of four muscle lineage blastomeres on each side of the midline. [(F) From Satoh et al. (1996a) Proc. Natl. Acad. Sci. USA **93**, 9315–9321; copyright National Academy of Sciences, USA; for further details and lineage, see Fig. 3.5A and text.] (G, H) Expression of HrMA4 muscle actin gene. (G) 64-cell stage; (H) 108-cell stage. The gene is expressed in muscle precursor blastomeres (red cells in (A) and (B), respectively. [(G, H) From Satou et al. (1995) Dev. Growth Differ. **37**, 319–327.] (B–H) are embryos of Halocynthia roretzi. (I) Expression of Cititf1, an endoderm-specific transcription factor, in all endoderm lineage progenitors in Ciona intestinalis at 76-cell stage (i.e., the A7.1, A7.2, A7.5, B7.1, and B7.2 blastomere pairs). [(I) From Ristoratore et al. (1999) Development **126**, 5149–5159 and The Company of Biologists Ltd.]

founder cells (Nakatani and Nishida, 1994, 1999); of mesenchyme lineage founder cells (Kim and Nishida, 1999); of trunk lateral cells (Kawaminani and Nishida, 1999); and of anterior and posterior epidermal fates (Wada *et al.*, 1999). Conditional specification of cell fates within the larval nervous system has long been known to occur conditionally (Rose, 1939; Reverberi and Minganti, 1947; Nishida, 1991; Satoh, 1994; Inazawa *et al.*, 1998). The territorial lineages of the ascidian embryo display the essential feature of Type 1 embryonic process: they proceed immediately toward cell differentiation once the founder cells have segregated from other lineage elements and become specified, whether this occurs by autonomous or conditional means.

As in the sea urchin the character of the embryonic mechanisms at work in the ascidian embryo is shown by the early expression of territory-specific genes. This is illustrated in Fig. 3.4D–I. The territories are coincident with given lineage elements, since the relevant portions of the cleavage process are invariant. This makes it possible to define the exact cleavage and the specific cells in which these genes are activated. In Fig. 3.4D we see that a gene encoding a Brachyury transcription factor (AsT) is activated, exclusively in progenitors of the notochord territory (Yasuo and Satoh, 1994). Brachyury is a key regulator of notochord development (Chapter 5). Transcription of this gene begins at the 64-cell stage, when two lineage founder cells for notochord have been specified on either side of the plane of symmetry. After the following cleavage this regulatory gene is expressed in the eight progeny of those cells and in two additional blastomeres which have now also been specified as notochord (Fig. 3.4E). The location of these cells at the respective stages can be seen in the context of the overall territorial maps in Fig. 3.4A and B, respectively. The expression of the *brachyury* gene is a direct output of the notochord specification mechanism as discussed below. Figure 3.4F–H illustrates expression of genes specific to the territory that gives rise to the tail muscles: at the 76-cell stage (Fig. 3.4F) a gene encoding a MyoD class transcription factor (Satoh *et al.*, 1996a) is expressed in four muscle precursors on each side (muscle specification is discussed in detail below); and Fig. 3.4G, H shows expression of one of the actin genes (HrMA4) in muscle progenitor blastomeres at 64– and 108-cell stages (Satou *et al.*, 1995). At the risk of unnecessary repetition, this is another example of the activation of a gene encoding a differentiation protein in cleavage, in which transcription begins long in advance of the morphogenetic events that formulate the tail muscle structures. Finally, in Fig. 3.4I is illustrated territorial expression during cleavage of a gene encoding a specific homeodomain transcription factor (Cititf1), here seen in five pairs of endoderm precursor cells (Ristoratore *et al.*, 1999). This gene encodes an endoderm-specific factor which continues to be expressed in all endoderm cells into gastrulation, and in postgastrular and larval stages it is utilized in the head endoderm. The colors in the diagrams of Fig. 3.4A and B can be taken to symbolize the institution of territory-specific domains of differential gene expression, all of which have been installed by midcleavage.

Mechanisms and Pathways in Mesoderm Specification

Just how direct can be the "direct cell type specification" of this discussion is evident in the process by which the muscle territory of ascidian embryos is specified. To appreciate this story a glance at the lineage map of the progeny of the 3rd cleavage B4.1 blastomere pair is necessary (Fig. 3.5A; a view of the 3rd cleavage embryo is shown in Fig. 3.5E5). From this pair of 3rd cleavage cells (one on each side) derive all the primary tail muscles and some mesenchyme cells of the larva (red and light green territories in Fig. 3.4A, B). Progeny of the B4.1 blastomeres produce exactly 14 primary tail muscle cells on each side of the larva, in both *Ciona intestinalis* and *Halocynthia roretzi*. A few additional "secondary" tail muscle cells positioned at the caudal end of the larva arise from different areas (i.e., ten from the B4.2 pair of blastomeres and four from the A4.1 cell pair in *Halocynthia;* in *C. intestinalis* these same two blastomere pairs instead produce a total of eight secondary muscle cells; Nishida, 1987; Meedel *et al.*, 1987). While the secondary muscle cells are conditionally specified, the primary muscle lineages stemming from B4.1 are autonomously specified. The experimental evidence for their autonomous specification includes all of the sorts of demonstrations listed above (for reviews see Satoh, 1994; Nishida, 1997; Davidson 1990). Molecules that activate muscle specific genes are localized in the portion of the cytoplasm inherited by the B4.1 blastomeres. The key observations (in *H. roretzi*) come from fusing vesicles that contain sectors of maternal cytoplasm to blastomeres that are normally fated to produce only epidermis. This was done by use of an electric field in the presence of polyethylene glycol. Fusion of cytoplasm from B4.1 causes the progeny of epidermal blastomeres to undergo muscle differentiation and express muscle markers, and the same is true if the cytoplasm is obtained from those sectors of unfertilized and fertilized eggs that are later incorporated in the B4.1 blastomeres (Nishida, 1992). Similarly, in *Ciona savigny*, certain centrifugal fractions of egg cytoplasm confer ability to activate muscle genes if fused with nucleated egg fragments that otherwise produce only epidermis when cultured (Marikawa *et al.*, 1994).

The diagram in Fig. 3.5A shows the origin of the seven tail muscle progenators that arise from the B4.1 lineage on each side, and traces the segregation of cell fates that occurs between 4th and 6th cleavages (the color codes are the same as in the territorial diagrams of Fig. 3.4A–C). Note that in the top half of the lineage there emerges at 5th cleavage an endoderm founder cell, B6.1 (that is, a cell all of whose progeny will become endoderm). A muscle founder cell, B7.4, is segregated at 6th cleavage. The sister cell, B7.3, produces both a mesenchyme and a notochord founder blastomere at the following cleavage. In the bottom half the final segregation of cell fates all occur at 6th cleavage: B7.5 and B7.8 give rise only to muscle; B7.7 to mesenchyme and B7.6 to endodermal strand. Note that two sublineages which have only mesodermal fates separate out at 5th cleavage. These are the B6.2 lineage, which produces muscle, mesenchyme, and notochord cell types; and the B6.4 lineage, which produces mesenchyme and muscle.

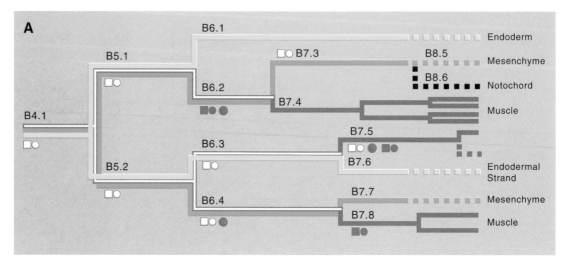

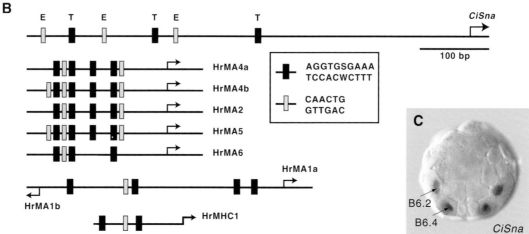

FIGURE 3.5   Autonomous specification of mesodermal precursors in ascidian eggs. (A) Partial B4.1 half lineage (i.e., lineage of one side of the bilaterally symmetrical embryo) displaying origins of primary tail muscle cells and of mesenchyme cells. Color coding is as in Fig. 3.4A, i.e., red, muscle; yellow, endoderm; light green, mesenchyme. The geometry of the 8-cell embryo, including the B4.1 cell pair, viewed from the side (anterior right, vegetal down) can be seen in (E5). The lineage is from *Halocynthia roretzi* (Satoh, 1994; Satou et al., 1995), but the portion shown is the same for *Ciona intestinalis*. Muscle and mesenchyme are mesodermal cell types. Not shown are some minor additional contributions: B7.5 produces adult muscle precursors (dashed red line) as well as two tail muscle cells. Also, note that a descendant of B7.3, viz, B8.6, produces four notochord cells on each side (black dashed line), while its sister cell, B8.5, produces mesenchyme. (Most notochord cells, 16 on each side, derive

FIGURE 3.5 *(Continued)*

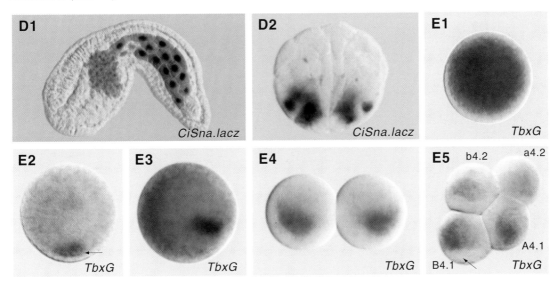

from A4.1.) There are 28 primary tail muscle cells in this species, 14 of which derive from each B4.1 blastomere. The diagram stops one cycle prior to the last in the muscle sublineages, so only seven tail muscle cells are shown. Restriction to single states of specification is indicated by singly colored lines; dashed lines indicate outcome of sublineage but not individual cell divisions. The solid symbols indicate gene activations and open symbols lack of expression of the same genes: red squares, muscle actin; red circles, myosin heavy chain (Satou *et al.*, 1995); green and red circles, timing of expression of *snail* gene: these data are superimposed from *Ciona* (Erives *et al.*, 1998). In *H. roretzi snail* gene expression has also been detected by *in situ* hybridization, but one cycle later (also in mesodermal lineages; Wada and Saiga, 1999). [(A) From Satou *et al.* (1995) *Dev. Growth Differ.* **37**, 319–327.] (B) Location of sites for bHLH factor (gray bars labeled E), and for the VegTr factor discussed in text (black bars labeled T) in the *cis*-regulatory elements of the indicated genes: *CiSna*, *Ciona snail* gene B4.1 element; *HrMA*, *H. roretzi* muscle actin genes; *HrMHCI*, myosin heavy chain gene. The sequences of the target sites indicated by T and E bars are shown in box at right; note that this bHLH factor utilizes sites in these genes that have an AC core. [(B) From Erives and Levine (2000a) *Dev. Biol.* **225**, 169–178.] (C) Expression of *CiSna* gene in 5th cleavage embryos, visualized by *in situ* hybridization. The gene is activated in B6.2 and B6.4 mesodermal precursors (cf. A). [(C) From Erives *et al.* (1998) *Dev. Biol.* **194**, 213–225.] (D) Expression of *CiSna.lacz* construct electroporated into eggs, monitored by β-galactosidase staining. The construct is driven by a 262 bp minimal *cis*-regulatory element that confers embryonic expression in B4.1 mesodermal lineages: (D1), larval

(Continues)

Observations on *Ciona* embryos show that this key event is marked by activation of a gene encoding a snail transcription factor (the *CiSna* gene) in the B6.2 and B6.4 cells on either side (Erives *et al.*, 1998; green surrounding red circles in Fig. 3.5A). Similarly, the *CiSna* gene is also activated at the next cleavage in the B7.5 pair, when this mesodermal cell, which gives rise only to muscle, separates from B7.6, its endodermal sister cell (see Fig. 3.5A). Expression of *CiSna* in the mesodermal B6.2 and B6.4 blastomere pairs is illustrated in Fig. 3.5C. In addition genes encoding contractile cell type specific proteins, *viz*, muscle actin (cf. Fig. 3.4G, H) and myosin heavy chain are activated in the muscle lineages either immediately before or just after final segregation of muscle lineages as shown in Fig. 3.5A (red symbols); (Satou *et al.*, 1995): specification has now occurred. In order for mesenchyme fate to be expressed an inductive signal from adjacent endoderm cells is required in the 5th–6th cleavage interval (Kim and Nishida, 1999). The effects of this signal are to suppress muscle-specific differentiation and permit mesenchyme differentiation.

One aspect of the mechanism underlying autonomous specification of muscle cell fates emerges from an analysis of the *cis*-regulatory element that controls expression of the *CiSna* gene, in those blastomeres of the B4.1 lineage that are fated to express mesodermal fates (Erives *et al.*, 1998). Other modules of this *cis*-regulatory system control larval expression of *CiSna* in secondary tail muscles and in the nervous system (later in development the *snail* gene is expressed in many mesodermal and neuronal domains; Corbo *et al.*, 1997; Erives *et al.*, 1998; Wada and Saiga, 1999). An exogenous reporter construct driven by the B4.1 lineage regulatory element of the *CiSna* gene is expressed in larvae in both tail muscle and mesenchymal cells (Fig. 3.5D1). In the 32-cell embryo expression occurs specifically in the B6.2 and B6.4 blastomeres (Fig. 3.5D2). The striking results of a comparison of *cis*-regulatory sequences between the B4.1 *CiSna* minimal enhancer and *cis*-regulatory elements from several muscle-specific contractile protein genes are shown in Fig. 3.5B (Erives *et al.*, 1998; Araki and Satoh, 1996; Kusakabe *et al.*, 1995; Satou and Satoh, 1996). The *CiSna* regulatory element shares two classes of target site with this battery of muscle genes: E-box sites, which include an AC core (rather than the usual GC core preferred by bHLH factors of the MyoD family), and sites for a Brachyury or T-box class transcription factor.

---

expression, in mesenchyme and tail muscle cells; (D2), expression at 5th cleavage in B6.2 and B6.4 cells. (E) Localization of maternal transcripts encoding *Ciona* VegTr (TbxG) transcription factor, visualized by *in situ* hybridization. (E1) Unfertilized oocyte, animal pole up; (E2) zygote, 30 min past fertilization; staining in cortical regions of the vegetal cytoplasm; (E3) 10 min later; staining now partially localized to cytoplasm that will be inherited by B4.1 blastomeres; (E4) two-cell stage, posterior view; (E5) eight-cell stage, lateral view. Transcripts encoding CiTbxG factor are present in B4.1 and also A4.1 cell pairs. Arrow indicates a region of B4.1 cytoplasm from which TbxG mRNA is excluded. [(D, E) From Erives and Levine (2000a) *Dev. Biol.* **225**, 169–178]

In general, the simplest mechanism that could underlie autonomous specification in early development is that localized maternal gene regulatory factors are sequestered in the region of the egg cytoarchitecture inherited by given blastomeres; and that this cytoplasm is necessary and sufficient for specification of the blastomeres because the factors it contains turn on the genes required for that state of specification (Davidson and Britten, 1971; Davidson, 1990). Such seems indeed to be the case here. A maternal T-box factor (CiVegTr) fills the bill: this factor binds the T-box sites in the B4.1 *CiSna cis*-regulatory module; and when these sites are mutated the expression vector fails to activate in the B4.1 lineage (Erives and Levine, 2000a). Most importantly, the VegTr factor is localized during the ooplasmic movements that follow fertilization so that it is mainly inherited by the B4.1 blastomeres. The progressive localization of this maternal regulator after fertilization is shown in Fig. 3.5E1–5. The "myoplasm" of the ascidian egg (Conklin, 1905) thus probably includes this factor, and muscle specificity may depend as well on interaction with a bHLH factor together with the VegTr factor, given that the sites for these factors are invariably adjacent in the muscle genes of Fig. 3.5B.

To summarize, autonomous specification of muscle lineages follows from the localization of at least one (and probably several) positively acting maternal transcription factors. These are evidently inherited by those blastomeres of the B4.1 lineage that generate progeny specified as mesoderm. We see that the same gene regulators may activate genes encoding cell-type specific contractile proteins as control expression of the Snail transcription factor. The *snail* gene begins to function as soon as the autonomously specified sister blastomeres of endoderm fate have segregated from the mesodermal precursors. Within the mesodermal progenitors of the B4.1 lineage, muscle fate is dominant, so a negative signal from adjacent endoderm cells is needed to suppress muscle fate and permit mesenchyme fate in the *snail*-expressing B6.2 and B6.4 lineages. These transcriptional regulatory events all occur during cleavage, essentially in the 3rd–6th cleavage interval. As predicted, this is a very "shallow" regulatory system: maternal factors define two successive levels of regulatory specification, i.e., the more general mesoderm specification, as defined by *Snail* gene expression; and muscle specification, marked by expression of muscle differentiation genes, including both transcription regulators such as the *myoD* gene and genes encoding contractile proteins. Downstream, of course, many additional muscle-specific genes might be expected to be activated, including those responsible for further tail muscle differentiation and the additional transcriptional regulators required, e.g., another T-box gene which is expressed specifically in muscle cells (Mitani *et al.*, 1999).

The notochord as well as mesenchyme lineages are conditionally specified in ascidian embryos, and we may ask whether this requires additional "layers" of regulatory interaction. But in fact the process is again one of immediacy and directness, and there are no intermediate stages between the initial specification and the institution of a differentiated regulatory state that causes expression of notochord differentiation genes. The notochord of the larva consists of a column of 40-cells that forms by the median intercalation of 20 notochord cells that arise on each side. Thirty-two of these notochord cells derive from the A7.3 and A7.4 blastomere pairs, i.e., those shown in Fig. 3.4D to be expressing the *brachyury* gene (or in *H. roretzi*, the *AsT* gene): and eight derive from the B8.6 blastomeres. The two B8.6 blastomeres and what are now four A-lineage blastomeres on each side are seen expressing the *AsT brachyury* gene in the 110-cell stage embryo of Fig. 3.4E. Expression of *AsT* occurs exactly when each of these six notochord founder blastomeres becomes uniquely specified (Yasuo and Satoh, 1998).

Specification of the A- or primary lineage blastomeres, and activation of the *AsT brachyury* gene, depend on signaling from vegetal blastomeres that include endoderm progenitors (Nakatani and Nishida, 1994, 1999; Nakatani *et al.*, 1996). The interaction takes place precisely during the 32-cell stage, and if the A-lineage blastomeres are separated from their vegetal neighbors prior to this their progeny never express vegetal markers; conversely, if the separated cells are recombined, the *AsT brachyury* gene is activated and notochord specification occurs. The signal may be a bFGF ligand, since human bFGF added to the medium substitutes for intercellular contact at the 32-cell stage. Expression of the *AsT* gene is also induced in isolated 32-cell stage A-lineage blastomeres by bFGF (Nakatani *et al.*, 1996). That this is a key event in notochord specification is shown by the observation that forced expression of Brachyury protein by injection of *AsT* mRNA into the egg substitutes for the inductive interaction: A-lineage blastomeres isolated at 32-cell stage from embryos containing exogenous *AsT* mRNA express notochord markers in the absence of any contact with vegetal cells (Yasuo and Satoh, 1998). Furthermore, while the 5th cleavage A-lineage blastomeres A6.2 and A6.4 both produce progeny one of which is a notochord progenitor (i.e., A7.3 and A7.5) while the sister cell is a spinal cord progenitor (i.e., A7.4 and A7.6), either bFGF treatment or *AsT* overexpression causes both sisters to become notochord progenitors (Nakatani *et al.*, 1996; Yasuo and Satoh, 1998). These experiments establish what is probably a direct link between the specifying signal and activation of the *AsT* gene, since its expression is detected very soon after passage of the signal at 6th cleavage; and also a causal link between *AsT* expression and downstream differentiation of the notochord cell type. Notochord cells have a unique elongate shape, and synthesize various unique proteins that can be detected by immunological means.

There remains to consider the regulatory distance between *AsT* gene expression and expression of these differentiation markers. In fact there is no distance, at least in respect to the initial cohort of cell type-specific genes: we now know that *AsT* directly controls a battery of genes that encode proteins expressed by differentiated notochord cells, as well as transcription factors that probably activate further sets of genes. Many candidates have been recovered from an overexpression screen using a vector which ectopically expresses the *Ciona brachyury* gene (Takahashi *et al.*, 1999a; Hotta *et al.*, 1999). Among these is a gene encoding a notochord-specific tropomyosin. This gene is unequivocally a direct target of the Brachyury regulator (Di Gregorio and Levine, 1999). A *cis*-regulatory module of this gene has been characterized which confers notochord-specific expression, the activity of which depends explicitly on Brachyury target sites. Another likely direct target is a collagen gene expressed in several cell types, including notochord. A *cis*-regulatory element from this gene that drives notochord (plus endoderm) expression also contains many binding sites for the *brachyury* gene product (Erives and Levine, 2000b). These examples suffice to make the point that conditional specification of notochord lineages requires as shallow a gene regulatory network as does the autonomous specification of the muscle lineages. Notochord specification is immediately marked by the signal-mediated activation of the *brachyury* regulatory gene, and this in turn functions as a direct controller of differentiation genes. Among the targets of the *AsT brachyury* gene is itself (Takahashi *et al.*, 1999b), so once turned on the differentiation program it controls will function autonomously.

But a complex *cis*-regulatory system underlies the function of this neatly organized specification function. Control of *brachyury* expression in the embryo probably requires multiple elements, including one which transduces the inductive signal early in development (Takahashi *et al.*, 1999b), and one which maintains expression in notochord cells later on. A regulatory component from the *Ciona brachyury* gene that probably carries out the maintenance function confers notochord-specific expression on reporter constructs (Corbo *et al.*, 1997; Fujiwara *et al.*, 1998), but its deletion from the overall *cis*-regulatory sequence does not prevent activation of the constructs in the notochord lineages. This element responds synergistically to a Suppressor of Hairless [Su(H)] regulator plus a bHLH regulator (different from the early mesoderm regulators discussed above, as shown by their target site specificity). The Su(H) factor is activated by the Notch (N) pathway, here as elsewhere (Bailey and Posakony, 1995), and its regulatory input is required for notochord expression (Corbo *et al.*, 1998). In addition to all this the *Ciona brachyury cis*-regulatory system includes target sites for the Snail repressor expressed in tail muscle cells, and if these are mutated, expression of reporter constructs spreads to tail muscle cells. Thus the notochord/muscle boundary turns out to depend on Snail repression (Fujiwara *et al.*, 1998). Though not yet sufficiently well known, it is already clear that the notochord-specific maintenance element of the *brachyury cis*-regulatory system is a typical information processing device, of the sort that we dealt with in Chapter 2. This brings us to a thought regarding the evolution of Type 1 embryonic specification processes. These processes depend largely on control of regulatory genes that direct cell type-specific gene expression in the early embryo. In Type 1 systems such genes reside in relatively shallow regulatory networks. Therefore evolution of this form of embryogenesis must have depended largely on loading in the sets of *cis*-regulatory target sites needed to cause their blastomere-specific spatial expression.

CAENORHABDITIS ELEGANS: THE GENOMIC APPARATUS FOR ENDODERM SPECIFICATION

C. elegans is an ecdysozoan (Aguinaldo *et al.*, 1997), and a highly derived terrestrial one at that. Yet though the deuterostomes are far removed from ecdysozoans (Fig. 1.6), the embryonic specification mechanisms of *C. elegans* can be parsed into the same kinds of developmental process elements that the sea urchins and ascidians use. This is again a small embryo which works by specifying blastomeres or sublineages during cleavage. The first-stage larva consists of only 558 cells, including 45 blast cells (in the male), that are set aside to carry out postembryonic programs of differentiation from which parts of adult sexual structures and gametes derive; the adult male has only 1031 somatic cells and the adult female 959 somatic cells (Sulston *et al.*, 1983). The earliest blastomere to be specified such that all of its progeny execute the same differentiation functions is the 3rd cleavage E blastomere. This gives rise to the intestine, and is our subject in the following. Elsewhere in the embryo cell fate specification occurs at different cleavages, in general earlier for lineages arising in the posterior quadrant of the embryo. The embryonic lineage and the cell fate assignments of *C. elegans* are almost (though not quite) all invariant. Though its rigidity and immutability are somewhat peculiar with respect to some other nematodes (Voronoy and Panchin, 1998; Wiegner and Schierenberg, 1999), this provides an experimental advantage. Typically for a Type 1 embryo the genomes are activated very soon after fertilization, and a number of specific genes are known to be expressed zygotically before the eight-cell stage (Edgar *et al.*, 1994; Seydoux and Fire, 1994; Seydoux *et al.*, 1996).

The A/P axis is set up by a cytoarchitectural polarization that is brought about by sperm entry (Goldstein and Hird, 1996; reviewed by Bowerman, 1998). Localization of maternal transcription factors in the posterior end of the egg plays an important role in the initial specification of this region as cleavage gets underway (for reviews see Bowerman, 1998; Schnabel and Priess, 1997; Newman-Smith and Rothman, 1998). Autonomous specification has a unique meaning in the germline lineage, which arises in this posterior region. At the 4-cell stage a maternal protein, the product of the *pie1* gene, localizes to the nucleus of the P2 blastomere, from which the germline descends. In the following cleavages, at each of which a germline precursor and a somatic blastomere are given off, this protein remains in the germline cell nucleus (i.e., in P3, P4 and its progeny), and there it acts to shut off all transcriptional activity dependent on polymerase II (Seydoux *et al.*, 1996; Seydoux and Dunn, 1997; Batchelder *et al.*, 1999). This prevents the P2 lineage from responding to any maternal transcriptional activators present in the posterior region of the egg. With the possible exception of the blastomeres descendant from P2 (see Fig. 3.6A), cell fates in all of the remainder of the embryo are either known or inferred on good grounds to be specified conditionally. So also is the establishment of the partially asymmetric right/left pattern of sublineage fates (for reviews op. cit; and Hutter and Schnabel, 1994).

Endoderm Specification

The E blastomere is specified soon after its birth, when its fate as a precursor of intestine cells is fixed. In isolation from other blastomeres it will generate progeny that express gut marker genes, that divide the correct number of times, and that display cytoskeletal attributes of intestine cell differentiation (Leung *et al.*, 1999; Schroeder and McGhee, 1998). The first four to five cleavages in the *C. elegans* embryo and the cell fates ultimately expressed in each of the sublineages by then established are shown in the diagram of Fig. 3.6A. By the 26-cell stage, when the diagram terminates, only the E, P4, and D sublineages have been specified. That is, in the undisturbed embryo each subsequently generates but a single cell type (indicated by color in Fig. 3.6A). Most of the blastomeres of the E lineage divide four times, and two divide a fifth time. The resulting 20 cells constitute the intestine, a columnar structure in which the cells are arranged in nine rings. The first (i.e., most anterior) ring consists of four cells and the rest are in pairs; the lumen of the intestine forms down the middle of the column within the inter-cellular junctions (Leung *et al.*, 1999).

E blastomere specification depends on a signal from P2 to EMS, the conse-quence of which is to polarize the EMS cell so that E, its posterior daughter, becomes the intestine founder cell, while its anterior daughter, MS, produces mesodermal, neuronal, and other cell types (see Fig. 3.6B). The signal passes within a few minutes during the first half of the EMS cell cycle (Goldstein, 1992, 1993, 1995). Its striking effect is illustrated in Fig. 3.6C, an experiment where EMS was separated from P2 after (Fig. 3.6C1) or before (Fig. 3.6C2) passage of the signal had occurred, and the progeny of the EMS cell were cultured in isolation and then stained for a gut-specific enzyme (red color; Goldstein, 1995). In the absence of this signal, the blastomere in the position of E adopts the MS fate. This occurs whether signaling is prevented by early removal of the P2 blastomere or by genetic means, i.e., in maternal mutants lacking one of the components of the relevant signaling pathways (Goldstein, 1995; Thorpe *et al.*, 1997; Rocheleau *et al.*, 1997). Genetic analyses have demonstrated that P2 activates two different signal-ing pathways in EMS. It presents a Wnt ligand, which activates the β-catenin pathway within EMS, and it also causes activation of a MAP kinase pathway in the EMS cells (Meneghini *et al.*, 1999; Shin *et al.*, 1999). Both pathways converge on a maternal transcription factor of the Tcf/Lef1 family (here the product of the *pop1* gene). In Fig. 3.6D1 the Pop1 protein is displayed immunocytologically in nuclei of the four posterior blastomeres of an eight-cell embryo, i.e., E, MS, C and P3 (Lin *et al.*, 1995; compare the DAPI-stained eight-cell embryo in Fig. 3.6D2). The signal pathway which activates the Pop1 protein provides the input to a *cis*-regulatory switch that performs the key control function in the specification of the EMS daughters.

In addition to the *pop1* gene product there is another autonomous component of the EMS specification system in the form of a localized maternal transcription factor called Skn1 (Bowerman *et al.*, 1992, 1993). This is a homeodomain/bZIP

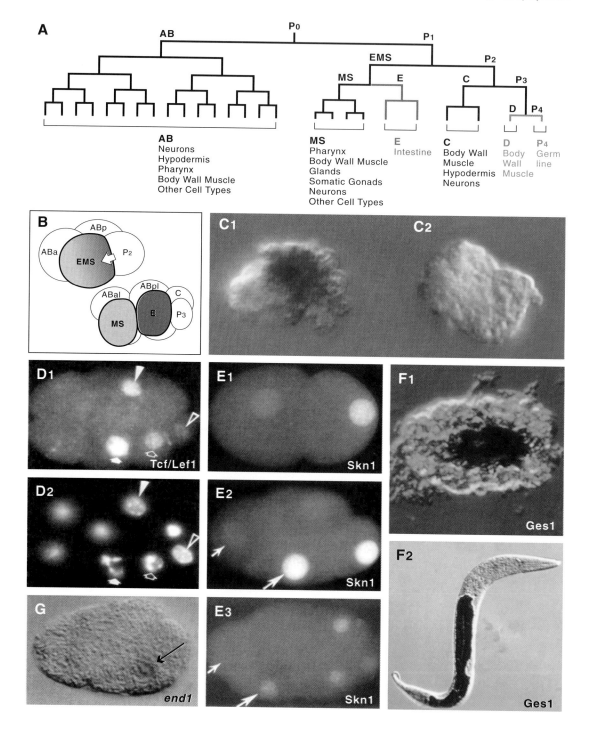

A

AB   P0   P1

EMS

MS   E   C   P2

D   P3

P4

AB
Neurons
Hypodermis
Pharynx
Body Wall Muscle
Other Cell Types

MS
Pharynx
Body Wall Muscle
Glands
Somatic Gonads
Neurons
Other Cell Types

E
Intestine

C
Body Wall
Muscle
Hypodermis
Neurons

D
Body
Wall
Muscle

P4
Germ
line

B
ABp
ABa   EMS   P2
ABal   ABpl
MS   E   C   P3

C1   **C2**

D1   Tcf/Lef1

D2

E1   Skn1

E2   Skn1

E3   Skn1

F1   Ges1

F2   Ges1

G   *end1*

FIGURE 3.6 (Continued)

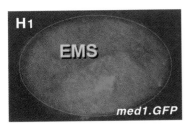
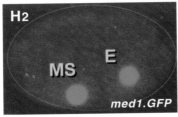
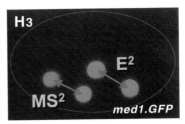

FIGURE 3.6 Specification of gut cell lineage in C. elegans. (A) Lineage of embryo to 26-cell stage. Names of some sister blastomeres are indicated from second cleavage on, but only in posterior half. Sublineages that have arisen in which all progeny give rise to only a single larval cell type are shown in color. Larval cell types that derive from the indicated lineage elements are shown below. Data from Sulston *et al.* (1983). (B) Diagrams of 4- and 8-cell embryos, viewed from ventral side. Signaling from P2 to its sister blastomere EMS is symbolized by the open arrow. The result is polarization of EMS (red shading) such that the blastomere adjacent to P2 is specified as the E founder cell. The E blastomere (red) arises at the 8-cell stage (see A), and from it derives all of the intestine cell lineage. Its sister cell MS gives rise to sublineages which produce mesodermal and other differentiated cell types as shown in (A). (C) Conditional specification of E lineage. The EMS blastomere was separated from P2, cultured, then stained for a gut-specific esterase (red). (C1) Separation of EMS from P2 after sufficient contact to permit passage of inducing signal (separation of EMS from P2 < 7 min before EMS division); (C2) earlier separation (>10 min prior to EMS division). The EMS cycle is about 15 min in length. [(A–C) From Goldstein (1995) *Development* **121**, 1227–1236 and The Company of Biologists, Ltd.] (D) Localization of maternal Tcf/Lef1 factor (product of *pop1* gene) in posterior cell nuclei. (D1) Immunocytological display, 8-cell embryo: the factor is present in the MS (closed white arrow) and E (open arrow) sister cells, as well as in C and P3 sister cells (arrowheads); cf. (B). (D2) DAPI stained 8-cell stage embryo to display location of all nuclei. Cells identified as in (D1). [(D) From Lin *et al.* (1995) *Cell* **83**, 599–609.] (E) Localization of maternal Skn1 transcription factor in early cleavage embryos, revealed by immunocytology. (E1) 2-cell embryo; Skn1 is present mainly in P1 nucleus; (E2) 4-cell embryo; (E3) 8-cell embryo (cf. B, D2). Short and long arrows indicate for comparison, respectively, a nucleus of an AB descendant and the EMS (E2) or MS (E3) nucleus. [(E) From Bowerman *et al.* (1993) *Cell* **74**, 443–452.] (F) Activity of a gut-specific esterase (product of the *ges1* gene); enzymatic stain. (F1) Ges1 esterase in embryo when there are about eight E lineage cells all expressing the gene; i.e., two cleavages beyond the end of the diagram in (A). Activity of *ges1* gene can be detected one cleavage before this (at the four E-cell stage). (F2) the Ges1 esterase is a terminal differentiation product of intestine cells, as seen in adult worm. [(F) From Edgar and McGhee (1986) *Dev. Biol.* **114**, 109–118.] (G) Expression of *end1* gene in E blastomere (arrow) by *in situ* hybridization (J. M.

(Continues)

transcription factor, which as illustrated in Fig. 3.6E, accumulates in the P1 nucleus and in its 2nd and 3rd cleavage progeny, but not in AB progeny. The Skn1 factor may be localized to posterior blastomere nuclei by differential translation (Bowerman et al., 1997). It is present only transiently and has disappeared by the 12-cell stage. In maternal skn1 mutants both E and MS specification functions partially fail. These two maternal transcription factors directly control the zygotic genes that mediate E and MS specification. The specification function of the Tcf/Lef1 factor is conditional on interblastomere signaling, and that of the Skn1 factor on its localization to posterior blastomere nuclei; as For the veg_2 endomesodermal precursors of the sea urchin embryo there is both a conditional and an autonomous component of the specification process.

The Network of Zygotic cis-Regulatory Interactions Required for Endoderm Specification

To take us where we want to go, we need to understand the network of interactions that link the cis-regulatory systems of the zygotic genes which control endoderm specification. The architecture of a specification network reveals how the genomic regulatory program for this developmental event works. It is the most important thing that we can know about mechanisms. The first problem is of course to identify the zygotic players.

The immediate target of Tcf/Lef1 is a pair of linked genes called end1 and end3, both of which encode transcription factors of the Gata family (Zhu et al., 1997). Gut specification does not occur in embryos which carry a large chromosomal deletion that includes these genes, and if expression of end1 and end3 is knocked out with "interference RNA" (RNAi), no intestine cells are formed (Kasmir et al., 2000). Furthermore, ectopic expression of end1 under heat shock control causes ectopic endoderm specification even in anterior (i.e., AB) lineages (Zhu et al., 1998). Forced end1 expression thus bypasses the need for the signal-activated Tcf/Lef1 factor and Skn1, since as illustrated in Fig. 3.6D, E, neither is present in AB descendants. The ges1 gene, which encodes a gut esterase, serves as a marker of downstream intestine cell differentiation. Expression of the ges1 gene normally begins when there are only four cells in the E lineage (Edgar and McGhee, 1986; its expression is shown at about the eight E-cell stage in Fig. 3.6F1, and in an adult worm in Fig. 3.6F2). When global end1 expression is induced by use of the exogenous heat shock construct the Ges1 esterase is synthesized everywhere (Zhu et al., 1998). These and other results show that the end1 and end3 genes are in the pathway that "reads" and transduces the

---

Rothman, personal communication). (H) Expression of med1.GFP construct. (H1) 4-cell embryo, expression in EMS; (H2) 8-cell embryo, GFP expressed in MS and E; (H3) the following cleavage (cf. A) when there are two E and two MS progeny, as indicated (J. M. Rothman, personal communication).

maternal Skn1 and Tcf/Lef1 inputs, and also that the *end* genes are upstream of the gut differentiation program (M.F. Maduro and J.M. Rothman, personal communication; Maduro *et al.*, 2000; Kasmir *et al.*, 2000). Either one of this pair of similar genes suffices. They are expressed in the E cell as illustrated in Fig. 3.6G, and in its immediate progeny. A *cis*-regulatory control system that mediates this early blastomere-specific pattern of transcription has been recovered from the *end1* gene (Kasmir *et al.*, 2000).

An MS vs. E switch operates within this *cis*-regulatory element. Some protein binding sites identified in the *cis*-regulatory DNA sequence of the *end1* and *end3* genes, and in the *end1* gene of another *Caenorhabditis* species, are indicated in Fig. 3.7A. Here we see that these regulatory genes are direct targets of the maternal Skn1 and Tcf/Lef1 factors (blue and green symbols in Fig. 3.7A). The effect of the P2→EMS signal is to convert the Tcf/Lef1 factor (Pop1) from a repressor of *end1* and *end3* to an activator (Maduro *et al.*, 2000; Kasmir *et al.*, 2000). Thus in MS, in the absence of the signal, the *end* genes are repressed; their activity is restored in *pop1* mutants (Thorpe *et al.*, 1997; Rocheleau *et al.*, 1997). In a minimal *end1* enhancer, deletion of a consensus Tcf/Lef1 site at which the maternal Pop1 factor probably binds, destroys the ability of this enhancer to direct expression in the E sublineage. Interactions in the *end1* *cis*-regulatory element with the maternal Skn1 factor are also essential for all but low activity levels. The Tcf/Lef1 and Skn1 interactions are synergistic: this is another example of *cis*-regulatory "and" logic (cf. Chapter 2), and double *skn1/pop1* mutations produce a nearly complete absence of endoderm specification, which neither single mutation does.

Two other nearly identical though unlinked genes also encoding Gata factors are "wired" in parallel to the *end* genes, in that they too respond directly to the maternal Skn1 factor. These are the *med1* and *med2* genes (Maduro and *et al.*, 2000). The relationship between the *med* and the *end* genes is shown at the top of the network diagram in Fig. 3.7C (direct interactions are shown here by solid line and inferences are indicated by dashed lines; see legend for evidence and references). Key sites in the *cis*-regulatory elements of the *med* genes include targets for both the Skn1 regulator and for Gata factors, as shown in Fig. 3.7B (Maduro *et al.*, 2000). Probably these two genes auto- and cross-regulate, so as to generate an autonomous amplification of output once they are turned on. The *med* genes are activated even earlier than are the *end* genes, in the EMS cell itself (Fig. 3.6H). The *med* gene products are required for MS as well as E specification, and *med* RNAi knockouts not only are missing structures usually derived from MS, but also sometimes lack the intestine or display other defects in endoderm development. The *med* genes operate in the initial zygotic specification of endomesoderm (i.e., EMS), and their own *cis*-regulatory elements do not in themselves mediate the E vs. MS choice. But they contribute to that function as executed by the *cis*-regulatory apparatus of the *end* genes: binding of the *med* gene Gata factors (probably at the Gata target sites shown in Fig. 3.7A) is necessary for full expression of the *end* genes, but only when the activated form of the Tcf/Lef1 regulator encoded by the

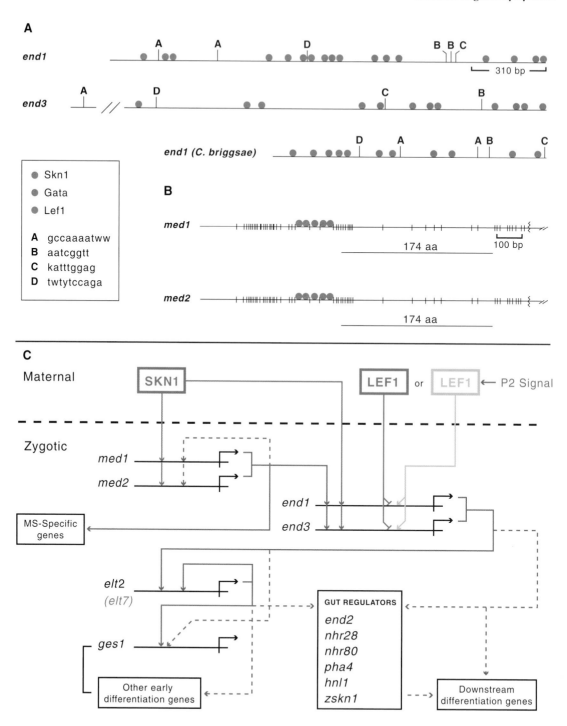

FIGURE 3.7 *Cis*-regulatory control of E lineage specification in *C. ele-gans*. (A) Observed target sites in *cis*-regulatory domains of the zygotic E lineage regulatory genes *end1* and *end3*, and the related *end1* gene of *C. briggsae*. All three *cis*-regulatory systems have multiple sites for Skn1, Tcf/Lef1, and Gata factors, plus a set of conserved sequence motifs (A, B, C, D) that may bind additional factors (see key). [(A) From Kasmir *et al.*, (2000), submitted for publication.] (B) Target sites identified immediately upstream of *med1* and *med2* genes, which also encode early zygotic Gata transcription regulators (see text). The two genes are almost identical in sequence though located on different chromosomes. Small vertical lines represent the only base pairs that are not identical between these genes. [(B) From Maduro *et al.*, (2000), submitted for publication.] (C) *Cis*-regulatory network for E lineage specification and immediately following events. For sources and evidence in addition to the following, see text. Maternal factors are shown in colored boxes. For the Tcf/Lef1 factor, green is used for repression and tan for activation functions. Red lines represent zygotic interactions. Where the red lines are solid there is evidence for direct inter-actions; dashed red lines are inferred interactions. *Cis*-regulatory elements of the *med* genes bind Skn1 protein *in vitro* and are also occupied by Skn1 factor *in vivo*, according to an *in situ* assay. Furthermore, *med1* expression constructs are dependent on presence of the Skn1 factor for activity, and ectopic expression of Skn1 causes a parallel ectopic expression of such constructs (Maduro *et al.*, 2000; M.F. Maduro and J.M. Rothman, personal communication). Presence of Gata sites (cf. B) suggest an autostimulatory feedback. As described in text the Tcf/Lef1 factor represses *end1* and *end3* (in MS and MS progeny), unless it is altered by the signal from P2, which converts it into an activator (in E lineage) (Kasmir et al., 2000). Binding of maternal Skn1 protein to *end1* and *end3* target sites (cf. A) is observed *in vitro*, and binding of Med1 protein at Gata target sites is found *in vivo* as well (J. M. Rothman, personal communication). The activation function of the Tcf/Lef1 factor which is the product of the *pop1* gene is required for expression. The normal level of output depends on the synergistic behavior of bound Skn1 and Med proteins (Kasmir *et al.*, 2000; Maduro *et al.*, 2000.). The *end1* and *end3* genes are expressed only transiently, during the initial few cleavages of the E cell. Therefore they are unlikely to be equipped with a self-sufficient autoregulatory device, depending rather on simultaneous inputs of the factors shown in (A), which are also present transiently. Autoregulation of the *elt2* gene was reported by Fukushige *et al.* (1998), and binding of the factor to its own *cis*-regulatory element has been visualized *in vivo* (Fukushige *et al.*, 1999). Genetic and molecular evidence require that *elt2* is downstream of the *end1/end3* genes: while ectopic expression of *elt2* causes ectopic gut-specific gene expression, including of the *ges1* gene (Fukushige *et al.*, 1998); elimination of *end1* and *end3* expression by use of RNAi prevents all endoderm differentiation (Zhu *et al.*, 1998; Kasmir *et al.*, 2000). *elt7* may function like *elt2*. The *elt2* gene product evidently activates *ges1*, according both to *in vitro* and genetic evidence, by interactions at a crucial pair of Gata target sites in the *cis*-regulatory element of the *ges1* gene (Fukushige *et al.*, 1996, 1998, 1999; Egan *et al.*, 1995). Initially the *ges1* gene might respond to regulation by *end1/end3* gene products,

(*Continues*)

pop1 gene is present (Maduro *et al.*, 2000; Kasmir *et al.*, 2000; M.F. Maduro and J.M. Rothman, personal communication). When it is absent, as in the MS cell, the repressor form is dominant.

The *end* genes may directly initiate expression of early gut-specific differentiation genes such as *ges1*, but they are not the major activators that drive differentiation of E-lineage progeny. Their role is specification. Immediately downstream of the *end* genes is another Gata-type regulatory gene called *elt2* (there may again be a pair of such genes; *elt7* is a likely candidate). The *elt2* gene has a strongly autostimulatory regulatory system which binds its own product *in vivo* (Fukushige *et al.*, 1998, 1999). Its role in the network is clearly to produce an amplified, gut cell-specific regulatory output. It is activated once the E blastomeres have divided to produce the first two E lineage daughters. *cis*-Regulatory analysis of the *ges1* gene shows that its expression depends on two Gata factor target sites (Aamodt *et al.*, 1991; Egan *et al.*, 1995). The *ges1* gene is activated by an *elt2* gene product which binds at these sites, or at least can bind there, since ectopic *elt2* expression results directly in ectopic *ges1* expression (Fukushige *et al.*, 1998). As Fig. 3.7C further indicates, the full program of intestine cell differentiation depends on a suite of other regulatory genes that operate further downstream. These must control expression of additional gene batteries that encode all the functions required for morphogenetic organization of the intestine (cf. Leung *et al.*, 1999), as well as for digestive operations. Some of these regulatory genes are gut specific (*end2, nhr80, elt4*), others not; and expression in the gut of at least one, the *forkhead/hnf3*β gene, *pha4* (Mango *et al.*, 1994), appears to be controlled by the factor encoded by the *elt2* gene (Kalb *et al.*, 1998).

The *cis*-regulatory network shown in Fig. 3.7C is simple, direct, and one might say, very nicely designed. It uses a pair of very similar *cis*-regulatory systems, those of the *end1* and *end3* genes, to transduce the signal that distinguishes E from MS and output a zygotic specification function. Another system, that of the *med* genes, transduces the localization pattern of the maternal Skn1 factor into a zygotic output that defines the EMS cell as endomesodermal. The two are combined in the synergistic *end1/end3 cis*-regulatory elements. And as soon as E is specified, the *end1/end3* products turn on *elt2*, a gene encoding an endoderm-specific regulator which is sharply stimulated by its own product. By the time

---

i.e., before institution of the *elt2* autoamplification circuit. Later differentiation of intestine cells involves other transcriptional regulators, some endoderm-specific, some not (for Nhr28, Miyabayashi *et al.*, 1999; for Pha4, Kalb *et al.*, 1998; for zSkn1, i.e., zygotically expressed Skn1, which is also required for normal gut development, Bowerman *et al.*, 1992; for others listed, J.M. Rothman, unpublished data). The dashed arrows from the *elt2* gene to the box containing the names of downstream transcriptional regulatory genes expressed in gut, indicates that for some genes such as *pha4* there is evidence for direct control by *elt2* (Kalb *et al.*, 1998), while for others it is not known whether control is direct or indirect.

there are only two E-lineage cells, a self-reinforcing gut-specific regulatory state has been generated. A downstream differentiation gene, *ges1* (and no doubt others), begins to be expressed when there are only four E-lineage cells, followed by all the other specific gene batteries required for differentiation of the intestine. The network shown in Fig. 3.7 is again a direct cell-type specification system that within a few divisions produces differentiated cell types from the progeny of unspecified early cleavage blastomeres.

SHORT SUMMARY: QUALITY OF TYPE 1 REGULATORY NETWORKS

In this chapter we have traversed regulatory mechanisms underlying the simplest and most direct processes of bilaterian embryogenesis, from the cytoplasmic interactions that mediate the initial specification events to the internal programs wired into the genome. Type 1 embryos display common phenomenological features, such as the early activation of embryonic genomes, and territorial specification during cleavage leading at once to differentiation. These features are general because they utilize a common regulatory "deep structure."

The most important feature of the architecture of the genomic regulatory networks that initiate Type 1 embryonic processes is that they are remarkably shallow. They do not have a deeply hierarchical or multilayered structure. The regulatory transactions that these networks mediate occur during cleavage. Essentially they depend on the information processing capabilities of key *cis*-regulatory elements, the job of which is to transduce the spatial inputs available in the early embryo into blastomere (and hence lineage- and territory-specific) transcriptional regulatory states. Thereby they execute what we call blastomere specification. Spatial inputs for specification derive from maternal cytoarchitectural location or from the regional activation of maternal components, sometimes caused by early interblastomere signaling. As specification states are established they are often supported by autostimulatory loops. The output of the specification process soon results in direct activation of differentiation gene batteries. The relation between specification state and the cellular organization of the Type 1 embryo is simplified by use of lineage relations: specification occurs in founder cells, and in later cleavage, these are divided up into polyclonal territories of differentiated progeny, all without net growth. From these territories the structures of the embryo arise. The complexity of the process of embryogenesis in Type 1 systems lies in the organization and operation of the cytoplasmic mechanisms that regionalize the activities of maternal regulatory molecules in the very early embryo; and in the internal organization of the key genomic *cis*-regulatory elements which respond to these regulators (cf. Chapter 2). But at least in relative terms the complexity is not in the number of layers in the network of intergenic regulatory connections, particularly in comparison to what we are to encounter in the next chapter.

Our task now is to approach the development of adult body parts. Here the answer to the same questions are quite different. Egg cytoarchitecture and sub-lineages of blastomeres that occupy fixed spatial positions are no longer relevant. The developmental process is itself far more complex, and it depends on far more elaborate forms of intergenic regulatory architecture.

CHAPTER 4

The Secret of the Bilaterians: Abstract Regulatory Design in Building Adult Body Parts

THE EVOLUTIONARY SIGNIFICANCE OF "PATTERN FORMATION"

The bilaterians share principles of developmental regulatory design that are uncannily similar in the most diverse examples, for instance flies and ourselves.

Yet at the DNA level these designs also hold the specific solutions to the general question with which this book begins: what accounts causally for morphological diversity in the bilaterians? The networks of gene regulatory interactions that direct the developmental construction of body parts in bilaterian development vary enormously in detail and outcome, but nonetheless they have common organizational properties; or one might say, they have a common sort of architecture. This is now to be our subject. From either an evolutionist's or a developmental biologist's point of view there may be no more generally important subject. For in their developmental regulatory architecture lies the real "secret of the bilaterians." That is so in two senses, one developmental and one evolutionary. The regulatory architecture explains how bilaterian genomes encode their body parts; but it follows also that the evolutionary appearance of the bilaterians had to have depended on the genomic assembly of such regulatory architecture in their common ancestors (a theory of bilaterian origins based in part on this concept is discussed elsewhere: Davidson *et al.*, 1995; Peterson *et al.*, 2000b; Peterson and Davidson, 2000). When conditions permitted, bilaterian diversification followed, powered not only by the capacities of their large complement of genes or gene families (Chapter 1), but by a shared heritage of regulatory "know how," that enables these genes to be deployed in an endless variety of deep regulatory formats.

"Deep" is a useful word here, because it connotes the stepwise and sequential nature of the necessary regulatory transactions, resulting in an architecture of multiple layers of control. At each level the functional object is to specify sets of transient spatial domains, or regions, and to control their growth, until all the subparts of an element of the body plan have been laid out. Within these subparts the batteries of differentiation genes that endow each with its tissue functions are finally activated. This is the process of "pattern formation" touched on in Chapter 1. The key mechanism in regional specification is the installation of new transcriptional regulatory states within each element of space. As indicated in Chapter 1, a transcriptional regulatory state denotes a region of a developing system in the cells of which given transcription factors are present and active. In the following we take up where this initial description leaves off: among other things we need to know what these transcriptional states accomplish; the nature of the genomic regulatory wiring needed to set them up; and how they are linked into a progressive series.

The outlines of the regulatory mechanism for body part formation seem clear, or are at least rapidly clarifying, but with respect to any particular body part, e.g., appendages or heads or glands, we can as yet only see through the glass darkly. Often we know relatively much about genomic mechanisms for the early stages, but almost nothing about the final steps of cell type diversification, and sometimes it is just the reverse. So the treatment that follows is synthetic. By use of many examples, drawn either from amniotes or from flies, we can see the elements of mechanism, the common principles of the bilaterian way of regulating adult body plan formation. Through these examples the elegance, the richness, and the variety of these mechanisms will hopefully be appreciated. But this is not the place to learn, from A to Z, how any particular piece of any particular animal develops: the focus of this chapter, like the object of its focus, is basically abstract. The mechanisms we want understood are abstract in that they utilize complex programs of genetic interactions to specify spatial domains, one after another, which do not in themselves resemble or function like parts of the structure that will emerge, until the final stages. To the observer taking a snapshot in developmental time, these domains appear as patterns, abstract patterns of regulatory gene expression.

THE FIRST STEP: TRANSCRIPTIONAL DEFINITION OF THE DOMAIN OF THE BODY PART

Consider body parts from an evolutionary perspective. The Cambrian lobopods lacked appendages which had the pattern elements of modern arthropod legs; the modern cephalochordates lack the organ level glands, and have at best only the simplest rudiments of the anterior brain organization of vertebrates (see Lacalli, 1996); modern agnathans and fossil conodonts lack the paired limbs of the jawed vertebrates, and so forth. Specific genomic regulatory apparatus is of course required for developmental formulation of every individual body part, and in each case the genomic apparatus must have been assembled stepwise during evolution, building upon the apparatus required to generate the relevant region of the body plan of the latest ancestor lacking that body part. Whatever its stage of evolutionary elaboration, a given body part is produced by a regulatory system that has a unit quality: it is a subprogram that starts with a particular developmental regulatory step, and proceeds from there. The following is about this first step.

The first regulatory step in body part formation is so general that it can be considered a law of bilaterian developmental biology. It is to set up a discrete regulatory state by expression of a gene or genes encoding transcription factors, specifically within a growing field of cells from which the body part will derive. A convenience term for this is the "progenitor field" for that part (Davidson, 1993). The *cis*-regulatory system activating each such gene in the progenitor field must read and integrate spatial inputs that define this field in terms of the preexisting regulatory coordinates of the body plan (here preexisting is used in the developmental sense, but it may often be true in an evolutionary sense as well). These coordinates specify position with respect to the midline, the anterior/posterior axis, the dorsal/ventral axis, the right/left location, or whatever. The following steps, by which the growing progenitor field is divided into subregions that will form the subparts of the future organ or structure, depend mechanistically on this first step. The observable earmarks of the first-step process are the appearance at a certain point in development of a pattern of regulatory gene expression such as can be detected by *in situ* hybridization or immunocytology, and which conforms to the cellular progenitor field for the body part; the conservation of this pattern within the clade of animals for which this body part is a specific shared character (i.e., the pattern is a synapomorphy); failure of the body part to develop if expression of the first-step regulatory genes can be knocked out or altered; and sometimes ectopic formation of elements of the body part if these genes are expressed ectopically, at least within those regions of the body plan wherein the additional necessary inputs are to be found. Of course this does not mean that expression of these first-step regulatory genes is both necessary and sufficient to make the whole body part; no one gene makes a structure (a strangely simple-minded view which for some reason has died hard). What it does mean is that the first step is necessary and sometimes sufficient to entrain all the subsequent processes, which do control the developmental assembly of the body part.

As an initial illustration of generality with respect to body part, Fig. 4.1 includes localized patterns of transcription of regulatory genes for four different organs of the mouse. In each case these patterns are evident in advance of morphogenesis of these structures. For example, the first known step in the transcriptional specification of the progenitor cells for the heart is activation of the *nkx2.5* gene (Fig. 4.1A, B; Harvey, 1996). There are several closely related members of the same gene family which have overlapping functions (reviewed by Evans, 1999), and orthologous genes are expressed at the beginning of heart development in the heart forming regions of all vertebrate classes (Harvey, 1996). As discussed below, *nkx2.5* expression is known to lie upstream of other regulatory genes that execute subsequent steps in patterning the heart, and much evidence exists for the role of this gene and its close relatives in the initial specification of the heart progenitor field (for reviews, Lin *et al.*, 1997; Evans, 1999; cf. Tanaka *et al.*, 1999). For instance, the introduction into *Xenopus* eggs of *trans*-acting negative versions of Nkx2.5 and the related Nkx2.3 factors that bear the *engrailed* repressor domain can result in total loss of heart specification, heart marker gene expression, heart morphogenesis, and cardiocyte differentiation (Fu *et al.*, 1998; Grow and Krieg, 1998). The point is that the pattern illustrated in Fig. 4.1A, B has a purpose: the first-step regulators set in train a sequential genetic machine, which first subdivides and specifies the parts of the heart, and then builds it.

Transcription of regulatory genes in the anterior neural tube of mouse embryos, from which the forebrain forms (Fig. 4.1C), probably plays similar roles. It is not easy to ferret out these roles because single gene knock-out phenotypes are not simply related to the early gene expression domains illustrated (e.g., for the *emx* and *otx* genes, see Boncinelli, 1999; Mallamaci *et al.*, 2000). Interpretation is complicated by the existence of multiple gene family members with partially overlapping functions; by prior embryonic lethality when the gene is knocked out; and more generally, by the density and parallel wiring of the regulatory systems that in mammals typically surround such genes. For example, activation of *emx1* and *emx2* in the domains indicated in Fig. 4.1C4 requires prior expression in the dorsal neural tube of a gene encoding another transcription factor, *gli3*; but expression of neither *otx1*, *otx2* (Fig. 4.1C4), *nkx2.1* (Fig. 4.1C1), *bf1* (Fig. 4.1C2), nor *shh* (Fig. 4.1C3) requires *gli3* (Theil *et al.*, 1999). These regulators are evidently "wired" in parallel and are initially activated by factors other than *gli3*. A powerful index of the importance of the patterns of gene expression in brain progenitor fields such as shown in Fig. 4.1C, though of a different sort than a gene knock-out phenotype, is their very remarkable conservation: throughout the vertebrates similar relative spatial patterns of genes encoding certain transcription factors occur early in brain development (e.g., for forebrain see Bally-Cuif and Boncinelli, 1997; Smith Fernandez *et al.*, 1998; for the midbrain-hindbrain region Joyner, 1996; and for hindbrain see below).

The role of the *pax6* gene in the development of the eye is taken up in the next chapter, but the first stage of that process illustrates the present point too well to omit here (Fig. 4.1D). Bilateral expression of *pax6* in the neuroepithelium of the amniote head region long before there is an eye, or any discrete structure relating thereto, is required for development of eyes (*pax6* is also utilized in development of brain regions and nasal placodes). In a *pax6* mouse mutant (*sey, small eye*) the earliest morphological manifestations of the progenitor field for the eye lens, the lens placode, fails to form, as shown in Fig. 4.1D2, 3); (Grindley *et al.*, 1995).

Figure 4.1E is included here because it provides a beautiful visual display of the kind of process by which the initial transcriptional regulatory states for body parts are established. They are located in the body plan by signaling. The spatial specification of these regulatory states demands integrative processing of signal inputs within individual *cis*-regulatory elements of genes that encode transcription factors (cf. Chapter 2). In the case considered in Fig. 4.1E, the body parts that will arise are the wings, halteres, and legs of *Drosophila*, which derive from the thoracic imaginal discs; the observation in the figure concerns the origins of these discs. The dorsal discs which give rise to wings and halteres and the ventral discs from which the legs emerge have a common origin in three patches of cells on each side in the postgastrular thoracic epithelium (reviewed by Cohen, 1993). These imaginal progenitor cells are marked by expression of the *distal-less* (*dll*) gene, or of a *lacz* gene insertion in the enhancer that directs *dll* expression in the early (5 h) postgastrular embryo. The expression of the *dll* gene is required for the developmental outcome of the ventral discs of the thorax, i.e., for formation of the legs (and also for the antennal discs of the head). In the absence of *dll* parts of the leg extending out from the body fail to form, while ectopic *dll* expression induces ectopic leg structures (Cohen and Jurgens, 1989; Gorfinkiel *et al.*, 1997). Figure 4.1E shows that the imaginal primordia are located precisely at the intersections between the parasegmental D/V stripes of *wg* gene expression and the bilateral longitudinal stripes of *dpp* gene expression. Expression of the *dll* Lacz marker is shown here superimposed on the ladder-like pattern of expression formed by these two genes, both of which encode signaling ligands. The pattern shown in Fig. 4.1E is causal as well as pretty, Wg acting as inducer of *dll*, and Dpp limiting its expression dorsally (Cohen *et al.*, 1993; Goto and Hayashi, 1997).

In summary, Fig. 4.1 shows us that development of very different body parts begins in just the same way, with the establishment of a patch of progenitor cells expressing certain genes encoding transcription factors. For each case we know

FIGURE 4.1   Initial stages of transcriptional progenitor field specifica-tion. (A), (B) Expression of *nkx2.5* gene marks the cardiac crescent and heart progenitor field, as visualized by *in situ* hybridization. This gene encodes a home-odomain transcription factor. (A) Mouse: (A1) 7.5 dpc, anterior aspect; (A2) 8.75 dpc, ventral view. (B) For comparison, expression in chick: (B1) stage 6, (B2) stage 10, ventral views. [(A, B) From Harvey (1996) *Dev. Biol.* **178**, 203–216.] (C) Expression of various transcription factors visualized by *in situ* hybridization in anterior neural tube of mouse. The expression domains demarcate future regions of the brain. (C1) *nkx2.1* expression in central region of developing forebrain, dorsoanterior view (he, heart; f, forebrain; m, midbrain). (C2) Expression of gene encoding a forkhead transcription factor, Bf1, at anterior end of neural plate; anterior view. (C3, C4) Two diagrammatic views of early gene expression domains: (C3) Expression of *bf1*, *otx1*, *nkx2.2*, *shh*, and *emx2* genes in anterior neural tube at 7–somite stage, dorsal view. (oc, optic cup; pr, prosencephalon; me, mesencephalon; rh, rhombencephalon; ap, alar plate; bp, basal plate; fp, floor plate; pos, postotic sulcus.) [(C1–C3) From Shimamura *et al.* (1995) *Development* **121**, 3923–3933 and The Company of Biologists Ltd.] (C4) Mouse embryo, 10 dpc, showing domains of expression of *otx1*, *otx2*, *emx2*, and *emx1* genes; telencephalon (Te), diencephalon (Di), mesencephalon (Mes). These genes are not expressed in the midbrain or hindbrain, i.e., metencephalon (Met), and myelencephalon (My). The *hox* genes are expressed in the hindbrain and more caudally. The domains of expression of the four genes indicated appear to be nested within one another such that *otx2* is expressed in the most broad pattern and *emx1* in the most confined pattern. [(C4) From Simeone *et al.* (1992) *Nature* **258**, 687–690, copyright Macmillan Magazines Ltd.] (D) Expression of *pax6* gene establishes the eye progenitor field in amniotes. (D1) Expression of *pax6* gene in chicken embryo, viewed in section after whole mount *in situ* hybridization. The embryo is at 2-somite stage (stage 7); the domain of expression shown is lateral to the V-shaped neural plate at the forebrain level, and includes the progenitors of eye lens and corneal epithelium. [(D1) From Li *et al.* (1994) *Dev. Biol.* **162**, 181–194.] (D2, D3) Consequences of mutation in *pax6* gene (*sey*, small eye mutation) in mouse: (D2) normal histology of early optic vesicle, in 7.5 dpc embryo. The lens placode (lp) arises as a local thickening of the surface epithelium (se). (D3) In *sey* homozygotes the lens placode fails to form. [(D2, D3) From Grindley *et al.* (1995) *Development* **121**, 1433–1442 and The Company of Biologists Ltd.] (E) Location of thoracic imaginal disc progenitors in *Drosophila*, with respect to intersec-tion of stripes of *wg* and *dpp* expressing cells. The embryo has been dissected and spread out, anterior left, ventral surface toward viewer; (t1), first thoracic segment. Longitudinal stripes of *dpp* expression, and metameric stripes of *wg* expression, both visualized by *in situ* hybridization, form a ladder-like pattern. This is shown together with six Lacz expression domains from a transgene driven by the early expression *cis*-regulatory module of the *distal-less* gene (dark blue). The cells expressing the transgene indicate the common progenitors of wing and leg imaginal discs [(E) From Cohen *et al.* (1993) *Development* **117**, 597–608 and The Company of Biologists Ltd.]

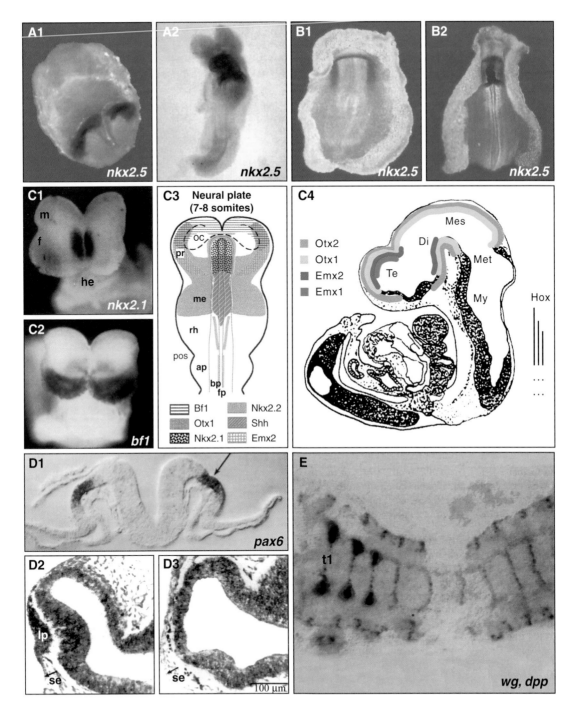

that these patterns of expression are an important first step: what happens next depends upon it. But the following steps are much more complicated. We first consider the ensuing process from the outside, i.e., we shall look at examples of more regulatory gene expression patterns, the function of which is to subdivide the initial field into the future subparts of that piece of the body plan. Then in the following section we turn to the underlying regulatory circuitry.

MORPHOLOGICAL PIECES AND REGULATORY SUBPATTERNS

The processes which occur next in the formation of adult body parts reflect the shape of the genomic regulatory networks that control bilaterian morphogenesis. This is our real objective, and though it is just becoming possible to perceive the architecture of such networks, the process itself is of tremendous interest. The initial progenitor field is transformed into a mosaic of regulatory subdomains, and, remarkably, these prefigure the morphological pieces of the body part. Many sequential steps are sometimes required before the process is complete. Setting up the developmental components of the morphology is to be distinguished from setting up particular differentiation programs, for the need for differentiation programs is frequently overlapping among the diverse pieces of which the body part is composed: cartilage, osteoclasts, and the various skeletal muscle fiber types are required within all the morphologically distinct components of the vertebrate limb; and the cell types of peripheral nervous system elements, veins, and inter-vein regions are generated alike in the diverse pattern elements to which the *Drosophila* wing disc gives rise. The last chapter dealt with regulatory processes that are aimed at direct cell type specification, but not so these. Furthermore, and most importantly, because we are large animals (i.e., compared to eggs or Type 1 embryos), the spatial components of the forming body part must also function during development as growth control units.

Here is where the term "abstract," as in the title of this chapter, is most needed. Genetic regulatory units hardwired in the genome produce as their output spatial transcriptional domains, which after many further patterning steps produce, as their output, the morphological elements of the body part. Five briefly discussed examples follow, to illustrate these causal, though abstract relations. For each case development of a particular morphological piece of the finished product, the body part, is seen to depend on an identifiable piece of the regulatory apparatus, that operates far upstream of the development of the final form.

Heart Parts

Upon specification of its progenitor field morphogenesis of the heart begins with the formation of a tube. Regions of this tube give rise to the different morpholo-gical and functional components of the heart (reviewed by Fishman and Chien,

1997; Christoffels *et al.*, 2000): in amniotes, these are the left and the right ventricles, and the left and right atria, plus the inflow and outflow tracts. Specification of the ventricular domains provides a particular case of the general process by which the subparts of an organ begin their development. The forming ventricles express two genes encoding bHLH transcription factors, the *dhand* and *ehand* genes (Srivastava *et al.*, 1997; Biben and Harvey, 1997), the expression patterns for which are shown in Fig. 4.2A and B. These patterns are quite remarkable: while the *dhand* gene is expressed strongly in the future right ventricle and more weakly in the future left ventricle (Fig. 4.2A2), *ehand* is expressed not at all in the future right ventricle but strongly in the future left ventricle (Fig. 4.2B2). If the *dhand* gene is knocked out, ventricular morphogenesis fails in the developing heart (Fig. 4.2C). The phenotype of the knockout includes failure to mobilize neural crest components that produce the aortic arches (in which the *dhand* gene is also expressed; Cserjesi *et al.*, 1995). The *hand* genes are downstream of the *nkx2.5* gene (Biben and Harvey, 1997; Srivastava *et al.*, 1997), and regional *dhand* expression also requires the *mef2c* gene, which together with the *nkx2.5* gene is expressed throughout the early heart tube and is required for its further morphogenesis (Lin *et al.*, 1997). Activation of *dhand* therefore depends on initial specification of the heart progenitor field. But obviously regulation of the *dhand* and *ehand* genes depends as well on additional spatial cues in the early embryo.

The *dhand* gene in turn controls other genes needed for ventricular development. Among these are *gata4* (Srivastava *et al.*, 1997) which is essential for further heart morphogenesis (Molkentin *et al.*, 1997). Another regulatory gene, *irx4*, encodes a homeodomain factor which in turn is needed for expression of the genes which encode the ventricle-specific myosin heavy chain isoform; but ventricular expression of *irx4* is not maintained in the absence of *dhand* expression (Bao *et al.*, 1999; Bruneau *et al.*, 2000). We can anticipate a *cis*-regulatory network that links the initial specification of the heart progenitor field (Fig. 4.1A, B) to the regionalization functions which specify the parts of the heart, and thence to the morphogenetic and functional development of these parts. A prominent aspect of the format of this network is that the genes which define each successive spatial and morphogenetic domain continue to be used to provide inputs for the following regulatory phase, together with additional new, regionally confined inputs. In this way the spatial achievement of each regulatory stage is integrated together with new cues so that specification is progressive. Some detailed examples of this principle appear in the following section.

Forelimb and Hindlimb Buds

The two bilateral pairs of limb buds arise in the lateral plate mesoderm of the amniote embryo. Their position is thought to depend on the prior A/P expression domains of *hox* genes (e.g., Gibson-Brown *et al.*, 1998; Cohn *et al.*, 1997), though the molecular linkages by which this is accomplished are unknown. At the beginning of limb bud outgrowth *hoxd9* is expressed in both leg and wing buds in

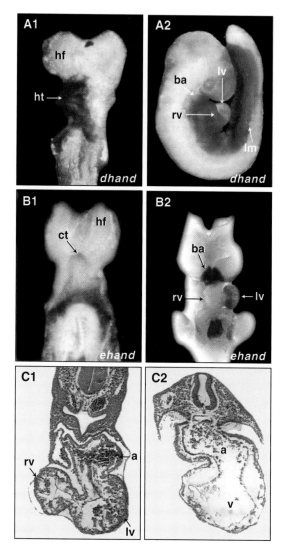

FIGURE 4.2   Transcriptional specification domains for the ventricular components of the heart in the mouse.   (A, B) Expression of *dhand* and *ehand* genes visualized by *in situ* hybridization. (A) Expression of *dhand* gene; (A1) 8.0 dpc; (A2) 9.5 dpc. (B) Expression of *ehand* gene: (B1) and (B2) same stages as (A1) and (A2). Abbreviations: ht, heart; hf, head fold; ba, branchial arch; lv, left ventricle; rv, proper ventricular, lm, lateral plate mesoderm; ct, conotrunchus. (C) *dhand* is required for right ventricle formation: (C1) Normal morphology, transverse section 9.5 dpc. (C2) morphology in embryo of same age in which the *dhand* gene has been knocked out: a, v indicate regions where atria and ventricles would have formed. [(C) Adapted from Srivastava *et al.* (1997) *Nature Genetics* **16**, 154–160, 410; copyright Macmillan Magazines Ltd.]

the chick (Nelson *et al.*, 1996; Cohn *et al.*, 1997), but not in the intervening lateral plate mesoderm. Application of beads which emit certain FGFs to this intervening region suffices to induce ectopic limb buds and limbs (Isaac *et al.*, 2000; whether or how these FGF family members function in the normal process of limb bud specification is not clear). If the beads are implanted in the anterior portion of this region, wing-like structures are formed, and if in the posterior region legs are formed (Cohn *et al.*, 1995). The extant positional inputs thus superimpose fore-limb or hindlimb identity on the ectopic limb outgrowth. The process by which "identity" is established with respect to the future morphogenetic character of a body part is now familiar: it requires institution of a specific transcriptional regulatory domain, as in all the cases considered in this chapter. Some of the regulatory players responsible for forelimb vs. hindlimb fate are shown in Fig. 4.3. The *pitx1* gene and the *tbx4* gene are expressed specifically in the hindlimb (Fig. 4.3A, C) in chicken (Logan *et al.*, 1998) and mouse (Szeto *et al.*, 1996); and the *tbx5* gene is expressed specifically in the forelimb (Fig. 4.3B; *ibid.*). The *tbx4* gene responds to expression of the *pitx1* gene, though the latter is activated in a broader area than the limb bud alone, while *tbx4* transcripts are here confined to the limb bud: *pitx1* is activated earlier than is *tbx4* in either normal or ectopi-cally induced limbs; and if *pitx1* expression is induced in the wing bud by introduction of a retroviral expression vector, the consequence is to activate *tbx4* expression there (Fig. 4.3D; Takeuchi *et al.*, 1999; Logan and Tabin, 1999). Furthermore, *tbx4* expression is exclusive of *tbx5* expression. A limb bud induced with an FGF bead in the anterior region of the flank expresses *tbx5* (Fig. 4.3E1) and as indicated above it produces an ectopic wing, illustrated in Fig. 4.3E3. But if expression of *tbx4* is forced within the ectopic limb by introduction of a retroviral expression vector, *tbx5* expression is turned off (Fig. 4.3E2), and the resulting ectopic limb now often displays morphological features of the hindlimb; e.g., it forms a claw (Fig. 4.3E4; Rodriguez-Esteban *et al.*, 1999; Takeuchi *et al.*, 1999).

Following the specification of forelimb or hindlimb identity for which the *tbx* genes are necessary and sufficient, other genes initiate the morphogenesis of the limb. Among those are sets of *hox* cluster genes, different members of which are expressed in specific patterns in the developing wing and leg (Nelson *et al.*, 1996; see Chapter 5). The *hoxd9* gene is upregulated in the wing bud and *hoxc9* down-regulated, while the opposite occurs in the hindlimb bud. Introduction of retroviral vectors encoding *tbx4* into wing bud causes the leg bud pattern of *hoxc9* and *hoxd9* expression; and introduction of a *tbx5* vector in the leg bud causes installa-tion of the wing bud pattern of activity of the *hox* genes (Takeuchi *et al.*, 1999).

All these results show that the *tbx* genes lie rather far upstream in the regulatory patterning network that establishes the limbs. Their regional activation follows initial specification of the limb buds in respect to the A/P axis of the embryo, while they themselves provide inputs that directly and/or indirectly set up the forelimb- and hindlimb-specific patterns of *hox* gene expression, and no doubt as well activate (or repress) many other target genes (in addition to affecting one another). But this is only an early stage of the abstract process of regional

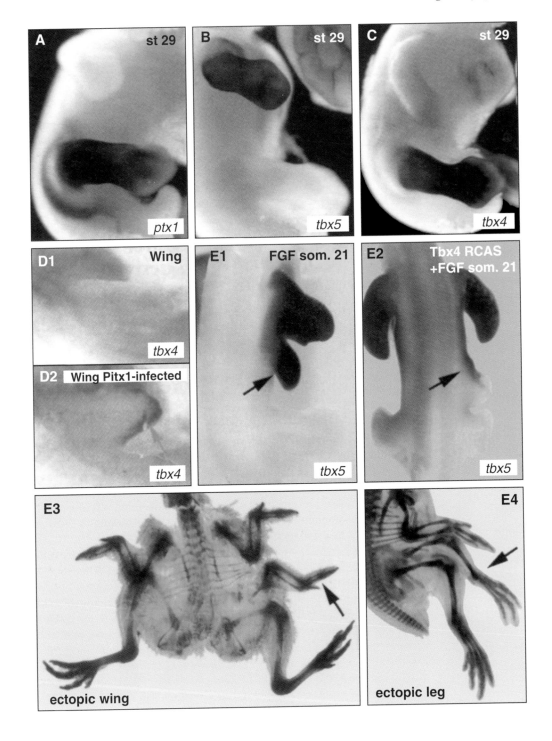

specification in the limb, which has many steps and stages and which is not yet very well known. Prominent elements of the control apparatus are the regulatory circuits that establish the A/P, D/V, and proximo-distal domains from which the different subparts of the limb will develop, and which later direct the actual morphogenesis of the wing.

Transcriptional Domains in the Gut Endoderm

In the two examples so far considered (i.e., in Figs. 4.2 and 4.3) we have looked at patterning within the progenitor fields for particular structures. But these kinds of mechanisms also apply to the overall body plan, and for this reason Fig. 4.4 is included. It serves as a reminder of the scope of the phenomenon we are discussing, before proceeding with better known and more detailed examples. Figure 4.4A shows in diagrammatic form the whole gut tube of a chick embryo: on the left is shown a series of transcriptional regulatory domains that have been imposed within the gut endoderm at the 15 somite stage; and in the center the organ that will form from each such domain (Grapin-Botton and Melton, 2000). The diagram shows that each organ arises from a region distinguished by a unique regulatory state at this stage. These patterns appear to be established by signaling from the surrounding mesoderm and ectoderm (Wells and Melton, 2000). Two examples of these patterns are also shown, *viz pdx1* expression in the dorsal and ventral pancreatic buds (Fig. 4.4B); and *hex* expression in the endoderm that will

---

FIGURE 4.3   Early transcriptional specification of hindlimb and forelimb identity in chick.   Domains of expression are visualized by *in situ* hybridization. (A) Expression of *pitx1* homeodomain regulatory gene in mouse hindlimb bud (this gene is also expressed in branchial arches and in the pituitary anlage). (B) Expression of *tbx5* gene in forelimb (wing) of chick; and (C) expression of *tbx4* in hindlimb. [(A–C) From Logan *et al.* (1998) *Development* **125**, 2825–2835 and The Company of Biologists Ltd.] (D) Demonstration that *pitx1* expression is upstream of *tbx4* expression: *pitx1* expression was forced in the wing bud by introduction of a retroviral expression vector, and *tbx4* expression can now be detected by *in situ* hybridization; (D1) control, (D2) ectopic *pitx1* wing bud, in which *tbx4* is expressed (red arrow). [(D) From Logan and Tabin (1999) *Science* **283**, 1736–1739; copyright American Association for the Advancement of Science.] (E) Fate of induced limb bud altered by forced *tbx4* gene expression. (E1) Induction of ectopic limb bud (arrow) in chick by implantation of bead emitting *Fgf2* in lateral plate mesoderm at the level of somite 21; the induced ectopic bud expresses the *tbx5* gene. (E2) Expression of *tbx5* in induced ectopic limb bud (arrow) is nearly extinguished on introduction of a retroviral vector (RCAS) expressing *tbx4*. (E3) Limb bud induced as in (E1) produces an ectopic limb that displays skeletal character of wing (70% of cases; arrow). (E4) Induced limb bud in which *tbx4* expression is forced instead produces mainly ectopic legs (56% of cases; arrow). [(E) From Rodriguez-Esteban *et al.* (1999) *Nature* **398**, 814–818; copyright Macmillan Magazines Ltd.]

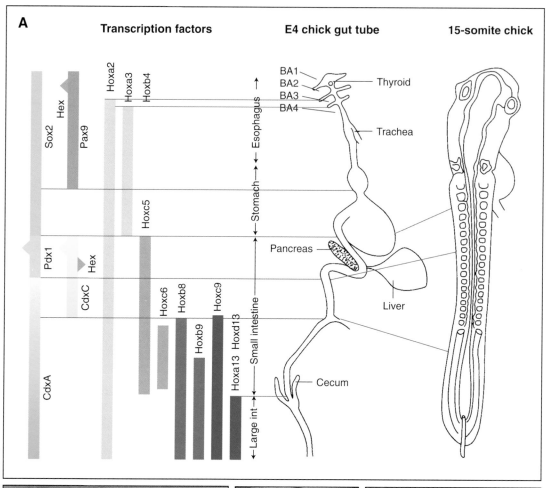

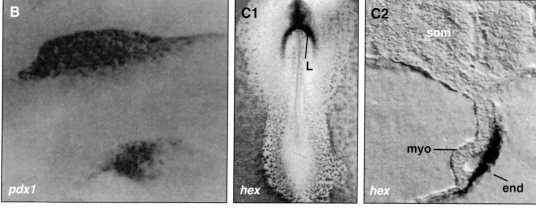

give rise to the liver (Fig. 4.4C). A *pdx1* knockout shows that this gene is indeed essential for proliferation and differentiation of pancreatic cell types within the bud (Offield *et al.*, 1996). We can be sure that Fig. 4.4A illustrates only a fraction of the regulatory transactions required to specify downstream morphogenetic fates in the endoderm. Even so, it shows that spatially confined transcriptional patterning is utilized throughout the whole extent of the endodermal germ layer.

Patterns in the Developing Hindbrain

The embryonic hindbrain of vertebrates transiently displays a segmental organization. In molecular and developmental terms the metameric units, or rhombomeres, are genetic regulatory units. That is, transcriptional regulatory states specific to each rhombomere are established early in hindbrain development, while each rhombomere can be considered the subunit of the hindbrain from which will ultimately derive specific facial and lower head ganglia, as well as specific populations of neural crest (Nieto *et al.*, 1995). The developmental outcome depends on both autonomous and nonautonomous (i.e., signaling) functions of the individual rhombomeres (reviewed by Lumsden and Krumlauf, 1996; Marin and Charnay, 2000). Rhombomere specification affords a clear causal link connecting early gene expression patterns with the later morphological output of the regulatory system.

Regional expression of many different genes has been recorded in the developing hindbrain. The domains of transcription of some genes which encode transcription factors, and others which encode components of signaling systems are shown against the metameric register of rhombomeres (r) 2–6 in Fig. 4.5A.

FIGURE 4.4 Regional expression of transcription factors foreshadows endodermal fate. (A) Summary of gene expression domains (left) mapped onto completed gut tube of chick (center) to illustrate the correlation between early transcriptional domains in the 15-somite embryo (right) and regional organogenesis in the endoderm (center). Pink triangles indicate *hex* gene expression in thyroid (top left) and in liver bud (center left); blue triangle (center left) indicates expression of *pdx1* gene in pancreatic bud; BA, positions of branchial arches. [(A) From Grapin-Botton and Melton (2000) *Trends Genet.* **16**, 124–130.] For additional *hox* cluster expression patterns in the endoderm see Beck *et al.* (2000). (B) Example of pattern shown in (A): expression of *pdx1* in pancreatic buds in mouse, visualized by immunocytology. Top domain is dorsal pancreatic epithelium, lower is ventral. [(B) From Ahlgren *et al.* (1996) *Development* **122**, 1409–1416 and The Company of Biologists Ltd.] (C) Expression of *hex* gene in endoderm that will give rise to liver bud (L) adjacent to heart epithelium in a stage 12 (18-somite) chick embryo, visualized by *in situ* hybridization. (C1) Whole mount; (C2) sagittal section showing that expression is confined to endoderm (end); myo, myocardium; som, somite. [(C) From Yatskievych *et al.* (1999) *Mech. Dev.* **80**, 107–109; copyright Elsevier Science.]

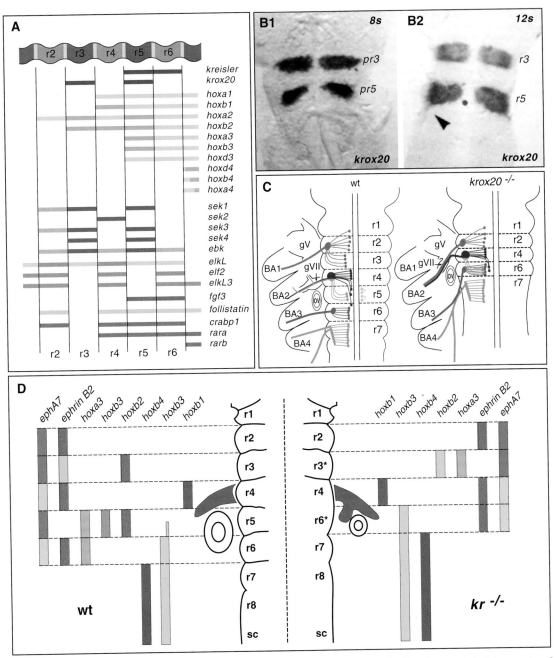

FIGURE 4.5 Rhombomere specification in hindbrain: establishment of essential spatial regulatory states. (A) Correlation of domains of spatial expression of indicated genes in the amniote hindbrain with respect to specific rhombomeres

From this diagram alone one can see that each rhombomere expresses a unique combination of developmentally important genes. The remainder of Fig. 4.5 concerns the functional significance of the two genes shown at the top of Fig. 4.5A, i.e., *krox20* and *kreisler*. The *krox20* gene encodes a Zn finger transcription factor, and its expression in the hindbrain is sharply confined to r3 and r5, as illustrated in Fig. 4.5B. The downstream consequences of knockout of this gene are illustrated in Fig. 4.5C: essentially the structures to which r3 and r5 normally give rise are just missing (Schneider-Maunoury *et al.*, 1997). Some neuronal populations are absent, others redirected, others respecified, a combination of

---

(r2–r6). These are the transient metameric units of the neuraxis in the hindbrain region. Genes encoding transcription factors are shown above and genes encoding signaling components below. Levels of expression are indicated by shading. Abbreviations for genes: *crabp1*, retinoic acid binding protein; *rar's*, retinoic acid receptors; *sek's* and *ebk*, Eph-related tyrosine kinase receptors; and *elk's*, *elf2*, ligands thereof. *kreisler* and *krox20* encode transcription factors. [(A) From Lumsden and Krumlauf (1996) *Science* **274**, 1109–1115; copyright American Association for the Advancement of Science.] (B) Patterns of expression of *krox20* transcription factor in developing mouse hindbrain. (B1) 8-somite (8s) stage; (B2) 12-somite (12s) stage. Expression is sharply confined to prospective rhombomeres (pr) (B1) and then rhombomeres (r) 3 and 5 (B2). [(B) From Mechta-Grigoriou *et al.* (2000) *Development* **127**, 119–128 and The Company of Biologists Ltd.] (C) Effects of *krox20* gene knockout on subsequent neural organization in the domains to which the rhombomeres give rise. The diagrams show mouse embryos at 10.5 dpc. Color coding indicates neuronal pools deriving from each rhombomere. The major components in normal embryos (left; wt, wild type) are the motor neuron pool, in red, from r2 and r3; facial motor neurons, dark blue, from r4; salivary neurons and neurons that enervate portions of the ear, light blue, from r5; IXth (green) and Xth (pink) motor nerves, from r6 and r7, respectively. In *krox20* knockout embryos most of the specific neuronal pools of r3 and r5 are missing or redirected, i.e., those shown in red in wild-type r3 and light blue in wild-type r5. BA, branchial arch; ov, otic vesicle, the developing ear. [(C) From Schneider-Maunoury *et al.* (1997) *Development* **124**, 1215–1226 and The Company of Biologists Ltd.] (D) Regional specification of posterior rhombomeres is controlled by expression of the *kreisler* (*kr*) gene. This gene encodes a leucine zipper transcription factor, normally expressed in r5 and r6 (Cordes and Barsh, 1994). *Kr* knockout embryos display altered spatial domains of several downstream regulatory genes, e.g., *hoxb3* (green) where rhombomeric expression is lost altogether; and *hoxb2* (red) where it is lost in r5, since r5 does not form. *Hoxa3* expression (orange) is ectopic as well. In addition, expression of Eph/ephrin receptors and ligands is affected (light and dark green and light and dark pink); levels of expression are represented by shades of color. The concentric circles are otic vesicles. Migrating neural crest patterns (purple) are also altered in the modified r2 which results from the mutation (r6*). The diagram represents embryos at 9.5 dpc. [(D) From Manzanares *et al.* (1999a) *Dev. Biol.* **211**, 220–237.]

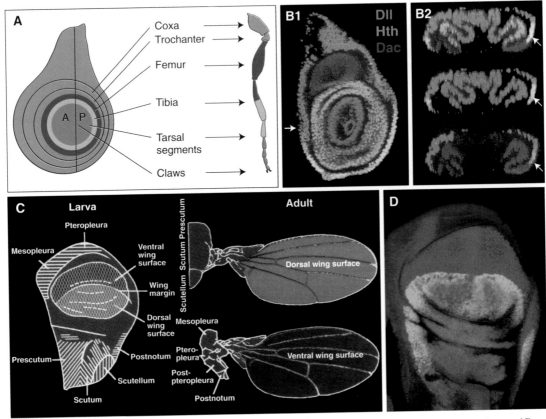

FIGURE 4.6 Transcriptional regulatory domains in the regional specification of thoracic imaginal discs in *Drosophila*. (A) Diagram of leg imaginal disc. Distal structures are generated from central regions of disc and more proximal structures from concentric peripheral domains. The coxa and adjoining regions are genetically defined by expression of *extradenticle* (*exd*) and *homothorax* (*hth*) genes (González-Crespo and Morata, 1996; Wu and Cohen, 1999), and the leg proper by expression of the *distal-less* (*dll*) regulatory gene together with other genes. [(A) From Lecuit and Cohen (1997) *Nature* **388**, 139–145; copyright Macmillan Magazines Ltd.] (B) Composite confocal images displaying domains of expression in 3rd instar discs of *hth* (red), *dll* (green), *hth* + *dll* overlap (yellow), *dachshund* (*dac*), dark blue; *dac* + *dll* (light blue) genes. Dac is a nuclear protein required for formation of femur, tibia, and the proximal three tarsi (Mardon *et al.*, 1994; Lecuit and Cohen, 1997). Images were obtained by immunocytology. (B1) Horizontal optical section; (B2) three images of an optical cross section (at arrow) of disc shown in (B1). The top image displays expression of all three factors; the center image of *Hth* and *Dll* only; the bottom image of *Hth* and *Dac* only. The convoluted form of the central (distal) region of the disc at this stage is here evident. Arrows indicate regions where expression domains overlap.

autonomous and indirect or nonautonomous developmental failures. In Fig. 4.5D the effects on rhombomeric gene expression patterns of a *kreisler* (*kr*) gene knockout are shown (Manzanares *et al.*, 1999a). The *kr* gene is normally expressed in r5 and r6, and the main effects of the knockout are in these domains, but again there are nonautonomous effects as well. Although not illustrated here, the subsequent consequence is again the development of defective neuronal architecture. In case there were any doubt, these examples show the essential role of the transient early transcriptional states by which specific downstream regulatory functions are brought into play in individual rhombomeres.

Appendage Parts and Transcriptional Patterns in *Drosophila* Imaginal Discs

Drosophila affords many excellent examples of the institution of spatial transcriptional domains at the beginning of a developmental process, the function of which is to specify subpopulations of cells that will give rise to particular components of a body part. But none so transparently reveal the nature of the mechanism as does transcriptional pattern formation in imaginal discs. This is partly because the cell populations of these anlagen are confined from the beginning; the imaginal disc cells are set aside from the rest of the epidermis as in Fig. 4.1E, and in terms of transcriptional state and morphology remain distinct. A major simplifying factor is the two-dimensional nature of their fate maps. Figure 4.6 illustrates such fate maps, and their relations to the three-dimensional structures to which the discs give rise, for leg (Fig. 4.6A) and wing (Fig. 4.6C). The centers of the discs evaginate

[(B) From Wu and Cohen (1999) *Development* **126**, 109–117 and The Company of Biologists Ltd.] (C) Diagrammatic display of 3rd instar wing disc (anterior left), and of the dorsal and ventral surfaces of the adult wing. [(C) From Williams and Carroll (1993) *BioEssays* **15**, 567–577, copyright Wiley-Liss, Inc., a subsidiary of John Wiley & Sons, Inc.] (D) Transcriptional regulatory states that mark the dorsal, ventral, and anterior domains of the wing disc, oriented as in (C), and displayed immunocytologically. The *vestigial* (*vg*) gene, which encodes a transcriptional cofactor (Halder *et al.*, 1998) is expressed throughout the wing blade (green), under control of the boundary and quadrant enhancers discussed in Chapter 2 (cf. Fig. 2.7); and the *apterous* (*ap*) gene, which encodes a Lim domain transcription factor, is expressed in the future dorsal domain of the wing blade and in more proximal regions (red). Thus the dorsal side of the future wing blade is denoted by *ap* + *vg* expression and the ventral side by expression of *vg* without *ap*. All cells of the anterior portion of the disc express the *cubitus interruptus* (*ci*) gene (blue), which is activated downstream of Hedgehog signaling on the anterior side of the A/P boundary (Jiang and Struhl, 1996). The A/P boundary itself is marked by a stripe of Dpp expression is not shown here. [(D) From Williams *et al.* (1994) *Nature* **368**, 299–305; copyright Macmillan Magazines Ltd.]

to form the distal structures of the appendages and the peripheral regions give rise to the structures that join the appendage to the body wall and the adjacent portions thereof. By comparison to the fate maps, the images of the transcriptional domains set up in these discs say it all: for leg discs see Fig. 4.6B (Lecuit and Cohen, 1997; Wu and Cohen, 1999); for wing discs, Fig. 4.6D (Williams *et al.*, 1994). The transcriptional domains mark out the regions of the discs from which the subparts of the emergent appendages will form, according to the fate map. As many mutational studies show (*ibid.* for references), expression of the genes is indeed essential for the development of the respective subparts, though many subsequent patterning steps intervene before institution of the final, detailed morphogenetic processes (see below).

In summary, in the modern Bilateria body parts are composed of subparts, and following the initial transcriptional specification of the progenitor field for the morphologically distinct body part, the subparts are each defined by institution of a particular regulatory state. This requires expression of a set of genes encoding transcription factors, and then the operation of an ensuing regulatory network, within the regions from which the subparts will form. The mechanisms by which the regional expression domains of such genes are set up; what these genes do; and what happens downstream, we take up in the following.

GLIMPSES OF HOW IT WORKS

For our purposes "how it works" means getting as close as possible to the way pattern formation is encoded in the genome. For that is of course where the whole process is programmed. As remarked above, thereby we can approach a direct DNA-level solution to the problem of understanding what has happened in evolution to generate diverse animal morphologies, i.e., the clade-specific parts and pieces of body plans. Someday, perhaps not too far off, we will be able to trace the genomic regulatory network for formation of a typical animal body part all the way from its inception, through the imposition of all the spatial transcriptional domains which specify the future subregions of the body part, to its terminal differentiation phase. This will require knowledge of the large number of functionally linked *cis*-regulatory elements which mediate expression of the key genes in the network. We will need to know the relevant inputs and outputs of these *cis*-regulatory systems, most importantly the signal pathway termini and the connections among the many genes encoding transcription factors. But nowhere does current knowledge yet encompass a large fraction of any pattern formation regulatory network for an adult body part. We do have some reasonably well illuminated pieces and bits of such networks, for a number of different developmental systems. Five examples follow. In discussing these only the bare bones of the developmental processes themselves are given: rather the intent is to focus on regulatory properties that these systems illustrate, and that can be regarded as intrinsic, even canonical aspects of genomic pattern formation architecture.

These general regulatory properties are as follows: (1) modularity in *cis*-regulatory organization underlying the specific subparts of the future structure; (2) use of integrative *cis*-regulatory information processing for establishing spatial transcriptional domains; (3) depth in the regulatory network, i.e., many layers of interaction between initial specification of the progenitor field and the ultimate activation of differentiation genes; and (4) the presence of inputs from intercellular signaling pathways at each level of spatial transcriptional specification. The reader will have to wade along, as these features cannot be taken up in a completely organized fashion. Each of the five systems that we shall touch on can be used to illustrate some, but none illustrate all of these aspects of pattern formation mechanism.

Transcriptional Domains and the Pattern Program for the *Drosophila* Wing Disc: Modularity and *cis*-Regulatory Inputs

The main pattern elements on the adult *Drosophila* wing are its veins and a large array of exactly positioned peripheral nervous system elements. These are stereotypic, species-specific features (for comparative consideration of veins see Biehs *et al.*, 1998; and of the relevant sensory elements of the peripheral nervous system, Wülbeck and Simpson, 2000). Each of the peripheral nervous system elements begins as a localized patch of cells, or a "proneural cluster," which expresses genes of the *achaete-scute* complex (*AS-C*). These genes encode a special class of bHLH transcriptional regulators. As illustrated in Fig. 4.7A2 more than two dozen proneural clusters can be detected in the 3rd instar disc as patches of *AS-C* expressing cells. The locations of the sensory organs to which each gives rise are shown in Fig. 4.7A3. In these figures the color coding reflects the modular organization of the *AS-C cis*-regulatory system (Gómez-Skarmeta *et al.*, 1995), as indicated in Fig. 4.7A1. This fascinating relation means that the individual proneural domains are set up by the individual enhancers of the *AS-C* genes, each of which processes the locally incident spatial information available to it in the form of transcription factors expressed or activated there. There is no single prior proneural cluster pattern: the modular *AS-C cis*-regulatory system generates it. In this light one can see why such pattern formation systems are modular, since each location requires that a unique *cis*-regulatory information processing job be performed (cf. Chapter 2). Ultimately a single sensory organ precursor (SOP) cell is specified within each proneural cluster, as touched on later in this chapter. *AS-C* regulators play a role in the process by which the SOPs are chosen from amongst the other cells of the proneural clusters; but once specified all of the SOPs present the same information to the *AS-C* control system, and now a single *AS-C* module runs those genes in all of the SOPs (black box in Fig. 4.7A1). SOP specification occurs asynchronously, so at the stage illustrated in Fig 4.7B an enhancer trap in the SOP module produces its Lacz product in only a subset of the proneural clusters (greenish color). The proneural clusters are identified by endogenous *AS-C* expression (purple) under control of the remainder of the modules shown in Fig. 4.7A1, i.e., the pattern formation modules.

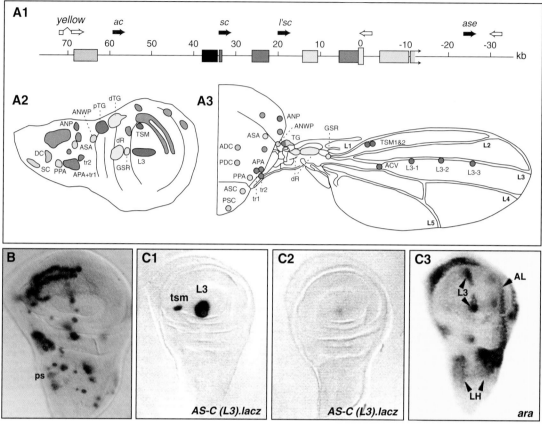

FIGURE 4.7 Transcription control of regulatory genes in the regional specification of the 3rd instar *Drosophila* wing disc.   (A) Modular enhancers of the *achaete-scute* gene complex (*AS-C*). (A1) Map of the locus; (A2) the proneural clusters controlled by each of the enhancers in the imaginal wing disc; and (A3) the location of the sensory structures to which they give rise in the adult wing and the dorsal surface of the adjoining thorax (heminotum). The enhancers, the proneural clusters, and the sensory organs are color coded. The *achaete* (*ac*) and *scute* (*sc*) genes encode bHLH transcription factors required for sensory organ development and are expressed in patches of cells within which the sensory organ precursor cell is subsequently specified (see following section of this chapter). The black box represents an enhancer required for expression in the sensory organ precursor (SOP) cells. The colored enhancer elements in (A1) promote expression of both *ac* and *sc*. Above the line representing the *AS-C* DNA, transcription units for the genes of the cluster are shown as arrows: the proneural genes, *ac*, *sc*, *lethal of scute* (*l'sc*) and *asense* (*ase*) solid; other genes, open. Distances are given in kb. In A2 and A3, sensory organ domains for which the enhancers are not indicated in this diagram are shown in gray. The positions

(Continues)

FIGURE 4.7 (*Continued*)

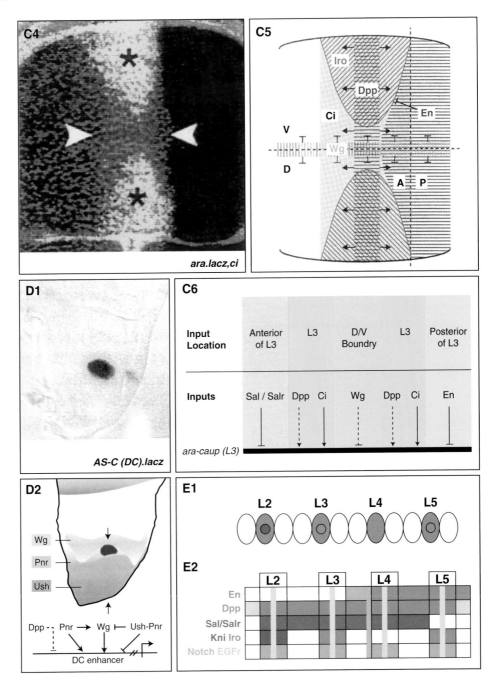

of the enhancer modules were determined from the phenotypic effects of chromosomal breakpoints, insertions and deletions, and by cis-regulatory analysis using lacz constructs, as illustrated for two examples below. [(A) From Gómez-Skarmeta et al. (1995) Genes Dev. **9**, 1869–1882.] (B) Wing imaginal disc stained for Lacz expression (green-blue stain) in an animal bearing a Lacz expressing insertion near the enhancer for the sensory organ precursor cells. Lacz expression reveals already specified SOPs. The disc was also hybridized with an sc probe (purple) to reveal activation of this gene in the proneural clusters. This disc was taken just prior to puparium formation; many of the prospective macrochaete SOP cells have been singled out from the clusters and have begun expressing Lacz. The ps cluster is an example of a proneural cluster which is expressing the proneural genes but has not yet activated the SOP lacz construct. The orientation of the disc is 90° rotated with respect to that in (A2), but most of the proneural clusters in (A2) can be recognized in (B). [(B) From Cubas et al. (1991) Genes Dev. **5**, 996–1008.] (C) Control of expression of arucan (ara) and caupolican (cau) genes. These genes encode homeodomain transcription factors which provide spatial inputs into the L3 cis-regulatory module of the AS-C. (C1) Expression of l3–lacz construct from the AS-C (cf. A2; initial pattern); (C2) same but after mutation of binding site for Ara/Caup factors. (C3) Expression of ara gene by in situ hybridization in various regions of the 3rd instar wing disc, including L3. [(C1–C3) From Gómez-Skarmeta et al. (1996) Cell **85**, 95–105; copyright Cell Press.] (C4) The L3 domain of ara and caup expression, visualized immunocytologically by use of an L3 ara-caup enhancer trap which expresses Lacz (red), shown together with domains of expression of cubitis interruptus (ci) gene (green). Ci encodes a transcription factor that mediates Hedgehog (Hh) signaling, which occurs throughout the anterior domain of the wing blade (see Fig. 4.6D); here anterior is left, and the black region is the posterior domain of engrailed (en) expression. The white arrows indicate the vertical stripe within which the highest level of ci expression occurs, centered on the Dpp stripe along the A/P boundary. Expression of ara and caup overlaps this stripe above and below the D/V boundary domain, and is bounded on the posterior side by the en expression domain. (C5) Relation between the transcriptional domains that set up the axes of the wing blade, and the L3 sensory organ progenitor field set up by expression of ara and caup: Iro, Iroquois, denotes the products of the ara-caup gene complex, red striped region; Dpp is secreted from the A/P boundary stripe and diffuses outward (blue striping); Ci domain, yellow; En domain (green striping); the wingless (Wg) signaling molecule is expressed along the D/V boundary (magenta striping). [(C4, C5 From Gómez-Skarmeta and Modolell (1996) Genes Dev. **10**, 2935–2945.] (C6) Summary of inputs into the cis-regulatory system controlling ara and caup expression in AS-C (L3), based on analysis of ara and caup activity in clones expressing or failing to express the ci, dpp, sall/salr, wg, hh, and Hh signal transduction intermediate genes (Gómez-Skarmeta and Modolell, 1996; Biehs et al., 1998; de Celis and Barrio, 2000). Transcription factor inputs are shown as solid arrows; signal system inputs are dashed lines. The actual order of target sites in the L3 cis-regulatory domain of the Iro-C is not known, and the diagram is oriented A (left) to P, as is the disc. (D) The DC enhancer of the AS-C. (D1) Expression of a DC.lacz enhancer construct (cf. A). (D2) Diagram of inputs

The individual pattern formation modules produce very sharply defined sub-sets of the overall pattern of *AS-C* expression, just as the diagrams in Fig. 4.7A indicate. The activity of the L3 module (green in Fig. 4.7A1) is demonstrated by expression of a Lacz reporter in Fig. 4.7C1. This enhancer is responsible for *AS-C* expression in a patch of cells that later separates into the several clusters along vein L3 from which arise the sensilla campaniformia associated with this vein (Fig. 4.7A3). The L3 enhancer also directs expression in a proneural cluster that pro-duces the twin sensilla (TSM) of the wing margin. The key positive input into the L3 module is provided by homeodomain transcription factors encoded by the *Iroquois Complex* (*Iro-C*). This contains three genes, *araucan* (*ara*), *caupolican* (*caup*), and *mirror* (Gómez-Skarmeta *et al.*, 1995, 1996; Kehl *et al.*, 1998), of which the first two, which are coregulated, are relevant to the L3 enhancer. If the *ara/caup* target sites within the L3 enhancer are destroyed, its activity is lost, as illustrated in Fig. 4.7C2 (Gómez-Skarmeta *et al.*, 1996). *Iro-C* expression (Fig. 4.7C3) includes the L3 domain at the stage shown. But institution of even this relatively simple-looking L3 pattern is in fact quite expensive in terms of regula-tory information, as summarized in Fig. 4.7C4–6. Figure 4.7C4 displays the loca-tions of some of the inputs which determine *ara* and *caup* expression in the L3 domain (see legend) and these and others are indicated spatially in Fig. 4.7C5 (Gómez-Skarmeta and Modolell, 1996) and more abstractly in Fig. 4.7C6.

into this enhancer. Direct evidence demonstrates activation at Gata sites where the product of the *pannier* (*pnr*) gene binds and for a negatively acting complex between the Pnr factor and a cofactor encoded by the *u-shaped* gene (*ush*); these two factors determine the dorsolateral boundaries of DC enhancer expression (Cubadda *et al.*, 1997; Haenlin *et al.*, 1997; Garcia-Garcia *et al.*, 1999). There is also indirect genetic evidence for DC enhancer repression by Dpp at the anterior boundary of the DC domain. Wg is necessary for DC enhancer expression but does not determine boundaries of expression. [(D) Adapted from Garcia-Garcia *et al.* (1999) *Development* **126**, 3523–3532 and The Company of Biologists Ltd.] (E) Transcription control of patterning in specification of longitudinal veins L2–L5 of the wing (see A3 for disposi-tion of these veins). (E1) Schematic representation of a transverse section of wing, veins indicated as magenta circles. The L2 vein field expresses *knirps* and *knirps-like* (blue), and L3 and L5 express *ara* and *caup* (green), (E2) Summary of spatial domains of gene products listed on left. Vertical orange line represents A/P boundary from which Dpp ligand diffuses; En is expressed only in posterior domain, except for region between L4 and L3. Within each vein progenitor field EGFR is activated in the central and flanking cells and N in adjacent cells, A and P (Sturtevant *et al.*, 1993). For L3, which contains peripheral nervous system elements, the key regulators are those of the *Iro-C* [as shown in (C); for the controlling inputs see (C5)]. This diagram suggests that there exists a distinct set of *cis*-regulatory inputs for each vein. The veins form at the borders of spatial transcriptional domains which specify the intervein regions (Bier, 2000; Mohler *et al.*, 2000; Sturtevant *et al.*, 1997). [(E) From de Celis and Barrio (2000) *Mech. Dev.* **91**, 31–41; copyright Elsevier Science.]

A very important aspect of the discovery of the role played by the *Iro-C* in positioning the proneural clusters is that as indicated in Fig. 4.7C, it links *AS-C* expression with the prior state of spatial specification. Thereby we can now perceive the depth of the genomic regulatory network which underlies wing patterning. Following initial establishment of the imaginal discs (Fig. 4.1E) a new set of interactions sets up the coordinate system of the wing disc, defining as transcriptional domains the wing blade as opposed to the hinge and more proximal (notum) structures (Casares and Mann, 2000); and within the wing blade, the A/P and D/V axes. The four quadrants thus formed within the wing blade, and the orthogonal axial stripes, are also defined as transcriptional domains in this process (see Fig. 4.6D; for reviews that summarize this layer of the patterning network and the signaling interactions by which the coordinates are installed, see Lawrence and Struhl, 1996; Williams and Carroll, 1993; Cohen, 1993; de Celis, 1998; Modolell and Campuzano, 1998). The next layer of the network is that determining expression of *ara* and *caup* in the L3 domain considered in Fig. 4.7C3–C6. As summarized in Fig. 4.7C6, the inputs are those generated by the preceding coordinate-making system: the boundaries are set negatively by the *en* domain (posterior), the *wg* domain (D/V boundary), and by the *spalt* (*sal*)/*spalt related* (*salr*) domain (anterior; de Celis and Barrio, 2000). Expression occurs within the A/P boundary domain of *dpp* gene expression, and also requires high levels of the *cubitis interruptus* (*ci*) gene product (expressed in consequence of the Hedgehog signaling set up with respect to the original A/P coordinates of the wing blade; Fig. 4.6D). The next level, after regulation of *ara* and *caup* expression, is that displayed in Fig. 4.7C1, i.e., the L3 *cis*-regulatory module of the *AS-C*, which is driven by the Ara/Caup factors. Following this are further levels where the downstream systems that control SOP specification within the proneural clusters operate. There follows the institution of cell type-specific sensory organ differentiation programs, as discussed in the next section of this chapter. The wing disc provides a remarkable illustration of the concept of regulatory network depth.

Furthermore, distinct sets of regulatory transactions take place at given levels in different regions of the disc, as represented by the *Iro-C cis*-regulatory inputs and the L3 *AS-C* enhancer. The operation of the DC enhancer of the AS-C, which is responsible for certain large sensory bristles, or macrochaetes, of the notum (Fig. 4.7A2, A3), is illustrated in Fig. 4.7D1. Spatial inputs which control this enhancer are indicated in Fig. 4.7D2 (see legend for brief description). Entirely different transcription factors are involved than are required for L3 (Garcia-Garcia *et al.*, 1999; Cubadda *et al.*, 1997). These are a Gata factor, the product of the *pannier* (*pnr*) gene, and a repressive cofactor of the Pnr protein, the product of the *u-shaped* (*ush*) gene (Haenlin *et al.*, 1997).

Transcriptional inputs required for the patterning system that sets up the veins illustrate the same general theme: each vein needs a different set of spatial inputs, which again represent prior pattern information. This is summarized in the diagrams of Fig. 4.7E (de Celis and Barrio, 2000), which indicate some of the transcriptional regulators and signaling systems necessary for the specification of veins 2–5 (cf. Fig. 4.7A3). Vein formation depends directly on a complex set of signaling interactions that define the positions where each vein will arise (for reviews and data Garcia-Bellido and de Celis, 1992, 1998; Biehs *et al.*, 1998; Sturtevant *et al.*, 1997; Mohler *et al.*, 2000; Milán and Cohen, 2000; Bier, 2000; Baonza *et al.*, 2000). Genes encoding transcription factors, *viz*, the *Iro-C* genes, *knirps/knirps-like*, *knot* and *sal/salr*, all play key roles, as does the original posterior expression of *en*. The outcome is a combinatorial specification of each vein position, defined in respect to the pattern of transcriptional domains in the wing disc.

Though knowledge is yet anything but complete, the regulatory mechanisms that drive pattern specification in the wing disc illustrate some very important, and very general points. This is an increasingly beautiful story: we can see how wing patterning is likely to be genetically organized at the *cis*-regulatory level, and can begin to think about how it evolved. Several of the cardinal features of bilaterian patterning networks are here clearly illustrated. These include the great depth of the network architecture, which layer by layer, controls a progression of transcriptional patterning states; the need for extensive *cis*-regulatory information processing which allows integration of the multiple spatial inputs presented at each stage; and the importance of *cis*-regulatory modularity. This last continues to surprise, as more and more examples are discovered in different systems. But given the complexity of the information processing job required for each pattern element, it is a pleasing outcome to find that each such job is done by a separate piece of regulatory DNA. In addition, though not considered explicitly here, note the direct inputs to the key transcriptional regulatory genes from signal systems; for example the *wg* and *dpp* genes are major players in the initial specification of the imaginal discs (Fig. 4.1E); in the organization of the wing blade coordinate system (Fig. 4.6D); and in the specification of both the *Iro-C* L3 domain (Fig. 4.7C5, C6) and the *AS-C* DC domain (Fig. 4.7D2).

Patterning the Heart Progenitor Field in *Drosophila*

The tubular heart of *Drosophila* forms by fusion of two bilateral, metameric arrays of precursor cells that arise in the dorsal mesoderm of the postgastrular embryo (reviewed by Bodmer and Frasch, 1999). A key transcriptional regulator required for this specification function is Tinman, a homeodomain transcription factor orthologous to Nkx2.5, which as we have seen (Fig. 4.1A, B) is also utilized for specification and subsequent development of the heart progenitor field in vertebrates. However, the *tinman* (*tin*) gene in *Drosophila* is expressed more broadly in the mesoderm than the heart progenitor field, and is required for development of all dorsal mesoderm derivatives, *viz.* dorsal muscles, visceral mesoderm, and heart, as well as for certain ventral muscles (Azpiazu and Frasch, 1993; Bodmer, 1993). The genomic apparatus controlling *tin* expression in the postgastrular embryo, the inputs to this apparatus, and its position in the overall network that

controls heart formation provide another excellent example with which to illustrate the major themes in this discussion. Several aspects are summarized in Fig.
4.8 (see legend for details).

The *tin* gene is initially activated in the invaginated mesoderm and its
expression thereafter goes through several developmental phases (Fig. 4.8A, C).
Its initial transcription is mediated by the mesodermal *twist* (*twi*) bHLH factor
(stage 9). The *tin* expression pattern becomes metameric in stage 10, as its
transcription is turned down within the domains defined by the *evenskipped*
(*eve*) stripes, and here the *bagpipe* regulatory gene is instead activated (Fig.
4.8A). The clusters of cells occupying these regions of *bagpipe* expression become
visceral rather than cardiac mesoderm. In the adjacent stripes where the *sloppy-
paired* (*slp*) forkhead family gene is expressed, *tin* transcription is stimulated and
these regions give rise to cardiac cells. Expression of *slp* promotes *wg* gene
expression, which is essential for heart development (Park *et al.*, 1996; Wu *et
al.*, 1995); a role of Wg factor secreted from the ectodermal cells could be to
stimulate high levels of *twi* expression in the prospective cardiac cells (Reichmann
et al., 1997). As shown in Fig. 4.8B, the originally metameric units of cardiac and
visceral mesoderm soon generate continuous columns. After gastrulation expression of *tin* becomes dependent on Dpp signaling (Fig. 4.8A, stage 9), and on its
own product. Later in development *tin* is expressed under Wg control, in the
two rows of cardiac cells from which the heart forms. These successive phases of
tin expression are illustrated in Fig. 4.8C1–C3. Still further on, cardiac cell types
are specified. A new set of transcription factors now become involved, particularly
Mef2, a Mads box factor; and the homeodomain factors Zfh1, Ladybird, and
Eve (Lilly *et al.*, 1995; Jagla *et al.*, 1997; Gajewski *et al.*, 1997; Su *et al.*, 1999).
Just as in the wing disc patterning process, the heart patterning regulatory network
is multilayered. At each stage the spatial information presented in the form of
regional transcriptional states is utilized by the *cis*-regulatory elements then
active to generate a finer pattern. Signaling interactions provide essential spatial
input in every phase of the process. As indicated in Fig. 4.8A, both Dpp and Wg
are required from early on, and at the subsequent stage of cardiac cell type
specification both of these same signal systems as well as Notch are utilized
(Su *et al.*, 1999).

The stepwise regulation of the *tin* gene is an essential driver of the heart
patterning process. As shown in Fig. 4.8E the *tin cis*-regulatory system is modular
in organization. It is particularly revealing that each of the phases of *tin* expression
is controlled by a specific enhancer element. Four such elements have been found
(Yin *et al.*, 1997), and the pattern of expression that each generates in isolation is
illustrated in Fig. 4.8D1–D4. These correspond remarkably to the successive
patterns shown in Fig. 4.8C. Module D operates at stage 11 (Fig. 4.8D2) and
confers control of *tin* expression to Dpp signaling and to its own product. Thereby
it integrates the result of the previous *tin* expression pattern set up by Module B,
the early acting *tin cis*-regulatory element which activates the *tin* gene in the
mesoderm (Fig. 4.8D1). The extent of mesodermal invagination can be decreased

by narrowing the domain of *snail* expression in the prospective mesoderm just before gastrulation, with the result that contact between the migrating mesoderm and the dorsal ectodermal Dpp domain is restricted (Maggert *et al.*, 1995). In consequence Dpp-driven *tin* expression is severely depressed, and heart formation is much reduced. Module C functions later, exclusively in cardial cells (Yin *et al.*, 1997). Some inputs into these modules are shown in Fig. 4.8E. Module B has been studied in most detail. Like many of the elements that carry out specification jobs discussed in Chapter 2 it utilizes both positive and negative inputs: Twi turns it on, and Eve, directly or indirectly prevents its expression in the *eve* stripe domains, while (probably indirectly) the *buttonhead* (*bth*) gene product represses Module B in the domain of the head where hemocytes form. In Fig. 4.7A we saw a strikingly modular *cis*-regulatory organization the function of which was to organize expression in the many individual proneural spatial domains; here we see one the function of which is to organize successive phases of expression. What the two strategies have in common is that in both, every individual module has to process different sets of inputs. In order for these alternative transcriptional states to be imposed on a given gene a rigidly effective "traffic control system" must exist, so that in any given cell at any given time only the relevant module communicates with the basal transcription apparatus.

At the stage when differentiation programs are called in, particular *cis*-regulatory modules control expression of *eve* (Su *et al.*, 1999) and of the *mef2* gene (Nguyen and Xu, 1998; Cripps *et al.*, 1999). These genes drive the specification of particular cardiac cell types. Figure 4.8F shows an example of a cardiac *cis*-regulatory element, from the *mef2* gene. This element integrates a spatial input from the preceding stage, i.e., Tin, with a different input, *viz.* the Gata factor Pnr (Gajewski *et al.*, 1998, 1999). The element uses "and" logic, and together these two factors suffice to activate the *mef2* enhancer and produce cardial cell fates even in ectopic locations (Gajewski *et al.*, 1999).

Though they operate in a very different context from the imaginal disc, the mechanisms of transcriptional control in embryonic heart formation reveal organizational similarities. We again encounter modular *cis*-regulatory programming that provides the execution of diverse information processing jobs; depth in the regulatory network, with multiple layers prior to cell type specification; and the ubiquitous use of signaling inputs to provide spatial cues at each layer of the regulatory system.

Encoding Hindbrain Regulatory Patterns

We move now to rhombomeric transcriptional patterning in the hindbrain of the mouse embryo, briefly touched on earlier (Fig. 4.5). Our subject is the genomic apparatus of which the function is to distinguish each rhombomere as a unique domain. This example is of particular value because many of the linkages in the fragments of the control network which are known have been established by direct *cis*-regulatory analysis.

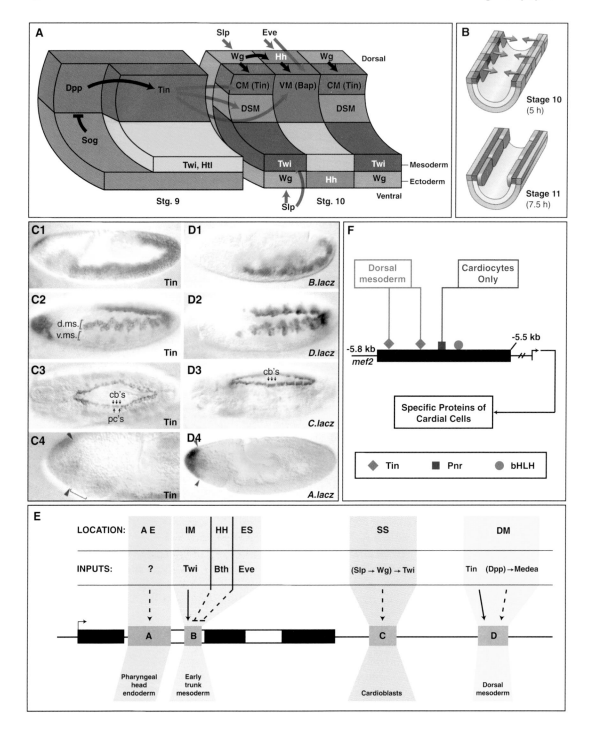

FIGURE 4.8   Transcriptional control of genes required for heart specification in *Drosophila*. (A–E) *tinman* (*tin*). This gene encodes a homeodomain regulator required for specification of dorsal mesoderm derivatives including cardiac mesoderm (CM), visceral mesoderm (VM), and dorsal muscles (Azpiazu and Frasch, 1993; Bodmer, 1993; Yin *et al.*, 1997). (A) Diagram of morphological gene expression domains in ectoderm and mesoderm of postgastrular embryos. At stage 9 the Dpp ligand is expressed in dorsal ectoderm, to which it is confined by its short gastrulation (Sog) antagonist. Transcription of the *sog* gene is controlled by the initial gradient of the Dorsal transcription factor (cf. Fig. 2.6). Twist (Twi) is an early mesodermal regulator that also responds to Dorsal, and is responsible for the initial activation of the *tin* gene. Expression of the *tin* gene then becomes Dpp-dependent. This maintains *tin* expression in the dorsal mesoderm. Htl (Heartless) is an FGF receptor required for the migratory behavior of the invaginated mesoderm, in the absence of which no dorsal mesoderm appears (Gisselbrecht *et al.*, 1996). At stage 10 combinatorial patterning devices divide the *tin* mesodermal expression domain into metameric stripes within each parasegment (Park *et al.*, 1996; Su *et al.*, 1999). The *wingless* (*wg*) gene is activated downstream of *sloppy paired* (*slp*), and *hedgehog* (*hh*) is activated downstream of *evenskipped* (*eve*). The *slp* domains later give rise to cardiac cell types, and the *eve* domains, in which *tin* initially activates the gene encoding the Bagpipe (Bap) transcription factor, become visceral mesoderm; DSM, dorsal somatic mesoderm. Black arrows and bars indicate signaling interactions; blue arrows indicate transcriptional interactions. (B) Metameric distribution of cardiac and visceral mesoderm at stage 10, followed by inward migration of visceral mesoderm to produce apposed mesodermal layers. [(A, B) From Bodmer and Frasch (1999) *In* "Heart Development" (R. P. Harvey and N. Rosenthal, Eds.), pp. 65–90, Academic Press, San Diego.] (C) Location of transcription factor encoded by *tin* gene in normal embryos visualized by immunocytology: (C1) Gastrulation, Tin factor in all cells of the trunk mesoderm; (C2) stage 11, *tin* product in dorsal mesoderm (d.ms.), not in ventral mesoderm (v. ms.); expression also in head cap. (C3) Stage 14, Tin factor in cardioblasts (cb's) and in pericardial cells (pc's); (C4) stage 8, high magnification view of head region; *tin* expression anterior of arrowheads but not in region indicated by bracket (hemocyte domain). (D) Expression of *lacz* constructs driven by *tin* cis-regulatory modules B (D1), D (D2), C (D3), and A (D4). Embryos in D1–3 are about the same stages, and are in same respective orientations as in (C1–3); D4 is stage 12. [(C, D) From Yin *et al.* (1997) *Development* **124**, 4971–4982 and The Company of Biologists Ltd.] (E) Organization of *tin* cis-regulatory sequences responsible for patterns of expression shown in (C) and (D). Question mark indicates that identity of factor is not known; dotted lines mean that the factor listed may act indirectly in that no direct interaction in *tin* cis-regulatory DNA has been demonstrated. Proteins indicated in parentheses are present in ectoderm and act upstream of mesodermal signal response system. The factors constituting the inputs are present in different regions of the embryo: AE, anterior endoderm; IM, invaginated mesoderm; HH, head hemocyte domain; ES, *eve* stripes; SS, *slp* stripes; DM, dorsal mesoderm. Exons are black; the four *tin* cis-regulatory modules in the figure are located in the first intron (A, B) and 3′ of the coding region (C, D). [(E) Adapted from Yin *et al.* (1997)

(*Continues*)

The *cis*-regulatory systems that determine the locations in which individual *hox* genes are expressed in the rhombomeres (see Fig. 4.5) are again remarkably modular in organization. Though the context is about as different as could be imagined, this is the same prominent feature that we confronted in the patterning systems of the *Drosophila* wing and heart. The disposition of modular enhancers governing the domains of expression of *hoxa3* is shown in Fig. 4.9A (Manzanares *et al.*, 1999b), and the modular neural (N) and other enhancers (M, mesodermal) controlling expression of the paralogous *hoxa4*, *b4*, and *d4* genes along the dorsal axis of the embryo are shown in Fig. 4.9B (Morrison *et al.*, 1997). Typically for *hox* genes, *hoxa3* is expressed in many regions of the postgastrular embryo. Its complex pattern of utilization is the sum of the subpatterns mediated by the different enhancer modules, some of which may indeed be divisible into even finer subelements. A discrete enhancer among these mediates expression of *hoxa3* exclusively in r5/r6 of the hindbrain (Fig. 4.9A; purple element). Similarly, discrete enhancers control expression of *hoxa4*, *b4*, and *d4* in r6/r7 (Fig. 4.9B, N modules); others in the contiguous somites (M modules). The experimental demonstrations in Fig. 4.9C–F illustrate dramatically the precision with which these and other *cis*-regulatory modules of the anterior *hox* genes control rhombomeric expression pattern, and indicate some of the necessary inputs which activate each (see legend for details and references). Note, for example, that the solid A/P pattern illustrated in Fig. 4.9E1 for *hoxb2* is actually a composite of several discrete genomic patterning devices: expression in r3–r5 depends on two such, one that operates in r3 and r5, and one that operates in r4. The explicit nature of these we see below.

Within less than a day (E8.0–E8.5) the rhombomeres achieve their distinct transcriptional identities. Figure 4.9G shows the rhombomeric expression domains just before and immediately after establishment of the respective transcriptional patterns for those genes considered in the following, *viz*, the *hox* group *1*, *2*, and *3* genes; *krox20* (*k20*); and *kreisler* (*kr*). Some relevant pieces of the control network are schematized in Fig. 4.9H, which is based on

---

Development **124**, 4971–4982 and The Company of Biologists Ltd, including additional information from Reichmann *et al.* (1997); Bodmer and Frasch (1999).] (F) Combined spatial inputs to the cardial cell expression module of the *mef2* differentiation regulator (Gajewski *et al.*, 1997; Cripps *et al.*, 1998; Nguyen and Xu, 1998). Expression occurs in cardial cells, a subset of the mesodermal cells expressing the *tin* gene. The Tin factor is essential for *mef2* expression (Gajewski *et al.*, 1997, 1998; Cripps *et al.*, 1999) and so is Pannier (Pnr), a Gata class transcription factor expressed in heart cells; ectopic expression of the *pnr* gene causes cardiac gene expression in all domains in which *tin* is also expressed (Gajewski *et al.*, 1998, 1999). The bHLH factor is unknown. Factors binding at the Pnr target site repress the *mef2* enhancer in pericardial cells (Gajewski *et al.*, 1998; another module of the *tin* gene generates expression in additional cardial cells independently of the Tin factor).

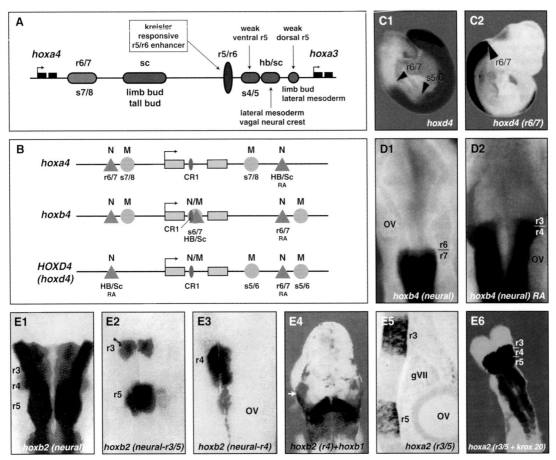

FIGURE 4.9   *cis*-Regulatory control of rhombomere specification during hindbrain development in mouse.   (A) Modular regulatory elements of mouse *hoxa3-hoxa4* intergenic control region: r, rhombomere; s, somite; hb, hindbrain; sc, spinal cord. [(A) From Manzanares *et al.* (1999b) *Development* **126**, 759–769 and The Company of Biologists Ltd.] (B) Modular regulatory elements surrounding mouse *hoxa4, hoxb4,* and human *hoxd4* (HOXD4) genes. CRI is a conserved intron sequence element the function of which is not known: M, mesoderm expression module; N, neural expression module; RA, retinoic acid response module. Each orthologue has a distinct *cis*-regulatory organization, though they all share the same set of elements. The gray oval in (A) denotes the S7/8 enhancer 3′ of the *hoxa4* gene in (B). (C) The *hoxd4* r6/7 module: (C1) expression of endogenous *hoxd4* gene, by *in situ* hybridization, 9.5 dpc; (C2) expression of *lacz* transgene driven by neural *hoxd4* r6/7 module, 10.5 dpc; no expression occurs in somites, in contrast to (C1). (D) Retinoic acid (RA) sensitivity of neural expression module of *hoxb4* gene (cf. B; for review of RA effects on pattern formation in hindbrain, Gavalas and Krumlauf, 2000): (D1) expression of *lacz*

(Continues)

FIGURE 4.9 *(Continued)*

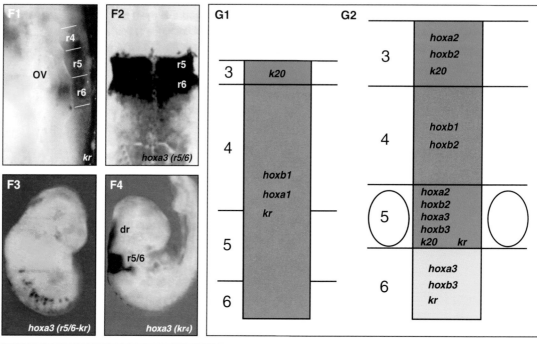

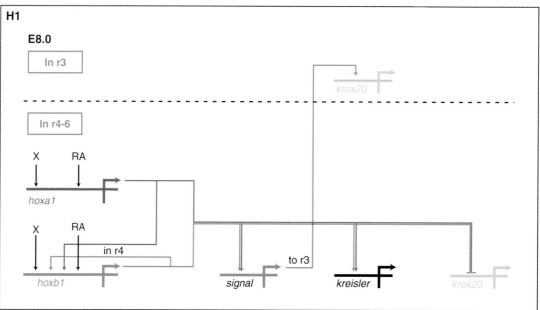

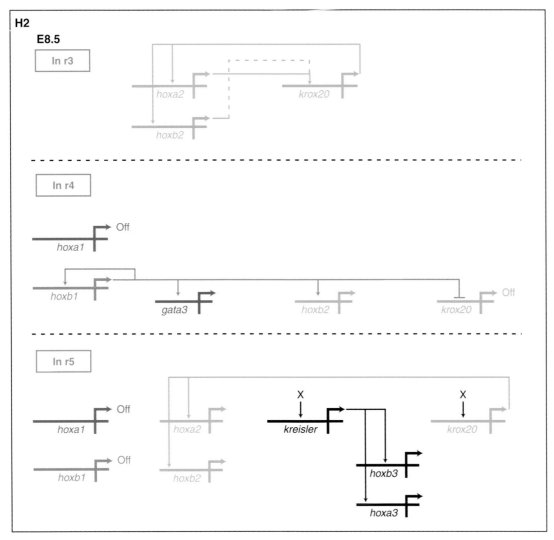

transgene, driven by downstream *hoxb4* neural (r6/7) module, dorsal view, 10.5 dpc; (D2) expression of same transgene in embryos that had been treated with retinoic acid one day earlier. [(B–D) From Morrison *et al.* (1997) *Development* **124**, 3135–3146 and The Company of Biologists Ltd.] (E) Control of *hoxb2* r3/5 *cis*-regulatory module by *hoxb1* and *krox20* transcription factors, dorsal views, 9.5 dpc. (E1) Expression of *lacz* transgene driven by 2.1 kb neural control region of *hoxb2* gene; r3, r4, and r5 as well as more caudal regions express this construct. (E2) Expression of same transgene after deletion of target sites for Hoxb1 protein and for cooperatively binding factors of Extradenticle and Homothorax (Hth) classes (Pöpperl *et al.*, 1995; Jacobs *et al.*, 1999; Ferretti *et al.*, 2000); only expression in r3 and r5 remain. If the *krox20* sites are destroyed in the r3/5 enhancer all expression in these rhombomeres is lost (Sham et al., 1993). (E3) Expression of *lacz* under control of an oligonucleotide consisting of three copies of the Hoxb1/Exd/Hth target site. This is sufficient to cause expression in r4. (E4) Expression of *lacz* construct driven by the *hoxb2* r4 module but in animal in which *hoxb1* is ectopically expressed under control of a cytoplasmic actin *cis*-regulatory element. The construct is now expressed in a much broader region including anterior to r4 (white arrow), while as shown in (E3) in normal hosts its expression is confined to r4. [(E1–E4) From Maconochie et al. (1997) *Genes Dev.* **11**, 1885–1895.] (E5) Expression of *lacz* transgene driven by r3/5 module of *hoxa2* gene, seen in section; ov, otic vesicle; gVII, facial ganglion, 9.0 dpc. (E6) Ectopic expression of construct in (E5) when introduced into a mouse in which the *krox20* gene is ectopically expressed under control of a *hoxb4* neural enhancer; expression has spread to r4. [(E5, E6) From Nonchev et al. (1996) *Development* **122**, 543–554 and The Company of Biologists Ltd.] (F) Control of *hoxa3* expression in r5 and 6 by Kreisler (Kr), a mafB class transcription factor: (F1) *kr* gene expression, 9.5 dpc, visualized by *in situ* hybridization; (F2) Expression of *lacz* transgene driven by *hoxa3* r5/6 module, 9.5 dpc; (F3) loss of expression if key Kr site is mutated; (F4) expression of a *lacz* construct in r5 and 6 as well as dorsal roof (dr) driven by an oligonucleotide containing four copies of this same Kr site. [(F) From Manzanares et al. (1999b) *Development* **126**, 759–769 and The Company of Biologists Ltd.] (G) Diagram of gene expression domains in r3–r6 domain of hindbrain; (G1) 8.0 dpc; (G2) after 8.5 dpc. Within each pre-rhombomeric (G1) or rhombomeric (G2) segment, genes expressed there are listed for which some direct *cis*-regulatory interactions have been determined (as e.g., in (F)): *k20, krox20; kr, kreisler.* [(G) From Barrow et al. (2000) *Development* **127**, 933–944 and The Company of Biologists Ltd.] (H) Network elements in rhombomere specification, demonstrated by *cis*-regulatory analysis, or inferred from genetic consequences of *hox* gene knockouts. (H1) 8.0 dpc. In r3, the *krox20* gene is activated by a signal from the prospective r4 region (Barrow et al., 2000). This signal is generated in consequence of *hoxa1* and *hoxb1* expression in the prospective r4–r6 region (Barrow et al., 2000). The *hoxb1* gene autoregulates in r4–r6 (Pöpperl et al., 1995). In addition *hoxa1* activates *hoxb1* (Studer et al., 1998; Rossell and Capecchi, 1999), and both are driven by retinoic acid (RA) sensitive *cis*-regulatory elements as well (Dupé et al., 1997; Studer et al., 1998; Niederreither et al., 2000). "x" here and in the following diagrams indicates additional, yet unknown spatial regulatory inputs, in this case those required to activate *hoxa1* and *hoxb1* throughout the neural plate up to the future r3/r4 boundary prior to E8.0 (Barrow et al., 2000). *hoxa1* and *hoxb1* expression also result in activation of the *kr* gene and repression of the *krox20* gene in the pre r4–r6 domain (Barrow et al., 2000); these interactions could be indirect. (H2) 8.5 dpc. In r3 the *krox20* gene product directly activates *hoxa2* and *hoxb2* [Sham et al., 1993; Nonchev et al., 1996; Vesque et al., 1996; cf. (E2), (E6), and Fig. 4.5B, C]. Its progressive expression in r3 depends in turn on that of *hoxa2* and possibly *hoxb2* (Barrow et al., 2000). It also may activate itself (not shown; reviewed by Barrow et al., 2000). Ultimately *krox20* activates other downstream targets as well, e.g., the *eph4A* gene (Theil et al., 1998). Expression of *krox20* is essential for subsequent development of r3 and r5 (Schneider-Maunoury et al., 1997, 1993; cf. Fig. 4.5B, C).

(Continues)

cis-regulatory analyses, together with the results of gene knockouts (see legend for key studies, and sources for earlier work). There are many interesting and illuminating features in these scraps of the regulatory system, although its overall design cannot yet be perceived. The most general point is that here we see explicitly what it is all about: every gene but one on the diagrams of Fig. 4.9H encodes a transcription factor, and the transactions these network elements mediate have as their sole outcome the imposition of rhombomere-specific transcriptional states. We can only imagine how many pages would be filled up were such diagrams to be extended forward in time, beyond the brief period represented in these regulatory snapshots.

Several special design devices are employed: for example in r4 expression of *hoxb1* depends on a powerful autoregulatory loop, on which r4 identity depends (Pöpperl *et al.*, 1995) in that in r4 *hoxb1* specifically both activates *hoxb2* (Maconochie *et al.*, 1997; Ferretti *et al.*, 2000; Jacobs *et al.*, 1999) and represses *krox20* (Barrow *et al.*, 2000; see Fig. 4.9H2). In r3 a self-reinforcing loop runs *krox20* via *hoxa2* (and perhaps *hoxb2*), but different inputs account for *krox20* activity in r5, though *krox20* activity is required in turn for *hoxa2* and *b2* expression in both. Another special device is a retinoic acid response system which sets the early limits of *hoxa1* and *b1* expression in the neural plate (e.g., Studer *et al.*, 1994; Dupé *et al.*, 1997; Niederreither *et al.*, 2000) as also for other *hox* genes (e.g., Gould *et al.*, 1998; Morrison *et al.*, 1996; Gavalas and Krumlauf, 2000). In addition, the diagrams in Fig. 4.9H show how important is the function of the *krox20* and *kr* genes: one of their roles is to transmit to the *cis*-regulatory apparatus of several *hox* genes the output of the immediately preceding state of regional specification, and by activating *hox* genes immediately downstream (cf. Fig. 4.5D) to send the patterning process forward to the next stage.

In summary, we see what animals have to do to make their body parts, far upstream of morphogenesis and differentiation. They utilize a library of *cis*-regulatory circuit devices with which they define beautifully precise patterns of regulatory gene expression. Each lasts only for a brief time in the life cycle. But in informational terms the job of installing successive spatial regulatory states is theessence of the developmental process.

Meanwhile in r4 *hoxa1* expression ceases, but *hoxb1* expression continues, driven by its autoregulatory loop. *hoxb1* activates *hoxb2* in r4 [Maconochie *et al.*, 1997; Ferretti *et al.*, 2000; cf. (E4)] and represses *krox20* (Barrow *et al.*, 2000). *Hoxb1* also activates a gene encoding the Gata3 factor (Pata *et al.*, 1999). In r5 *krox20* is activated by intrinsic spatial factor(s) denoted "x" (Barrow *et al.*, 2000). The *kr* gene was initially activated by *hoxa1* and *hoxb1* (H1), and continues to be active after disappearance of *hoxa1* and *hoxb1* products. The *kreisler* product is required for *hoxa3* and *hoxb3* expression [Manzanares *et al.*, 1999a,b, as illustrated above in (F); effects of *kr* knockout were summarized in Fig. 4.5D]. All of the genes included here are essential for subsequent hindbrain development (for current summaries, Rossel and Capecchi, 1999; Barrow *et al.*, 2000; Manzanares *et al.*, 1999a; Studer *et al.*, 1998).

The Role of Signaling

Signaling interactions are essential in all of the patterning processes mentioned in this chapter. Signaling is utilized to position the transcriptional domains that are the output of the regional specification process at every stage. That is, signaling interactions are invariably upstream of the activation of patterning genes encoding transcription factors (except for special cases such as the syncytial blastoderm of the *Drosophila* embryo). Signaling pathways are activated, repressed, or included in autostimulatory loops in the course of pattern formation, and genes encoding signaling components are also to be found downstream of patterning genes encoding transcription factors. In a crude sense it could be said that patterning proceeds by means of series of the form signal → transcriptional regulatory domain → more signals → more transcriptional regulatory domains (e.g., Davidson, 1993); but such a simple-minded description completely bypasses the architecture of the underlying genomic control network. As always, it takes specific cases to confer reality. We have already encountered many situations in which signals provide the essential spatial inputs in the establishment of transcriptional patterns. The two examples which follow specifically illustrate the position of signaling systems within patterning networks, both upstream and downstream of the establishment of regional transcriptional states.

The first of these examples concerns the specification and early stages of morphogenesis of the tracheal system of *Drosophila*. The tracheal progenitor fields consist of ten metameric "tracheal placodes" on each side of the postgastrular embryo, which later invaginate, producing cells that undergo complex A/P and D/V migrations. In their terminal phase of development the tracheal cells produce a fine network of branched tubules which invade the tissues of the larva in response to their need for oxygen (reviewed by Metzger and Krasnow, 1999). We are here concerned only with the earlier stages of tracheal pattern formation, just after the tracheal placodes have been initially specified, and up to the onset of migratory behavior in the tracheal cells. The tracheal placodes are initially defined by the expression of two genes encoding transcription factors, *trachealess* (*trh*) and *drifter* (*dfr*; also known as *ventral veinless*). The early pattern of expression of *trh* in the placodes is illustrated in Fig. 4.10A1. The *trh* gene here plays the role of a typical transcriptional identity factor, defining the tracheal progenitor fields. Thus *trh* mutants fail to develop trachea and ectopic *trh* expression generates ectopic tracheal placodes (Isaac and Andrew, 1996; Wilk *et al.*, 1996). The *dfr* gene is required for tracheal morphogenesis and *dfr* mutations cause presumptive tracheal cells to fail to differentiate and to display defective migratory behavior (Anderson *et al.*, 1995). Expression of a *dfr-lacz cis*-regulatory construct in all tracheal cells soon after they begin migrating is shown in Fig. 4.10B1. The *trh* and *dfr* genes are activated in parallel in response to the A/P and D/V coordinates of the postgastrular embryo (Wilk *et al.*, 1996; Zelzer and Shilo, 2000a). The specific inputs required for initial activation of *trh* and *dfr* in the placodes are not known, except that D/V positioning depends on *dpp*, that some aspect of the metameric

A/P transcriptional organization of the epidermis is likely to be involved, including the *wg* gene (de Celis *et al.*, 1995); and that transcription of these genes is specifically repressed in the anterior and posterior terminal domains by the *spalt* (*sal*) gene (Kühnlein and Schuh, 1996; Boube *et al.*, 2000). The *knirps* (*kni*) and *knirps-related* (*knrl*) genes are also activated in tracheal progenitor fields, in parallel, independently of *trh* and *dfr* (Chen *et al.*, 1998; Boube *et al.*, 2000).

By stage 12 the initial spatial inputs have been supplanted by an auto- and cross-regulatory system that maintains *trh* and *dfr* expression in an enforcing activation loop. Evidence for this is shown in Fig. 4.10A2, A3, B2, and B3; and Fig. 4.10C provides a diagram displaying the interactions of *trh* and *dfr* based on *cis*-regulatory as well as genetic analysis (see legend for references). For present purposes the main point in Fig. 4.10C is the nature of the genes downstream of *dfr* and *trh* (Boube *et al.*, 2000). Those genes shown in green in Fig. 4.10C all encode molecules essential for the operation of different signaling systems in the tracheal cells. Those shown in blue encode additional transcription factors required for the differentiation and further morphogenesis of the tracheal tubules (e.g., Wilk *et al.*, 2000). The signal systems controlled by the *dfr/trh* transcriptional state are used for separation of cell fates within the tracheal placodes, manifested by the distinct migratory behavior of cells giving rise to the A/P tracheal trunk and the D/V branches. The cells producing these parts of the tracheal system respond to different cues, express different downstream genes and carry out different morphogenetic roles (reviewed by Zelzer and Shilo, 2000b). Three signaling systems, FGF, EGF, and Dpp, are utilized, and components of each are downstream of the tracheal identity genes *trh* and *dfr* (Fig. 4.10D-F; for details see legend). As a primary example of control of a signaling pathway by the prior transcriptional specification function, expression of the *breathless* (*btl*) gene encoding the FGF receptor active in the placode cells is directly regulated by Trh and Dfr (Anderson *et al.*, 1996; Oshiro and Saigo, 1997; Fig. 4.10C). Cells migrating dorsoventrally do so in response to Dpp, by means of their Tkv receptor; note that activation of the *tkv* gene also requires the *dfr* gene product (Fig. 4.10C). In the Dpp response domains from which the dorsal and ventral bound tracheal cells emerge, a specific transcriptional state is established under control of *kni/knrl*, as shown in Fig. 4.10F. On the other hand the domain which gives rise to the cells migrating in A/P directions, which build the dorsal trunk, instead activate an EGF signaling pathway. This pathway utilizes the products of the *rho* and *dof* genes which are also downstream of *trh* and *dfr* (Fig. 4.10C). Activation of the EGF pathway in these cells turns on the *sal/salr* regulatory genes (Fig. 4.10F; see legend for references).

To summarize, Fig. 4.10 shows that once an autonomously operating transcriptional state is installed in the tracheal placodes, one of its major objects is to activate genes encoding signal response systems. In the next stage of patterning, in response to dorsal and ventral sources of the Dpp signaling ligand, the placodes are subdivided further, into *kni* and *sal* transcriptional domains. These in turn

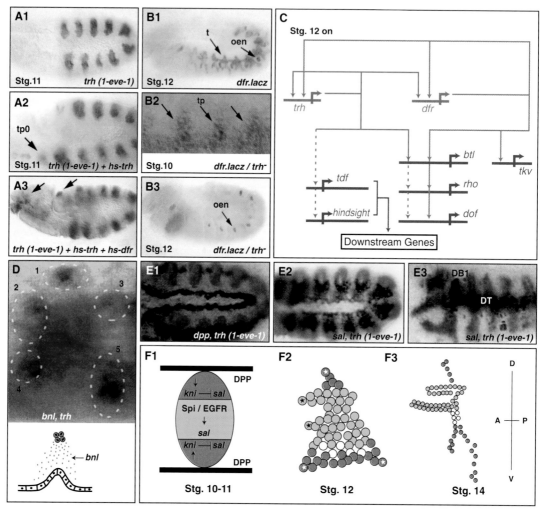

FIGURE 4.10   Early development of trachea in *Drosophila*.   (A) Expression of *tracheless* (*trh*) gene, which encodes a bHLH-PAS transcription factor, and acts as a determinant of tracheal identity. The locus contains a β-gal expression enhancer trap (*1-eve-1*) and expression of the *trh* gene is conveniently monitored by use of an αβ-gal antibody. (A1) *trh(1-eve-1)* expression in ten tracheal placodes; normal pattern, stage 11. (A2) *trh* autoregulates: expression is extended anteriorly to generate an ectopic tracheal placode, (tp₀), in embryos expressing *trh* under heat shock control (*hs-trh*). (A3) *trh* cross-regulation by *drifter* (*dfr*) gene product. Extensive ectopic *trh* (*1-eve-1*) expression is seen in an embryo expressing both *dfr* and *trh* under heat shock control (arrows). In (A2) ectopic expression driven by *hs-trh* occurs in tp₀ because *dfr* is already being expressed in that location. (B) Expression of *dfr* gene, which encodes a POU domain transcription factor. (B1) Expression of a *dfr.lacz* construct (visualized

as in A1–A3 i.e., by immunocytological detection of the Lacz gene product) at stage 12, as tracheal cells (t) begin migration; oen, oenocytes, where this gene is also expressed. (B2) Expression of *dfr.lacz* in *trh⁻* embryo at stage 10: the initial activation of *dfr* in tracheal placodes (tp) does not depend on *trh* expression. (B3) Same as (B2) but at stage 12; expression of *dfr.lacz* has now disappeared from the tracheal placodes, in trunk remaining only in oenocytes. [(A, B) From Zelzer and Shilo (2000a) *Mech. Dev.* **91**, 163–173; copyright Elsevier Science.] (C) Regulatory interactions of the *trh* and *dfr* specification genes. At stage 9 the A/P and D/V patterning systems of the embryo set up the ten bilateral *trh* and *dfr* expression domains. By stage 12 the initial spatial inputs have been supplanted by an auto- and cross-regulatory system that maintains *trh* and *dfr* expression in an enforcing activation loop, as shown in (A) and (B). Downstream interactions of *trh* (red) and *dfr* (purple) are shown: target genes require activity of either *trh*, *dfr*, or both. These gene products interact physically (Zelzer and Shilo, 2000a), and the bHLH-PAS product of *trh* also requires a second factor, encoded by the *arnt* gene (Ohshiro and Saigo, 1997). Target genes which encode components of signal transduction systems are shown in green (for evidence and references, Boube *et al.*, 2000): *btl*, *breathless*, an FGF receptor; *rho*, *rhomboid*, a transmembrane facilitator of EGF signaling; *dof*, which encodes a component of the EGF signal transduction system; *tkv*, *thickveins*, which encodes a Dpp receptor. Two genes encoding transcription factors are targets of Trh, shown in blue: *tracheal defective* (*tdf*), which encodes a bZip factor; and *hindsight*, encoding a Zn finger factor required for further tracheal morphogenesis (Wilk *et al.*, 2000). The interactions with the *btl* cis-regulatory system are known to be direct (Anderson *et al.*, 1996; Ohshiro and Saigo, 1997); the other interactions are inferred from genetic evidence, i.e., observations on expression in *trh⁻* and *dfr⁻* embryos. However, while necessary, *trh* and *dfr* gene products are not sufficient to induce expression of any but the *btl* gene when those factors are expressed ectopically. (D) Expression of gene encoding the FGF ligand Branchless (Bnl; dark spots) in five domains surrounding a *trh(1-eve-1)* tracheal placode (red). The cells closest to the *bnl* expressing patches migrate in their direction, as indicated in the cartoon below. Control of migration and branching of tracheal tubules is mediated by subsequently expressed genes within the tubule cells (see review of Metzger and Krasnow, 1999). [(D) From Sutherland *et al.* (1996) *Cell* **87**, 1091–1101; and Metzger and Krasnow (1999) *Science* **284**, 1635–1639; copyright American Association for the Advancement of Science.] (E) Signaling domains which determine tracheal cell fates. (E1) Expression of *dpp* (purple) dorsally and ventrally, by *in situ* hybridization, about stage 11 just prior to migration. Domains of *dpp* expression border the tracheal placodes, which are revealed by expression of *trh(1-eve-1)* (Lacz product as above, red). [(E1) From Vincent *et al.* (1997) *Development* **124**, 2741–2750 and The Company of Biologists Ltd.] (E2) Expression of *sal* gene within a subset of tracheal cells destined to form specific components of the tracheal system; *sal* expression (violet) superimposed on *trh(1-eve-1)* expression (red) at stage 11; (E3) *sal* expression in extending tracheal system (brown) superimposed on *trh(1-eve-1)* expression (violet), which continues to mark all tracheal cells. The *sal* expressing cells constitute the dorsal trunk (DT); (*sal* expression transiently marks the dorsal branches; DB1, dorsal branch 1). [(E2, E3) From Kühnlein

(Continues)

express the Dpp and EGF signal response systems, respectively, and use them to carry out different aspects of migratory tracheal morphogenesis.

Our second example is drawn from the chick limb, the morphogenesis of which depends on a series of regional signaling interactions (reviewed by Johnson and Tabin, 1997; Dudley and Tabin, 2000). The specific aspect shown in Fig. 4.11 concerns dorsoventral specification in the wing bud. The evidence comes from a series of forced ectopic expression experiments using retroviral vectors. The ectodermal signaling ligand, Wnt7a, causes expression of a mesenchymal regulatory gene of the *lim* class, *lmx1*, which defines the subpart of the limb bud that will form dorsal structures. Spatial control, i.e., the function of excluding *lmx1* expression from the ventral side, is the responsibility of the *engrailed1* (*en1*) gene, which encodes a transcription factor that is expressed only in cells of ventral ectodermal origin (Altabef *et al.*, 2000). The *en1* gene performs this function by preventing expression of the *wnt7a* gene on the ventral side. The normal disposition of these domains of gene expression are shown in the diagram of Fig. 4.11A (Johnson and Tabin, 1997), and some of the experimental results supporting the relations shown there are illustrated in Fig. 4.11B1–B10. The outcome is that if the *lmx1* transcriptional domain is permitted to extend to the ventral side of the limb bud, by ectopic expression of *wnt7a*, the limb develops as if it had two dorsal sides (Fig. 4.11B7–B10). The *lmx1* expression domain is a defining transcriptional

---

and Schuh (1996) *Development* **122**, 2215–2223 and The Company of Biologist Ltd.] (F) Subdivision of tracheal placode into dorsal and ventral "Dpp response" domains (red) and central "EGFR" domain (yellow). (F1) *sal* and *kni* domains. The *sal* gene product serves as a determinant of anteroposterior tracheal migration and morphogenesis, controlling the activity of certain signaling pathways. In the central domain where *sal* is expressed, *rho* expression results in processing of the EGF ligand precursor (Spitz, Spi), activating the EGFR, and maintaining *sal* expression. These cells produce the dorsal trunk, as seen in (E3). In the regions subject to the Dpp signal, *kni* and *knrl* expression are required to mediate Dpp signal effects following reception of the signal by the Tkv receptor. These cells migrate dorsoventrally and ectopic expression of these genes in tracheal cells causes them to migrate in dorsoventral directions, i.e., toward the Dpp stripes. The *sal* gene is directly repressed by Kni (Chen *et al.*, 1998). (F2) Migration begins, driven by the Bnl/Btl (FGF) system controlled by the *trh* and *dfr* genes [(C, D)]. Migration tips will form at leading cells marked by stars. The red and yellow colors mark the Dpp response domain cells and the EGF signaling domains, as in (F1). (F3) Resulting migration patterns. The differences in migratory behavior are apparently mediated by different downstream genes that control diverse signal systems, cell adhesion molecules, special guidance cues, etc. (see reviews of Zelzer and Shilo, 2000b; Metzger and Krasnow, 1999). [(F1) From Zelzer and Shilo (2000b) *BioEssays* **22**, 219–226, copyright Wiley-Liss, Inc., a subsidiary of John Wiley & Sons, Inc.]; [(F2, F3) From Wappner *et al.* (1997) *Development* **124**, 4707–4716 and The Company of Biologists Ltd.]

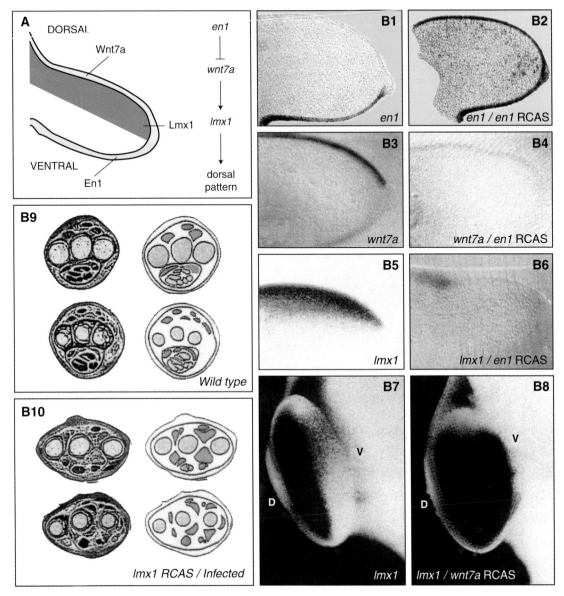

FIGURE 4.11   A typical relation: transcriptional control of a gene encoding a signal ligand determines the dorsal transcriptional domain in chick limb bud. (A) Diagram of limb bud and early regulatory interactions required for D/V specification. The *engrailed* (*en1*) gene is expressed in ventral ectoderm; a Lim class homeodomain gene (*lmx1*) is expressed in dorsal mesoderm; and the *wnt7a* gene is expressed in dorsal ectoderm. [(A) From Johnson and Tabin (1997) *Cell* **90**, 979–990;

(*Continues*)

state upstream of the dorsal program of morphogenesis, and the example shows how its spatial localization is controlled by repression of the gene encoding the Wnt7a signaling ligand.

We have now briefly explored some of the forms of regulatory design that underlie bilaterian pattern formation, and the morphogenesis of body parts. There is not nearly complete enough knowledge of any patterning system, and so we have been forced to extract elements of mechanism where they can be found, from a variety of different developmental processes. But there is an advantage. The complete generality of the mechanisms underlying the pattern formation process emerges strongly from these comparisons, disparate as they are. *cis*-Regulatory modularity, the requirement for spatial *cis*-regulatory information processing, the depth of the regulatory networks that control first the definition and then the subdivision of future body parts and subparts, and the intimate sequential linkage between installation of regional transcriptional states and signaling events: these are all entirely general features.

THE LAST ROUTINES: CALLING IN DIFFERENTIATION PROGRAMS

There remains the culmination of the process of building body parts. After the successive transcriptional patterns have been generated, and the right morphogenetic processes set in train in the right places, differentiated effector cells must be produced. Regional specification continues until the final fine-scale pattern is achieved and the transcriptional mechanisms which determine the spacing of

---

copyright Cell Press.] (B) Experimental demonstrations supporting genetic interactions indicated in (A); visualization of gene products in sections (B1–B6) or whole mounts (B7, B8) by *in situ* hybridization. (B1) Normal ventral expression of *enI* gene in hindlimb of stage 21 chick embryo; (B2) Transcription of *enI* gene extended to dorsal ectoderm by introduction of a retroviral vector expressing the mouse *enI* gene (*enI* RCAS); stage 10; (B3) normal expression of *wnt7a* gene, stage 21; (B4) loss of *wnt7a* expression in hindlimb bud injected with *enI* RCAS; (B5) *lmxI* expression in dorsal mesoderm, stage 21; (B6) loss of almost all *lmxI* expression in limb bud injected with *enI* RCAS. [(B1–B6) From Logan *et al.* (1997) *Development* **124**, 2317–2324 and The Company of Biologists Ltd.] (B7) Normal expression of *lmxI*, stage 24; (B8) ectopic expression of *lmxI* in limb bud injected with *wnt7a*RCAS. (B9) Histological cross-sections of normal hindlimb, day 11, dorsal up, at two successively more distal levels in foot. In the diagrams green indicates tendons; red, dorsal muscles; pink, ventral muscles; yellow, ventral tendons; light blue, bones. (B10) Similar sections and diagrams of foot deriving from limb bud that had been injected with *lmxI*RCAS at stage 8–10. Ventral structures are missing and the morphology approaches mirror image dorsal symmetry. Gray, tendons appearing in an abnormal position. [(B7–B10) From Riddle *et al.* (1995) *Cell* **83**, 631–640; copyright Cell Press.]

terminal morphological elements are in place. Terminal differentiation typically follows mitotic expansion of already specified precursors, and this process also has to be locked in to the underlying morphological pattern. From a programmatic point of view a lot of special regulatory functions are required to set in place the right patches, arrays, or masses of differentiated cells, and many or most of the differences that distinguish closely related species are controlled at this level.

A revealing aspect of the processes that control final events of cell type differentiation is how separate they are from what has gone before. For example, return for a moment to Fig. 4.7E: there we saw that a unique combination of patterning functions is required to specify each of veins L2-L5 of the *Drosophila* wing, but we notice that at the end all the vein territories are identically divided into initial EGFR and flanking Notch territories (Sturtevant *et al.*, 1993; Baonza *et al.*, 2000). That is, the diverse vein patterning processes terminate with the installation of the same differentiation programs. Of course this is a commonplace, true as well of the diverse patterning processes that end by calling in the same differentiation programs for muscle cell types, or bone forming cells, given types of neuron, and so forth, both within and amongst organisms.

Genomic differentiation programs are of a different, because more ancient, evolutionary origin than pattern formation processes (Davidson *et al.*, 1995; Peterson and Davidson, 2000). Bilaterian clades are distinguished by the specific design of the pattern formation processes by which their definitive morphologies are constructed. But many of the differentiation programs they use are completely panbilaterian, and others are shared by the very diverse forms that make up huge clades, such as those required for formation of the endoskeletal and dermal bone of all vertebrates or for the exoskeleton of ecdysozoans. That is, genomic programs for many forms of cell type differentiation were already present in the common ancestors that preceded the morphological diversification of the Bilateria, and some even preceded the bilaterian-cnidarian split. Differentiation programs have remained as separate subelements of the developmental regulatory network, that can be called into play and discretely "wired" into diverse regional specification systems. To link them into body part development means that their genomic controls must respond to a transcriptional output of the preceding pattern formation process. Two examples follow which illustrate the nature of the interfaces immediately upstream of differentiation gene batteries.

Specification of Peripheral Nervous System Elements in the *Drosophila* Wing

The wing disc control systems we last looked at determine the position of the proneural clusters. As precise and elegant as these systems are, further regulatory apparatus is still needed in order to refine the position at which SOPs are actually specified. Only thereafter does the terminal differentiation of peripheral nervous system cell types occur. The function of the additional positioning apparatus is to limit proneural fate both without and within the proneural clusters of *AS-C* expressing cells. For example, the *extramacrochaetae* (*emc*) gene provides an additional negative spatial input in the 3rd instar wing disc, the effect of which is to increase the accuracy of the proneural positioning system (Modolell, 1997; Garrell and Modolell, 1990; Van Doren *et al.*, 1992). Another transcriptional regulator which spatially controls *AS-C* expression is Hairy (Van Doren *et al.*, 1994), which in certain regions acts as a DNA-binding repressor that is required to prevent ectopic expression. Within the clusters proneural fate is ultimately confined to the cells that will actually become SOPs. In the case of the large macrochaete bristles of the notum the SOP which emerges is one of a patch of very few cells precisely positioned within the field of 20-30 cells that make up the proneural cluster. The SOP accumulates high levels of *AS-C* gene products, expresses downstream neurogenic genes, and gives rise to a single bristle. In large bristle clusters such as on the wing margin, the positions of the nervous system elements are less precise, except for their characteristic spacing (Simpson, 1990; reviewed by Garcia-Bellido, 1981; Modolell, 1997). SOP cells actively express Delta, the Notch (N) ligand (under direct *AS-C* control; Kunisch *et al.*, 1994), and in consequence of activation of their N receptors the surrounding cells of the cluster are diverted to epidermal fate (reviewed by Jan and Jan, 1994).

Many of the N effects on gene expression are mediated in the proneural cluster cells by genes of the *Enhancer of split* [*E(spl)*] cluster. A direct link between N and its target genes is a transcription factor, Suppressor of Hairless [Su(H)], that forms a transcriptionally active nuclear complex with the intracellular domain of N, which is released upon N activation. Figure 4.12A shows diagrams of the *cis*-regulatory elements for a number of *E(spl)* genes (Nellesen *et al.*, 1999). Some of these genes are critical elements of the spatial refinement system of the wing proneural clusters, in that they encode bHLH repressors of the proneural genes (see legend). The key point here is how the activity of the *E(spl)* genes is focused to the appropriate cells. The answer is to be seen in the target sites of the *E(spl)* cis-regulatory elements: Fig. 4.12A shows that each of these genes includes sites both for Su(H) and *AS-C* factors, i.e., these genes will run in prospective proneural cells, when these cells receive N signals. The result of their activation is suppression of neural differentiation potential, and execution of epidermal fate. Some *E(spl)* genes respond directly to ectopic Su(H) and *AS-C* expression *in vivo* (Cooper *et al.*, 2000) and in Fig. 4.12B two of the genes, *m4* and *mγ*, are shown to increase their expression dramatically on stimulation of N signaling under heat shock control (Nellesen *et al.*, 1999). But note that the expression still occurs in proneural clusters, in the case of *mγ* in clusters in which it is not normally expressed.

Once specified, the SOP undergoes the stereotypic divisions shown in Fig. 4.12C (Posakony, 1994; Gho *et al.*, 1999). The four cells of the bristle assemblage, i.e., shaft, sheath, socket, and neuron, express four distinct differentiated fates, and at each division the asymmetry in fate depends on N signaling (see legend). We are now finally at the level at which cell type-specific differentiation processes must be called into play. A transcriptional regulatory factor which executes that function is Pax2. The *pax2* gene is expressed initially in all the SOP

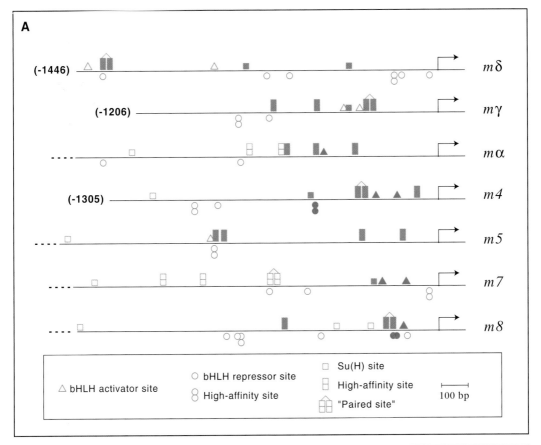

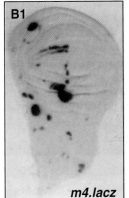

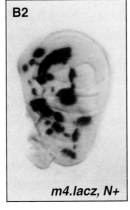

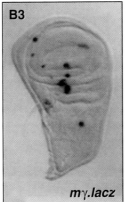

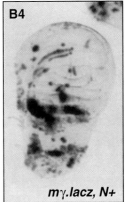

FIGURE 4.12 (Continues)

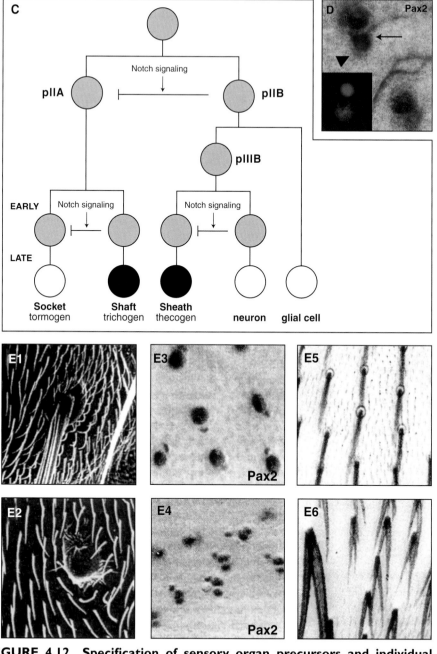

FIGURE 4.12   Specification of sensory organ precursors and individual cell types in the *Drosophila* wing imaginal disc.   (A) *cis*-Regulatory elements controlling expression of some *Enhancer of split* (*E(spl)*) *Complex* genes; solid symbols

(Continues)

indicate interactions established by direct DNA-protein binding studies; open symbols, probable sites based on their similar sequence (see color key for identity). Each gene has sites for *AS-C* proneural activators (green triangles); for bHLH repressors (red circles); and for the Suppressor of Hairless (Su(H)) factor (blue square). Together with the intracellular domain of Notch (N), which is released on ligand binding, the Su(H) protein acts as a transcriptional activator (Struhl and Adachi, 1998; Artavanis-Tsakonis et al., 1999; Lecourtois and Schweisguth, 1998). Other genes of the *E(spl)* Complex share the same sets of target sites in addition to those shown (Nellesen et al., 1999). The *E(spl)* genes themselves encode bHLH repressors (of those shown, *mδ*, *mγ*, *m5*, *m7*, *m8*) or small proteins probably involved in N signaling (*mα*, *m4*). [(A) Adapted from Nellesen et al. (1999) *Dev. Biol.* **213**, 33–58.] (B) Expression of individual *lacz* constructs driven by *E(spl)* *cis*-regulatory elements in wild-type wing discs and in discs in which an activated form of N is expressed ubiquitously. Despite these similar inputs some *E(spl)* genes must respond to additional cues as well, since they are expressed in several different patterns (Nellesen et al., 1999; Ligoxygakis et al., 1999; Cooper et al., 2000). (B1) *m4.lacz*, revealed by *in situ* hybridization against lacz mRNA; (B2) *m4.lacz* in ectopic N disc; (B3) *mγ.lacz*; (B4), *mγ.lacz* in ectopic N disc. [(B) From Nellesen et al. (1999) *Dev. Biol.* **213**, 33–58.] (C) Lineage of mechanosensory bristle cell types. In each of the divisions indicated N signaling differentiates between alternative fates, in that the cell on the left as shown by the barred horizontal lines is prevented by the signal from executing the fate of its sister cell. Signaling is bidirectional, but the PIIB cell and the shaft and neuron cells do not respond because of the activity of N pathway antagonists, *viz*, Numb and Hairless (H); furthermore, the effectors of N signaling are different in the socket and sheath cells (Posakony, 1994; Frise et al., 1996; Nagel et al., 2000). The shading indicates levels of accumulation of the Pax2 transcription factor. (D) Pax2 expression in microchaetes of the notum, 32 h after puparium formation, identified immunocytologically (red, green in inset) together with an antigen expressed on the neuron and shaft cell (blue stain). The two nuclei containing Pax2 protein are the small sheath cell nucleus (arrowhead) and the shaft cell nucleus. The neuron is identified by the arrow. The inset shows a macrochaete: Pax2 is here indicated in the sheath cell in green; the neuron, identified by a neuron-specific antibody, in red. Pax2 is also present in the shaft cell, but not the socket cell (cf. C). (E) Pax2 function in cell type specification. (E1) Scanning EM of wild-type macrochaete shaft surrounded with microchaetes; (E2) failure of shaft formation in a loss of function *pax2* mutant, resulting in an empty socket; (E3) normal expression of *pax2* gene in bristle lineage identified immunocytologically as above; (E4) *pax2* expression after repression of N signaling (by overexpression of the H antagonist of N signaling, driven by a heat shock *cis*-regulatory element). Three cells rather than two now express the *pax2* gene in each mechanosensory structure; these are the sheath, shaft, and what should have been the socket cells. (E5) Normal bristle shafts; (E6) bristles with double shafts, due to conversion of sheath cell to shaft fate after repression of N signaling. Other experiments show that this effect is indeed caused by the ectopic *pax2* expression. [(C–E) From Kavaler et al. (1999) *Development* **126**, 2261–2272 and The Company of Biologists Ltd.; in (C) information from Gho et al., 1999 has been added.]

descendants, but the gene is turned off in the prospective neuron and socket cells (Fig. 4.12C). Expression of Pax2 is seen cytologically in Fig. 4.12D, and the requirement for this factor in order for differentiation of the shaft to take place is dramatically illustrated in Fig. 4.12E1, E2 (Kavaler *et al.*, 1999). Without *pax2* gene expression a hole appears in place of the bristle shaft. Furthermore, *pax2* gene is a direct target of N signal transduction (Kavaler *et al.*, 1999). If the N pathway is repressed, the Pax2 protein remains present in both of the PIIA daughter cells (Fig. 4.12E3, E4); the result is that both become shaft cells, producing a double shaft phenotype (Fig. 4.12E6).

A considerable number of differentiation gene batteries must be expressed in the development of the whole wing, including its veins, its intervein epidermis, its hinge structures, its other peripheral nervous system elements, and so forth; there are four cell types in the macrochaete alone. This is scarcely atypical. Formation of an adult body part always involves the (more or less exact) spatial installation of multiple gene batteries encoding effector proteins, superimposed on all the patterning mechanisms that have gone before.

Installation of Cell Type-specific Differentiation Programs in the Pituitary

The transition between morphogenetic pattern formation processes and those detailed strategies used to call in cell type-specific differentiation programs can be seen anywhere one looks. The terminal differentiation programs themselves (the "routines" of the title of this section) are not our subject here (cf. Chapter 2); rather is their linkage into the prior spatial regulatory apparatus. Our final example in this chapter is the pituitary, in which many differentiation programs that are responsible for expression of this organ's endocrine genes have been defined (Simmons *et al.*, 1990). Key aspects of the pattern formation process leading to the morphogenesis of the structure from which the pituitary develops, Rathke's pouch, are also known. Figure 4.13A1 shows a diagram of the formation of Rathke's pouch in the mouse, and some prominent signaling interactions involved in setting up its regional transcriptional states. Rathke's pouch arises at the interface of the dorsal oral ectoderm and the ventral diencephalon. A large set of genes encoding transcription factors is expressed in different regions of the pouch, as shown in Fig. 4.13A2 (Treier *et al.*, 1998). Six different endocrine-secreting cell types arise, in specific spatial domains of the forming gland (Fig. 4.13A3). These are the corticotropes, which produce adrenocortico-stimulating hormone (ACTH); the melanotropes, which produce melanocyte-stimulating hormone (MSH); the thyrotropes, which produce thyroid-stimulating hormone (TSH); the somatotropes, which secrete growth hormone (GH); the lactotropes, which produce prolactin (PRL); and the gonadotropes, which produce luteinizing hormone (LH) and follicle-stimulating hormone (FSH). Each of these cell types arises within a domain defined earlier by a particular transcriptional state, as indicated in the diagram of Fig. 4.13B. As implied there, part of the mechanism we want to know about will emerge from detailed descriptions of the *cis*-regulatory transactions which account for the activation of the genes encoding these transcription factors. There seem always to be lots of special little mechanistic devices utilized at this level of the developmental control apparatus, and a closer look at the specification of the cell types that arise in the ventral region of the forming pituitary after about day 13 provides an illustration.

The ventral region is where gonadotropes, thyrotropes, and somatotropes plus lactotropes differentiate, in three adjacent domains. The key upstream transcriptional regulators which are required to install these respective differentiation programs in their respective regions are the POU homeodomain factor Pit1 and the Gata2 factor (Fig. 4.13B, C; see legend for details and references) The regulatory relations by which this occurs are indicated in Fig. 4.13C2, C3 (Dasen *et al.*, 1999). Essentially, a ventral-to-dorsal concentration gradient of *gata2* expression that is set up under the influence of a ventral source of BMP2, plus the interrelations between the *gata2* and *pit1* genes and their products establish the three different regulatory domains. This works as follows: high levels of Gata2 factor repress the *pit1* gene, but Pit1, off the DNA, interacts with the Gata2 factor so as to interfere with its ability to serve as a transcriptional activator, except in *cis*-regulatory elements where sites for both factors are contiguous. Given the constitution of the *cis*-regulatory target sites in genes controlling the final differentiated functions of gonadotropes, thyrotropes, and somatotropes plus lactotropes, this suffices to establish the three ventral regulatory states (Fig. 4.13C; see legend). Note that in this case as in the preceding example of bristle cell-type specification, signaling plays an essential role, right to the end of the process. But if we ask where lies the spatial "intelligence" of the regulatory system by which the differentiation programs are installed, the answer is of course in the genomic *cis*-regulatory elements. It is these elements which enable the genes of the different cell types to respond differentially to the Pit1 and Gata2 concentrations with which they are confronted (Fig. 4.13C3); and it is the relevant *cis*-regulatory modules of the *pit1* and *gata2* genes which determine their own sites of expression.

CONCLUDING REMARK

We end where this traverse of the regulatory mechanisms underlying adult body part formation began, with a thought about its evolutionary meaning. It should be obvious now how great is the distance between what it takes to carry out direct cell type specification, as in the Type 1 embryonic processes considered in Chapter 3, and what it takes to build the pieces of an adult bilaterian body plan. This distance must be considered in terms of the required genomic regulatory information: the number of regulatory linkages, the network depth, the specificity of *cis*-regulatory modules involved throughout, and the number of diverse differentiation gene batteries ultimately needed. In architectural terms comparing the regulatory networks underlying direct cell type specification with

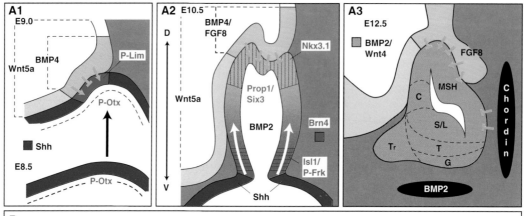

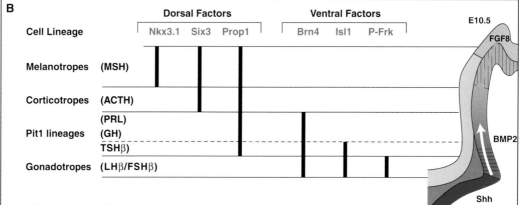

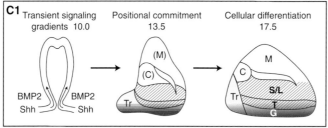

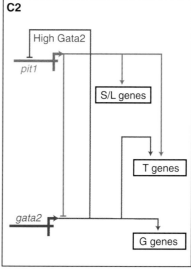

FIGURE 4.13   Cell-type specification late in pituitary morphogenesis. (A) Signaling interactions in mouse pituitary development; in A1, A2 signaling molecules are indicated in black; transcription factors in red. (A1) At 9.0 dpc a region of the oral epithelium from which the pituitary will form is distinguished by repression of sonic hedgehog (Shh; green) in response to a BMP4 signal (yellow) from the adjacent ventral diencephalon (tan). In the pituitary progenitor field, i.e., the cells that will give rise to Rathke's pouch (blue), the consequence is activation of a gene encoding a Lim domain transcription factor. The *p-otx* gene is expressed throughout the oral ectoderm. (A2) At 10.5 dpc, Rathke's pouch has invaginated, and a ventrodorsal gradient of BMP2 (white arrows) and a dorsoventral gradient of FGF8 (yellow arrows) are set up. The infundibulum is forming from the ventral diencephalon. The source of the BMP2 is the interface with the adjacent Shh expressing oral ectoderm, and the source of the FGF8 is the infundibulum. Genes encoding transcription factors that are required combinatorially for cell type specification are expressed in different spatial domains in consequence of these signal inputs. The factors are Nkx3.1 (pink), Six3, and Prop1 (black vertical stripes) at the dorsal end of Rathke's pouch, and Brn4 (blue), Isl1, and P-Frk (black horizontal stripes) toward the ventral end. (A3) At 12.5 dpc the *bmp2* and *wnt4* genes are expressed in Rathke's pouch with additional BMP2 input from the underlying mesenchyme. BMP2 signaling is counteracted by dorsal expression of FGF8 from the infundibulum, and posterior expression of the chordin BMP antagonist (yellow symbols). Within the pouch the domains of cell type specification are indicated: MSH melanotropes; C, corticotropes; S/L, somatotropes and lactotropes; T, thyrotropes (r, rostral); G, gonadotropes. (B) Correlation of dorsal and ventral transcription factor domains with cell types and their endocrine products, in parentheses (see text). [(A, B) From Treier *et al.* (1998) *Genes Dev.* **12**, 1691–1704.] (C) Transcriptional control of four ventral differentiation pathways in the pituitary by the POU domain factor Pit1 and by Gata2. (C1) Distribution of Pit1 and Gata2 factors, at 13.5 and 17.5 dpc. The *gata2* gene is activated ventrally by BMP2, and if expression of this gene is forced in the dorsal pituitary it converts the cells of this region to ventral cell types. Similarly if *pit1* expression is forced in the ventralmost regions it converts prospective gonadotropes to thyrotropes. Expression of the *pit1* gene also requires BMP2 (Treier *et al.*, 1998). Gata2 is present in graded concentration, highest in the most ventral region (purple shading), from which the gonadotropes will derive. Pit1 is present in a more central domain, from which thyrotropes, somatotropes, and lactotropes will derive (hatching). [(C1) From Dasen *et al.* (1999) *Cell* **97**, 587–598; copyright Cell Press.] (C2) Gene interactions, from data of Dasen *et al.* (1999). At high levels Gata2 is a transcriptional repressor of the early expression module of the *pit1* cis-regulatory system. The *pit1* gene product, where it is present in very high concentrations, interacts physically with the Gata2 factor, preventing it from activating target genes except where Gata2 and Pit1 target sites are adjacent. (C3) Pit1 and Gata2 functions in controlling expression of different sets of ventral differentiation genes of the pituitary. *cis*-Regulatory organization of the S/L, T, and G genes, and the relationships in (C2), account for regional activation of the four differentiation programs. G region: expression of the *pit1* gene is

(*Continues*)

those underlying adult body part formation may be a little like comparing a one-story factory to a city of enormous buildings, within which the density of machinery per cubic foot is the same as that in the one-story factory.

So what is the "secret of the bilaterians," with respect to building body parts? It is freeing many regulatory transactions from the job of running differentiation gene batteries, and instead using them, layer after layer, for progressive regional transcriptional specification: the use of abstract patterning mechanisms. When we can perceive the real regulatory architecture, in the DNA, that drives adult body part development from the beginning to the end of at least some components of the process, then we will be masters of the secret of the bilaterians. And then we will be able to think in more effective detail about how the bilaterians arose. As more and more fragments of bilaterian developmental regulatory architecture come into view, its overall dimensions and form look less a black box than they used to. Our great advantage is that the bilaterians all know the same secret, which they use to build their many and various morphological components. It is their regulatory heritage, and thus, as we have all learned, what we see in one developmental context illuminates how another works. But a take-home lesson of this chapter is intended to be that the significance, whether comparative, mechanistic, or otherwise, of any given observation on any piece of the developmental regulatory network depends on appreciating where it fits in; what part of the process it illuminates. Differentiation does not particularly illuminate the stepwise transactions of pattern formation, and shallow networks do not much illuminate deep ones. That is why it has seemed useful to divide this discussion into the successive regulatory processes by which adult body parts are generated.

---

excluded from the ventralmost region by Gata2, and only gonadotrope genes which require Gata2 but not Pit1 operate there. S/L region: similarly, somatotrope and lactotrope genes have no requirement for Gata2, but respond to Pit1. T region: thyrotrope genes operate where both factors are present. In this region there is insufficient free Gata2 to repress the *pit1* gene, nor can gonadotrope genes be activated because Pit1 inhibits binding of Gata2 to its DNA target site by a protein interaction off the DNA. [(C3) Adapted from Dasen *et al.* (1999) *Cell* **97**, 587–598; copyright Cell Press.]

CHAPTER 5

Changes that Make New Forms: Gene Regulatory Systems and the Evolution of Body Plans

Once we see something of how genomic programs for morphogenesis are designed, the nature of the basic mechanism by which bilaterian body plans change in evolution becomes obvious. Evolutionary change in the form of body parts requires evolutionary change in the gene regulatory network that controls pattern formation processes: this is the specific subject of this chapter. Our objective is to understand what underlies bilaterian diversity. Similar developmental regulatory principles are useful for thinking about the Precambrian evolutionary origins and the antecedents of the Bilateria as well, but the reader is referred elsewhere for discussion of that subject (Davidson *et al.*, 1995; Peterson *et al.*, 2000b; Peterson and Davidson, 2000).

Change in regulatory networks underlying pattern formation processes occurs at every level of the developmental programs that they encode. The relevant regulatory networks consist of the linkages amongst the genes engaged in setting up the morphogenesis of a body part, from the initial specification of its progenitor field to the terminal installation of differentiation pathways, as discussed in the last chapter. Change in terminal connections to regulators that control differentiation gene batteries results in detailed alterations in fine aspects of morphology; change in upstream connections amongst genes specifying spatial transcriptional state is required for more gross alterations in the form of the subparts of a structure. As we have seen, the architecture of these regulatory networks is encoded in the *cis*-regulatory systems of the genes that they include. These systems determine where and when each gene is expressed; which regulatory genes are functionally linked to which others at each stage; what are the downstream regulators; and to what signals they respond. Logic requires that changes in *cis*-regulatory systems which have the effect of altering spatial transcriptional patterns are the only kind of DNA level events which could cause novelty in organismal form at the level of body parts and subparts. But there is also plenty of direct supporting evidence from comparative studies of developmental processes. Indeed one could almost derive the principles of developmental mechanism from the nature of evolutionary change in gene use, rather than the reverse. The evolutionary processes by which regulatory genes are reassigned to different functions in a pattern formation network goes by the name of "cooption" or "recruitment," and examples are at hand wherever one chooses to look. This chapter is about cooption at all levels of patterning gene networks, and about the kinds of cooptive processes that have been responsible for the evolution of new body parts during the divergence of the Bilateria.

SOME EXAMPLES: EVOLUTIONARY COOPTION OF GENES TO NEW PATTERN FORMATION FUNCTIONS

Cooption of regulatory genes involved in developmental pattern formation to new roles means that the gene is performing a different function with respect to the role it had in the ancestor. The gene may have acquired target sites in downstream genes that it did not previously control but now does. Or it may have retained control of the same apparatus downstream, but operate this apparatus in a new domain of the developing animal, because its own locus of expression has changed. In either case the course of development is altered, and new spatial patterns of gene expression result. Comparative observations that reveal differences in gene expression patterns between animals can often be used to indicate cooption events, providing that the required phylogenetic information is available. That is, cooption is implied at a specific evolutionary branch point if there is a convincing cladistic argument that the ancestral form did not use the gene the way a particular clade of its descendants does: if the usage observed in a given clade of animals is truly evolutionarily novel with respect to the ancestor. Since the pattern of gene expression in the ancestor cannot be observed because it is extinct, the argument necessarily turns on a proper analysis of various of its living descendants.

Some impressive examples of cooption are illustrated in Fig. 5.1. These cases are drawn from the deuterostomes, which have the great advantage for this purpose that the phylogenetic relations amongst the major deuterostome clades are clear and incontrovertible (see legend to Fig. 5.1A for references). The deuterostome clades are distinguished by very different adult body plans, and yet there is no question but that the deuterostomes are monophyletic. They descend from a common ancestor (Adoutte *et al.*, 2000; Cameron *et al.*, 2000). Among the major features specific to given deuterostome clades are the notochord of the chordates; the paired anterior, middle, and posterior larval coeloms from which the mesoderm of the adult body plan arises in the echinoderm plus hemichordate clade; and the fivefold radial symmetry, calcite endoskeleton, and water vascular system of the echinoderms. A glance at Fig. 5.1A shows that fivefold radial symmetry (character 3) is a feature that is novel with respect to the evolutionary ancestors of echinoderms. The sister group of the echinoderms, the hemichordates, remain bilateral; the chordates are also bilateral; and essentially all protostomes are bilateral as well (here represented by the arthropod outgroup). Therefore the common deuterostome ancestor was bilaterally organized, as was the last common ancestor of the hemichordate plus echinoderm clade. Another character primitive for both hemichordates and echinoderms is indirect development by means of similarly organized embryos and ciliated feeding larvae (Peterson *et al.*, 1997, 2000b). The diverse uses of the *brachyury* gene in the various deuterostome clades (Fig. 5.1B–F) exemplifies the rather startling incidence of the evolutionary process of cooption.

The *brachyury* genes constitute a subfamily of genes encoding T-box transcription factors (Papaioannou and Silver, 1998). These genes are expressed during development in the posterior end of the gut in most bilaterian animals that have been examined. A *brachyury* gene is even expressed at the opening of the gut in a cnidarian (Technau and Bode, 1999), and cnidarians can be taken as an outgroup for the bilaterians. Presumably this very ancient gene functioned in gut development in the remote metazoan ancestor of both the cnidarians and the bilaterians. In bilaterians *brachyury* genes are expressed in the posterior mesoderm as well as in the hindgut during embryogenesis (reviewed by Herrmann and Kispert, 1994; Smith, 1997; Peterson *et al.*, 1999a). For example, in *Drosophila*, the *brachyenteron* gene (an orthologue of the mouse *T* gene and all the other *brachyury* genes included in Fig. 5.1B–F) is expressed at the posterior end of the syncytial blastoderm stage embryo, in a domain from which the hindgut endoderm will invaginate, and after cellularization in a band of cells that form the posterior visceral mesoderm (see Fig. 5.1G). The gene is required for specification of both of these components of the body plan (Kispert *et al.*, 1994; Singer *et al.*, 1996; Kusch and Reuter, 1999). The mouse *brachyury* gene is also spatially required for development of the posterior mesoderm (Herrmann and Kispert, 1994).

So the ancestral bilaterian use of the *brachyury* gene, conserved from ecdysozoans to deuterostomes, is expression in posterior mesoderm and hindgut. But beyond this there have been installed a remarkable variety of clade-specific,

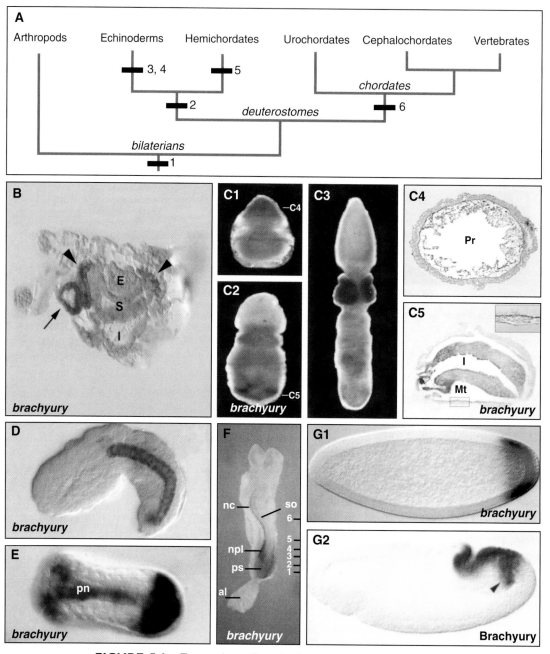

FIGURE 5.1 **Examples of evolutionary cooption of transcriptional regulatory genes that define morphogenetic patterns in deuterostomes.** (A) Deuterostome phylogeny, based on molecular evolution (Cameron *et al.*, 2000;

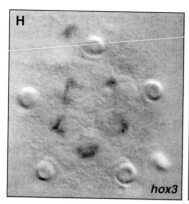

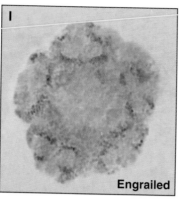

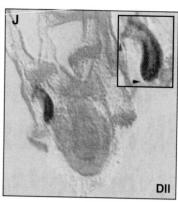

hox3

Engrailed

DII

Castresana *et al.*, 1998; Turbeville *et al.*, 1994; Wada and Satoh, 1994). The arthropods, which are ecdysozoans (see Fig. 1.6), serve as an outgroup. Numbers represent acquisition of characters relevant to the following figures: 1, bilateral organization; 2, paired anterior, middle, and posterior larval stage coeloms which give rise to the major mesodermal components of the adult body plan; 3, fivefold radially symmetric adult body plan of echinoderms; 4, calcite endoskeleton of echinoderms; 5, hemichordate proboscis, incorporating anterior coelom(s); 6, chordate notochord. (B–F) Cooptions of *brachyury* transcriptional regulators in deuterostome evolution. (B) Distribution of *Strongylocentrotus purpuratus brachyury* transcripts visualized by *in situ* hybridization in larva of *S. purpuratus*. Expression is observed in the invaginating vestibule (arrow) and the middle coelom (hydrocoel; arrowhead) with which the vestibule makes contact on the left side of the larva to form the rudiment of the adult. Expression is also observed in the right axohydrocoel (arrowhead; the anterior coelom or axocoel, and the middle coelom, the right hydrocoel, are fused on this side). E, esophagus; S, stomach; I, intestine. [(B) From Peterson *et al.* (1999b) *Dev. Biol.* **207**, 419–431.] (C) Expression of *brachyury* gene in metamorphosing larva of an enteropneust hemichordate, *Ptychodera flava*, by *in situ* hybridization. (C1–C3) Dorsal views, anterior up; (C4, C5) sections at planes indicated in (C1) and (C2). (C1) Expression in anterior coelom of the proboscis in a competent larva. (C2) Metamorphosing larva (one day after stimulus). Expression of *brachyury* gene is now also seen in middle coelom, and in caudal region. (C3) Three-day juvenile; expression of *brachyury* in coelomic mesoderm of all three body regions. (C4) Transverse section through proboscis, showing expression in mesodermal cells of protocoel (Pr; these cells are the progenitors of the proboscis muscle), but not in ectoderm; (C5) Transverse section through hindgut: both endoderm and posterior coelomic mesoderm express *brachyury* but not ectoderm as seen in inset (Mt, metacoel, posterior coelomic lining of body wall; I, intestine). [(C) From Peterson *et al.* (1999a) *Development* **126**, 85–95 and The Company of Biologists, Ltd.] (D) Expression of *brachyury* gene in presumptive notochord of a larvacean urochordate (*Oikopleura dioica*), visualized by *in situ* hybridization, lateral view, anterior left. Larvaceans retain the swimming tail in the adult form. An early hatched stage is shown. The *brachyury* gene is also expressed in the developing hindgut

(Continues)

additional *brachyury* functions. In both sea urchin and hemichordate embryos the *brachyury* gene is expressed at the blastoporal (prospective anal) and oral ends of the gut (as well as in the secondary mesenchyme in sea urchin embryos; for sea urchins, Harada *et al.*, 1995; Peterson *et al.*, 1999b; and unpublished data; for hemichordates, Tagawa *et al.*, 1998; Peterson *et al.*, 1999a). However, the embryonic phase of *brachyury* expression in the oral domains of these embryos has nothing to do with adult body plan formation, since the embryonic expressions are extinguished before the larval stage when the adult body is formulated, and since they occur in regions that are not incorporated in the adult body plan. The use of the *brachyury* gene in the oral regions of these embryos is a cooption particular to the echinoderm plus hemichordate clade (cf. Fig. 5.1A). However, each group has coopted the gene for different developmental purposes in respect to adult body plan formation.

(not visible here). [(D) From Bassham and Postlethwait (2000) *Dev. Biol.* **220**, 322–332.] (E) Expression of *brachyury* genes in amphioxus (*Branchiostoma floridae*); there are two similar *brachyury* genes in this species and the probe detects both. Expression shown by *in situ* hybridization in neurula stage embryo, dorsal view, anterior left, and in presumptive notochord (pn) plus caudal mesoderm. [(E) from Holland *et al.* (1995) *Development* **121**, 4283–4291 and The Company of Biologists, Ltd.] (F) Expression of *brachyury* (*T*) gene product, visualized by immunocytology in 8.5 dpc mouse embryo. Expression is seen in notochord (nc), notochordal plate (npl), and caudal primitive streak (ps) mesoderm; al, allantois; so, somites. [(F) From Kispert and Herrmann (1994) *Dev. Biol.* **161**, 179–193.] (G) Expression of orthologous *brachyury* gene of *Drosophila*, the *brachyenteron* gene. (G1) *brachyenteron* transcripts visualized by *in situ* hybridization at late syncytial blastoderm stage (cycle 14), anterior left. The gene is expressed in the region from which hindgut, anal pads, and posterior mesoderm will develop. [(G1) From Kispert *et al.* (1994) *Genes Dev.* **8**, 2137–2150.] (G2) Brachyenteron protein at gastrula stage visualized by immunocytology, same orientation as in (G1). The hindgut and the primordium of the posterior visceral mesoderm (arrowhead) display the protein. [(G2) From Kusch and Reuter (1999) *Development* **126**, 3991–4003 and The Company of Biologists, Ltd.] (H) Expression of *Sphox3* gene in larval rudiment of adult *S. purpuratus*, visualized by *in situ* hybridization; view from future oral surface. Expression occurs in a fivefold radially symmetric pattern, in mesodermal structures related to the dental sacs. [(H) From Arenas-Mena *et al.* (1998) *Proc. Natl. Acad. Sci. USA* **95**, 13062–13067, copyright National Academy of Sciences, USA]. (I) Detection of Engrailed protein in a juvenile brittle star (an ophiuroid, *Amphipholis squamata*) by immunocytological staining. Expression is observed in a fivefold radially symmetric pattern, in regions bounding skeletogenic domains in which the endoskeletal plates are at an early stage of formation. (J) Distribution of Distal-less protein observed by immunocytology in the invaginating vestibule (between arrowheads in inset), the neuroectodermal anlage of the adult body plan in larva of the echinoid *S. drobachiensis*. [(I–J) From Lowe and Wray (1997) *Nature* **389**, 718–721, copyright Macmillan Magazines, Ltd.]

A specific feature of echinoid echinoderms (i.e., sea urchins) is their mode of indirect development. The adult body plan forms within a feeding larva. The oral surface of the adult, including its radially symmetric water vascular and central nervous systems, arise within a multilayered imaginal structure, the "rudiment." This imaginal structure is constituted initially from the union of an invaginating neuroectodermal pouch, the "vestibule," and the left middle coelom of the larva, the "hydrocoel." As shown in Fig. 5.1B, in sea urchins the *brachyury* gene is expressed in both the invaginating vestibule and the hydrocoel to which it is apposed at the beginning of adult rudiment development. The hydrocoel is equivalent to the middle coelom of the hemichordate. But Fig. 5.1C shows that all three of the coeloms of the adult body plan of *Ptychodera flava* express the *brachyury* gene during metamorphosis. Expression in the posterior coelom is the ancestral bilaterian usage; expression in the middle coelom is an echinoderm plus hemichordate character; but expression in the mesoderm of the anterior coelom, which forms the hemichordate proboscis, is a cooption particular to this group, just as the proboscis is an anatomical feature particular to hemichordates.

The *brachyury* gene assumes its most famous function in chordates, as a transcription factor required to specify notochord (reviews *op. cit.*). Expression of *brachyury* genes in the notochord of a larvacean ascidian is shown in Fig. 5.1D; in the notochord of amphioxus in Fig. 5.1E; and in the notochord and notochord progenitors of a mouse embryo in Fig. 5.1F. In all of these examples expression can also be seen at the caudal end of the embryo (see legend for details and references). But the phylogeny in Fig. 5.1 shows that the famous notochordal role of *brachyury* is also a cooption, peculiar to the chordates, and distinct from its ancestral panbilaterian function at the caudal end of the embryo.

In summary, we see in Fig. 5.1 unique *brachyury* cooptions for every one of the deuterostome clades included in Fig. 5.1A. Some other examples, which unmistakably indicate cooptions to the unique developmental processes of echinoderms, are shown in Fig. 5.1H–J. Figure 5.1H displays transcripts of the *hox3* gene in the imaginal structure from which the radial body plan of the sea urchin arises. The expression domains are associated with the five developing dental sacs (Arenas-Mena *et al.*, 1998). In Fig. 5.1I the radially symmetric distribution of the *engrailed* gene product is seen in a developing brittle star (Lowe and Wray, 1997). Expression of this gene occurs at the boundaries of the skeletogenic domains within which the endoskeletal elements of the juvenile are being laid down, near the bases of the future arms. Since the radially organized body parts to which the patterns of Fig. 5.1H and I pertain are echinoderm-specific (characters 3 and 4 in Fig.5.1A) with respect to the bilateral ancestor (character 1), these transcriptional regulatory genes also must have been coopted to new patterning functions. In Fig. 5.1J the panbilaterian *distal-less* gene which we last met in *Drosophila* imaginal discs, is seen to be expressed in the invaginating vestibule of the sea urchin larva. This is an echinoid-specific cooption of the *distal-less* gene (Lowe and Wray, 1997), as only the echinoids (i.e., sea urchins) produce a vestibule. The cooptions illustrated in Fig. 5.1H–J may indeed have occurred during the evolution of the particular echinoderm classes represented in these examples (Lowe and Wray, 1997).

The examples here are mainly cases of evolutionarily novel gene use early in the transcriptional patterning of important body parts: the notochord in chordates; the proboscis in the hemichordate; and the major components of the imaginal rudiment in the sea urchin. But cooption happens at every level and affects all of the stages of body part formation sketched in the last chapter. It is the fundamental process by which evolutionary change in bilaterian form has occurred, from the major changes in body plan that define high level taxonomic groups (e.g., those of Fig. 5.1A) to the detailed changes that occur in speciation.

hox GENE FUNCTIONS AND COOPTIONS OF THE *hox* CLUSTER PATTERNING SYSTEM

The A/P *hox* gene patterning system is widely regarded as the definitive shared developmental mechanism of the bilaterians. This is because during development arthropods and chordates, about which we know by far the most, generally express genes of the *hox* cluster(s) in the same sequence along their A/P axes. There are relatively few observations on animals belonging to other groups, except for *Caenorhabditis elegans*, which has a degenerate and rearranged *hox* complex and in which only parts of the general A/P expression pattern can be discerned (Ruvkun and Hobert, 1998; Van Auken *et al.*, 2000; Brunschwig *et al.*, 1999.). The most important and unusual aspect of the A/P expression patterns of *hox* cluster genes is that they are "colinear" with the order in which these genes occur in the chromosome; that is, at least the anterior boundaries of the *hox* gene expression domains lie in the same posterior-to-anterior morphological sequence as the 5' to 3' sequence of the genes within the cluster(s) (reviewed by McGinnis and Krumlauf, 1992). Some exceptions are known, due to the dynamic quality of *hox* gene expression patterns, particularly in mammals where there are four clusters and in which individual *hox* genes have acquired special functions. Some genes are expressed a bit out of order at given stages (e.g., *hoxb1* in r4 of the mouse hindbrain at day 8.5; e.g., Fig. 4.9G). But these special situations are irrelevant to the generality of colinear A/P expression. Colinearity means that some aspect of the chromosomal gene arrangement is utilized in the spatial control mechanisms that determine the domains of *hox* gene transcription. The organization of the cluster could be important either in respect to setting up the initial domains of *hox* gene expression or in maintaining stable patterns that are initially installed by specialized *cis*-regulatory modules (as in the examples in Fig. 4.9). How colinearity works remains a subject of much interesting debate, which, however, does not concern us here. We shall simply take for granted that it does work (see, Gellon and McGinnis, 1998; Kmita *et al.*, 2000). The essential point is that the *hox* gene clusters are to be thought of as vectorial transcriptional patterning systems: they produce an ordered sequence of expression domains in embryonic space. Such systems are potentially very useful anywhere a transcriptional specification sequence is required to be arranged directionally along some kind of an axis.

Numerous experiments in both *Drosophila* and mouse have demonstrated regional developmental failures in specification along the A/P axis if given *hox* genes are knocked out or mutated (for general review of the observations that provide this foundation see McGinnis and Krumlauf, 1992). However, *hox* genes do not "make" the anterior or middle or posterior domains of the forming body plan. The genes of the *hox* clusters are activated after the primary spatial regulatory systems of the embryo have set up transcriptional domains that denote its major future components, such as its head, tail, gut, central nervous system, body wall, and so forth. In indirectly developing animals the *hox* gene cluster is not used at all for formation of the completed embryo, and in postembryonic larval development the genes of the cluster are all mobilized only in tissues from which elements of the adult body will form (e.g., in a sea urchin, Arenas-Mena *et al.*, 1998; in an indirectly developing polychaete annelid, Peterson *et al.*, 2000a). The general view is not that individual *hox* genes directly act to produce heads or tails, but that the vectorial expression pattern of the *hox* gene cluster is essential for installing the correct developmental pattern along the A/P axis of the body plan. And that A/P axial patterning by genes of the *hox* cluster is a deeply conserved and fundamental bilaterian feature.

But when we come to mechanism there is something paradoxical about this last conclusion. Consider the example of rhombomere specification, on which we expended some attention in Chapter 4. This is a specific and clear case of axial A/P patterning (cf. Figs. 4.5 and 4.9), in which the rhombomeric expression domains are in general colinear with gene order. However, this cannot be an example of an ancient, conserved patterning mechanism: animals that are not vertebrates do not generate rhombomeres, so the detailed genetic regulatory network that sets up rhombomeric expression patterns cannot be conserved, except within the vertebrates. We saw parts of that network in Fig. 4.9H: its architecture, and the specific connections indicated, must (at least) in part represent the genomic anatomy of a vertebrate cooption. At minimum, key components of this network must be novel with respect to the antecedents of the vertebrates. Furthermore, many of the network connections link genes of different *hox* clusters in specific ways, and lower deuterostomes, i.e., amphioxus (Garcia-Fernandez and Holland, 1994) and echinoderms (Martinez *et al.*, 1999) have only single *hox* gene clusters. Therefore all intercluster regulatory linkages are necessarily cooptions. What we need to understand are the evolutionary pathways that have led mechanistically to novel *hox* cluster gene functions, which nonetheless still mediate axial A/P patterning. In the real terms of specific regulatory linkages these specific functions are not in detail "old," let alone panbilaterian.

Hox gene products act at many different levels of the developmental regulatory apparatus. Some even operate in single differentiation pathways, e.g., *hox7* in aboral ectoderm of sea urchin embryos (Angerer *et al.*, 1989); or *labial* in copper cells of *Drosophila* (Hoppler and Bienz, 1994; see below). However, by far the most important functions of *hox* genes are interventions in the *cis*-regulatory networks that effect transcriptional pattern formation. They do this by altering expression of genes that encode signaling components and transcription factors (for reviews see Carroll, 1995; Weatherbee and Carroll, 1999; Gellon and McGinnis, 1998; Bilder *et al.*, 1998; Bienz, 1994). The following example shows this in detail.

A Specific Case of A/P Patterning: How a *hox* Gene Does its Job

Among the now classic examples of A/P patterning mediated by *hox* genes is the diversification of segmental appendages in *Drosophila* (Lewis, 1978). Specific *hox* genes affect the identity of halteres, wings, and legs, and repress antennal fate, acting on developmental processes within the imaginal disc from which these structures derive (reviewed by Weatherbee and Carroll, 1999). The halteres form from the dorsal imaginal discs on the third thoracic segment (T3), and the wings from the corresponding discs on the second thoracic segment (T2). The halteres, which are small balancing organs, lack many of the specific pattern features of the wing (for latter, Fig. 4.7). One of the most impressive definitions of the "power of the *hox* genes" was the demonstration (Bender *et al.*, 1983) that a complete loss of the *ubx* gene product causes the haltere imaginal disc to give rise instead to a wing. In these mutants execution of the normal T2 imaginal disc program for wing development remains unaffected (*ubx* is of the same subclass as the vertebrate *hox6–8* genes; de Rosa *et al.*, 1999). The result is a four-winged fly, illustrated in Fig. 5.2A. The direct implication is that wing patterning functions in the haltere imaginal discs are repressed by the Ubx transcription factor, which at the key time is present in the T3 disc but not the T2 disc.

We now know a good deal about exactly what the Ubx factor does in the haltere disc to prevent wing patterning from taking place: it acts as a repressor for at least six different wing-specific patterning functions (Weatherbee *et al.*, 1998). Some examples are shown in Fig. 5.2B–E, and a summary of known Ubx effects on the wing patterning network is reproduced in Fig. 5.2F. Figure 5.2B concerns expression of one of the transcriptional regulators of wing disc patterning that we encountered earlier, *spalt-related* (*salr*), which acts as a controller of vein positioning (Fig. 4.7C6, E2). The *salr* gene is turned off by Ubx in the haltere, and is repressed in the wing if ectopic expression of the *ubx* gene is forced to occur there. Similarly, expression of the gene encoding DSRF, a transcription factor that is required to set in train the differentiation of the collapsed intervein epithelium of the wing surface, is repressed by Ubx (Fig. 5.2E). So also is expression of Wg in the posterior region of the haltere disc (Fig. 5.2D). These effects are all autonomous and they are likely to be direct effects on the respective *cis*-regulatory elements; and this is even more clearly so for the example in Fig. 5.2C. The *vestigial* (*vg*) gene quadrant enhancer (for which see Fig. 2.7), is normally silent in the pouch region of the haltere disc. But as illustrated in Fig. 5.2C, a quadrant enhancer *vg.lacz* construct springs to life within a Ubx⁻ clone made in a haltere disc (Weatherbee *et al.*, 1998). The summary diagram in Fig. 5.2F shows that Ubx represses regulatory functions needed to set up a number of different transcriptional patterning domains in the developing wing: it blocks expression of a factor that mediates growth in the wing blade (i.e., Vg); of a factor that controls vein designation (i.e., Salr); of factors that are necessary for peripheral nervous system development (i.e., the *AS-C* factors); of a differentiation factor for intervein epithelium (i.e., DSRF); and upstream of this it interferes with expression of signaling ligands (i.e., Ser and Wg) which are involved in the initial spatial specification of some of these transcriptional domains.

So we have here a clear explanation, at the level of the regulatory pattern formation network, of how an A/P specification function of a *hox* gene actually works. But this function is like the rhombomere patterning example discussed above, in that it also has to be considered an evolutionary cooption, and not a very ancient one at that. For though the *ubx* gene is expressed similarly throughout insects (for references, Abzhanov and Kaufman, 2000), less derived orders do not differentiate their hindwing and forewing appendages; in dragonflies, for example, the T2 and T3 appendages are similar and the adult has four wings. The most basal insects do not even have wings. Therefore the cooption of the *ubx* gene to repression of wing patterning in the haltere disc necessarily postdated the origin of the wing patterning system in which it intervenes. It must have occurred by steps, in which the precursor of the haltere progressively lost wing size and pattern elements. This would account for both the parallel sites of Ubx intervention, and the ultimate effectiveness of the patterning blockade (Weatherbee *et al.*, 1998).

Evolutionary Changes in *hox* Gene Expression in the Arthropods

The arthropods include among living animals insects plus crustaceans, myriapods (i.e., millipedes and centipedes), and chelicerates (spiders, mites, and horseshoe crabs). The basic anatomical similarities in the segmental organization of these animals provides a leg up in considering evolutionary homologies among their diverse body plans. In insects many of the segments are distinguished from one another by special sensory, feeding, and locomotory appendages and other structures; while at the opposite extreme, in myriapods most segments are alike. The arthropods offer illuminating comparisons of *hox* cluster functions in patterning segmental appendages along the A/P axis. In terms of *hox* cluster genes the major arthropod groups appear to share the same genomic equipment, so the differences in their body plans cannot be regarded as the consequence of differences in their *hox* gene complements (Akam, 1995; de Rosa *et al.*, 1999; Grenier *et al.*, 1997). To the extent that the segmental morphologies of these animals depend on *hox* gene function, the diversity of these morphologies is due to change in the regulatory linkages which affect *hox* gene targets, and by which their own domains of expression are controlled.

FIGURE 5.2   How Ubx blocks the wing transcriptional patterning system in the haltere disc of *Drosophila*. (A) Four-winged fly, resulting from total loss of *ubx* gene function. In this animal the imaginal discs of the third thoracic segment have executed the same spatial patterning functions as the wing discs of the second thoracic segment. [(A) From Bender *et al.* (1983) *Science* **221**, 23–29, copyright American Association for the Advancement of Science.] (B–E) Examples demonstrating Ubx repression of specific genes that participate in wing patterning, displayed by immunocytology. The initial A/P patterning systems of wing and haltere discs are the same, as marked by expression of *engrailed* in the posterior domains and by the stripe of Dpp along the A/P boundary of the distal pouch of the discs (not shown): Ubx does not affect this early A/P patterning system (Weatherbee *et al.*, 1998). (B) Effects of Ubx on *salr* expression (green). (B1) The *salr* gene is expressed in the central portion of the wing blade where it is necessary for spatial specification of certain veins (see Fig. 4.7E), but it is not normally expressed in this region in the haltere disc (right side of figure). (B2) Ectopic expression of *salr* in a *ubx⁻* clone (arrowhead) in a haltere disc: Ubx protein is shown in red. (B3) The converse experiment: ectopic expression of the *ubx* gene in the blade region of the wing disc (red; white arrows) blocks *salr* expression in the *ubx⁺* cells. (C) Repression of the quadrant enhancer of the *vestigial* gene (vg^Q). The figure shows a haltere disc from an animal carrying a vg^Q.*lacz* construct. Ubx protein, shown in red, is present throughout, except in clones (arrow) of *ubx⁻* cells, and there the vg^Q.*lacz* construct is expressed; the lacz product is displayed immunocytologically in green. (D) Ubx effect on posterior expression of the *wg* gene, shown in green. (D1) Wg protein is seen along the D/V boundary of the wing blade region of the wing disc, and the Apterous transcription factor (cf. Fig. 4.6D) in the dorsal domain (purple). (D2) Same for the haltere disc, but *wg* expression does not extend to the posterior margin (white bracket). Since Wg is required for growth, this may account in part for the smaller size of the haltere (Weatherbee *et al.*, 1998). (D3) Haltere disc with several ubx⁻ clones (dark holes; Ubx in red). One of these clones lies at the posterior D/V boundary, and here ectopic Wg is detected (white arrow). (E) Ubx effects on expression of the DSRF transcription factor (green), which is required for the intervein epidermal fate. (E1) DSRF is expressed in the center of the wing blade region of the wing disc, left, between the vein domains (dark stripes), but not in the equivalent region of the haltere disc (right). (E2) Haltere disc bearing a large clone of *ubx⁻* cells in center of pouch region: here DSRF is expressed and the vein domains are formed as in the wing disc (Ubx in red). (E3) Wing disc, with ectopic stripe of Ubx expression (red). The arrow indicates a region of the normal intervein DSRF domain where, in consequence of Ubx expression, DSRF is absent. (F) Summary of Ubx effects on wing patterning system; post, posterior. These effects occur at genes shown in red boxes; solid red box denotes the Ubx effect on the vg^Q enhancer, assayed directly with a *cis*-regulatory construct. [(B–F) From Weatherbee *et al.* (1998) *Genes Dev.* **12**, 1474–1482.]

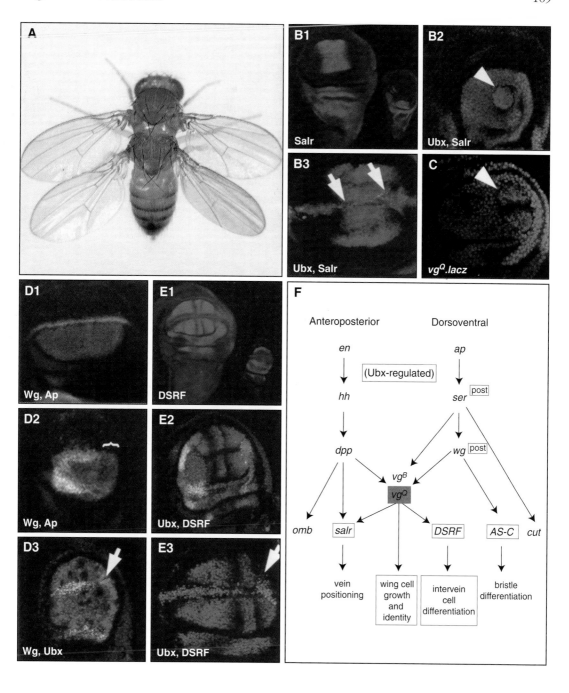

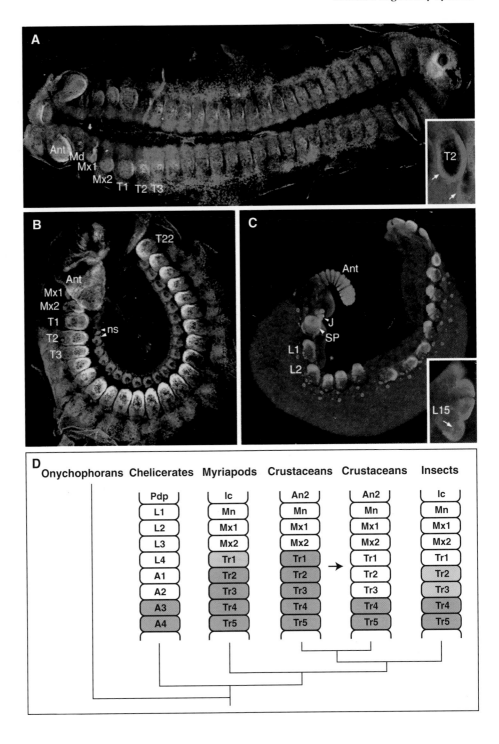

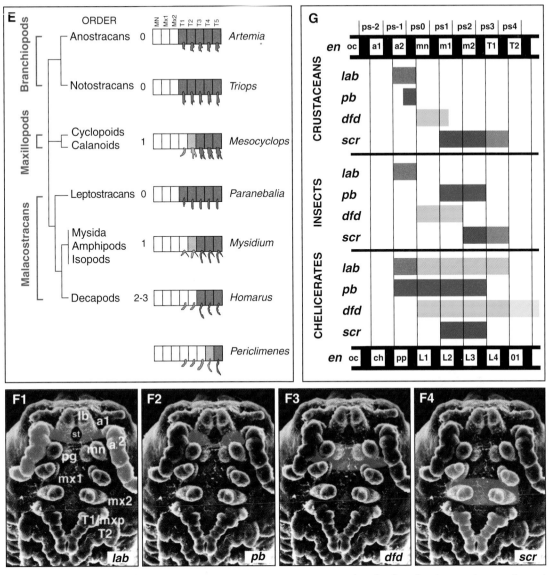

FIGURE 5.3 Domains of *hox* gene expression in developing arthropods and onychophorans. (A)–(C) Ubx and Abd-A proteins identified immunocytologically (red), together with Distal-less (Dll), (green), in embryos in which all trunk segments produce the same structures. (A) Centipede (*Ethmostigmus rubripes*). The Dll transcription factor is present in all appendages or outgrowths; anterior left. Ant, antenna; Md, mandible; Mx, maxillary segment; T1, maxilliped, a specialized poison claw; T2, T3, and more posterior appendages, walking legs. Ubx/Abd-A is present in trunk limb-bearing

(Continues)

segments up to the T1/T2 boundary, and in all the limb buds (yellow). Inset: Ubx/Abd-A staining around base of limb bud. (B) Slightly older centipede embryo, anterior top left; Ubx/Abd-A staining in body wall up to T1/T2 boundary, now strongly in appendages T2–T22 (yellow), and parasegmentally, also in the nervous system (ns). (C) Onychophoran embryo (*Acanthokara kaputensis*; anterior, top left). Indicated body parts are Ant, antenna; J, jaws,; Sp, slime papillae; L1–L15, lobopods. All of these structures express the *dll* gene. Ubx/Abd-A are detected only in the last pair of lobopods, in L15 (inset), partly overlapping with Dll (yellow). [(A–C) From Grenier *et al.* (1997) *Curr. Biol.* **7**, 547–553.] (D) Summary of differences in developmental domains in which Ubx/Abd-A factors are detected in representative arthropod species; onychophorans are a sister group of the true arthropods. In chelicerates (here spiders) the body is divided into prosomal and opisthosomal domains, and the anterior boundary of *ubx/abd-A* expression (Damen *et al.*, 1998; Abzhanov *et al.*, 1999) lies between the first and second opisthosomal segments (A1 and A2); i.e., following the segmental analysis of Damen *et al., op. cit.*). Pdp, pedipalps; L1–L4, legs; A1–A4, opisthosomal segments. In myriapods (centipedes and millipedes) the expression boundary is further anterior [see (A, B)]; lighter color in trunk segment 1 (Tr1) indicates an early expression domain in a centipede; Ic, intercalary segment; Mn, mandibles; Mx1 and 2, maxillary segments. In crustaceans, as shown in more detail in (E), the expression boundary varies; in the most basal group, the branchiopods, it is at the junction between head and trunk, and in more derived groups more posterior (arrow). The anterior segments shown are An2, second antenna segment; and mandibular and maxillary segments. In insects (i.e., here *Drosophila*) *ubx* expression is dynamic and initially extends up to the anterior boundary of T3 (i.e., ps5; ps, parasegment), while Ubx/Abd-A protein is detected up to a boundary within segment A1 (i.e, the ps6/7 boundary); but at the end of germ band extension *ubx* expression extends anteriorly into T2 and T3, symbolized by pink color. [(D) Slightly modified (by addition of onychophoran branch) from Abzhanov *et al.* (1999) *Evol. Dev.* **1**, 77–89.] (E) Summary of changes in expression of *ubx/abd-A* genes, correlated with thoracic appendage morphology in the different crustacean clades. Representatives of several crustacean orders are included. Red appendages are locomotory thoracic limbs; gray appendages are specialized for feeding (maxillipeds). Three gnathal segments (appendages omitted) are indicated in white, followed by five thoracic segments. Red indicates domain of Ubx/AbdA protein as detected by immunocytology; orange, weak expression. [(E) From Averof and Patel (1997) *Nature* **388**, 682–686, copyright Macmillan Magazines Ltd.] (F) Expression of anterior *hox* genes superimposed on scanning EMs of head surface structures in a crustacean, *Porcellio scabar*. Structures and segments are labeled as in (F1). mx1, mx2, T2 as above; T1/mxp, first thoracic segment/maxillipeds; pg, paragnaths; mn, mandibles; st, stomodaeum; lb, labrum; a1, a2, first and second antennal segments. (F1) *labial* (*lab; hox1*) expression domain; (F2) *proboscepedia* (*pb; hox2*) expression domain; (F3) *deformed* (*dfd; hox4*) expression domain; and (F4) *sex combs reduced* (*scr; hox5*) expression domain. (G) Summary of *lab, pb, dfd,* and *scr* expression domains in head regions of crustaceans, insects, and chelicerates. The *engrailed* (*en*) domains serve as *parasegmental* (*ps*) markers, identifying the posterior

The greater group to which the arthropods belong is termed the "panarthropods." These include, in addition to the arthropods, the tardigrades (or water bears), plus onychophorans (Nielsen, 1995). Spatial differences in *hox* gene expression within the homologous segmental domains of various panarthropod groups are illustrated in Fig. 5.3. A perfectly situated outgroup for the arthropods within the panarthropods is represented by the onychophorans (i.e., velvet worms). These animals provide a revealing side light on the process of panarthropod diversification. The onychophorans are living descendants of a lobopodian fauna that had diverged from the ancestors of the arthropods by the later Early Cambrian (Budd, 1996; Wheeler *et al.*, 1993). Lobopods differ from the arthropods in that their appendages are not jointed as are arthropod appendages, but like the arthropods their bodies are segmented, and they share many anatomical features with the arthropods (Nielsen, 1995). As in myriapods, the locomotory appendages on onychophoran trunk segments are all alike. In Fig. 5.3A–C the distribution of Ubx and Abd-A protein (assayed together with a single antibody) in a developing centipede is compared to that in a developing onychophoran (Grenier *et al.*, 1997). In both species the *distal-less* (*dll*) gene is expressed in all the appendages, as displayed by an antibody against the Distal-less protein (green). But the distribution of the Ubx/Abd-A protein (red) in these embryos is entirely different. In the onychophoran the *ubx/abd-A* genes apparently have little to do with the morphological plan of most of the appendage-bearing segments, since expression is confined to the very last of the 15 lobopod pairs (Fig. 5.3C). In contrast, in the centipede (Fig. 5.3A, B) Ubx/Abd-A is detected in every trunk segment (i.e., T2–T22), excepting only the first, which generates a special appendage. Expression of Ubx/Abd-A in the body segments also turns out to differ greatly within the arthropods. A summary is shown in Fig. 5.3D (Abzhanov *et al.*, 1999; see legend for details and further references).

Two conclusions strike one in staring at Fig. 5.3D: first that the same *hox* genes evidently intervene in many different patterning processes, since these animals generate diverse structures on their trunk segments (there are no wing blades on shrimp!); and second, that the expression domains of the *ubx/abd-A* genes with respect to these segments has been anything but immutable in evolution. This point is extended in Fig. 5.3E, where it can be seen that within crustaceans Ubx/Abd-A distribution is correlated with diversity in the organization of the appendages on the thoracic segments (Averof and Patel, 1997). The appendage assignments with respect to segment are diverse in these crustaceans, and the rule is that

---

part of each segment or anterior part of each parasegment. Segments are, at top, for crustaceans, oc, ocular; and the remainder as in (F). At bottom, for chelicerates, ch, chelicerate segment; pp, pedipalps, and the remainder as in (D). Columns indicate homologous segments. Horizontal bars represent domains of expression, and light colors denote weak, late, or transient expression. [(F–G) From Abzhanov and Kaufman (1999) *Proc. Natl. Acad. Sci. USA* **96**, 10224–10229, copyright National Academy of Sciences, USA.]

locomotory appendages form in domains where Ubx/Abd-A is present, while the specialized feeding appendages (maxillipeds) arise on segments lacking these factors. The conclusion from these and additional observations (Abzhanov and Kaufman, 2000) is that the various panarthropods have coopted the *ubx* and the *abd-A* genes for segmental pattern formation processes particular to each clade, and the same is true of the *antennapedia* gene. It follows that the insects and many of the crustaceans have independently reorganized the regulatory systems controlling segmental appendage development; in the common ancestor from which both derived the same appendages were probably present on all the trunk segments (Averof and Akam, 1995; Abzhanov and Kaufman, 2000).

Figure 5.3F, G deal with anterior *hox* gene expression in the head regions of arthropods from spiders to *Drosophila*. Each of the four *hox* genes included in Fig. 5.3G is expressed in a particular domain of the head of a crustacean in which certain structures arise. No doubt these *hox* gene expressions are required for some aspect of morphogenesis in each region. But when the domains of expression of these same four genes are compared in a spider, in a crustacean, and in *Drosophila* (Fig. 5.3G), we again see that, except for the *labial* (*hox1*) gene, they are all expressed in a unique fashion in each animal (Abzhanov and Kaufman, 1999). As Fig. 5.3G indicates, different combinations of *hox* genes are expressed in the homologous anterior segments of these three clades.

In considering Fig. 5.3 we must keep in mind that downstream of each of the *hox* gene expression domains will lie regulatory patterning networks that include multiple targets for their products, just as illustrated in Fig. 5.2 for Ubx in the *Drosophila* haltere. What is described as the general A/P patterning function of *hox* cluster genes consists actually of an amazing variety of different networks, different linkages, and different regulatory functions. These have evidently been rebuilt over and over, even within clades that construct their body plans in more or less similar ways.

Off the A/P Axis

Hox genes are expressed all over the bilaterian body, though not anterior of the antennal segments in arthropods nor anterior of the hindbrain in vertebrates (McGinnis and Krumlauf, 1992; van der Hoeven *et al.*, 1996; Akam, 2000). Colinear A/P patterns of expression occur in the endoderm as well as the trunk mesoderm and neuroectoderm (Grapin-Botton and Melton, 2000; Beck et al., 2000; Fig. 4.4A). But colinear expression of *hox* cluster genes also occurs off the A/P axis, in the limb buds of tetrapods, and as we see below, in much more remote contexts as well.

The posterior *hox* cluster genes are expressed in an interesting and dynamic way in developing amniote limb buds. A summary of *hoxa* and *hoxd* cluster expression patterns in chick wing and leg buds is shown in Fig. 5.4A (Nelson *et al.*, 1996; some genes of the other clusters are utilized in limb buds as well). There are three phases of expression. In phase 1, the *hoxd9* and *hoxd10* genes are initially expressed throughout the bud, and in its initial distal outgrowth (Fig. 5.4A1). Following this, in phase 2, a nested pattern of expression is set up such that *hoxd9, 10, 11,* and *12* are all expressed at the posterior margin of the wing bud but only the more 3′ of this group, i.e., *hoxd9* and *hoxd10*, are expressed toward the anterior side of the bud (Fig. 5.4A2). But in phase 3 a different pattern is installed in which the orientation of the phase 2 expression is completely reversed, so that the most 5′ gene (*hoxd13*) is now expressed toward the distal anterior margin of the bud (Fig. 5.4A3). Numerous comparative, mutational, and expression perturbation studies show that *hox* gene expression is essential for limb development. Phase 2 expression is a component of the patterning system that sets up the developmental domains from which derive major proximal and distal structural elements of the limb (e.g., our upper arms and forearms), while phase 3 expression is associated with patterning of the terminal portion of the limb (the autopod, e.g., our hand and its digits); (reviewed by Johnson and Tabin, 1997; Nelson *et al.*, 1996; Shubin *et al.*, 1997; Zákány *et al.*, 1997; Peichel *et al.*, 1997). The argument could be made that expression of the 5′ *hox* genes at the posterior margin of the limb buds (Fig. 5.4A2) is just another case of colinear A/P *hox* gene use. However, activation of these genes in this order in the limb bud depends on signaling from a site at the posterior margin of the limb bud. This signaling system is set in place through an independent mechanism operating completely within the growing limb bud (reviewed by Johnson and Tabin, 1997). The reversal of orientation in the phase 3 expression pattern clinches the matter in any case: here the vectoral *hox* gene expression system is used along a transient that is not only independent of the A/P axis of the trunk, but is nearly opposite to it. The anteriorly curved secondary axis which is marked by phase 3 *hox* gene expression serves as a mount for the secondary skeletal growths that produce the digits (Fig. 5.4B3); (Shubin *et al.*, 1997).

The expression of *hoxd11* in mouse and zebrafish forelimb limb buds is compared in Fig. 5.4B (Sordino *et al.*, 1995). Early on, the same posterior pattern of expression is seen in the two species (Fig. 5.4B1, B4). But then something different occurs in the mouse limb buds: the axis of expression curves anteriorly, while the axis of the developing fish appendage (i.e., its pectoral fin) remains straight (Fig. 5.4B2–6).

Ancestral chordates did not have paired appendages such as derive from the limb buds of tetrapods. These uses of the *hox* cluster patterning systems off the A/P body axis are obvious evolutionary cooptions, and the cladistic evidence for this is summarized in Fig. 5.4C (Shubin *et al.*, 1997). There are no appendages on modern invertebrate chordates, e.g., larvacean urochordates, or cephalochordates such as amphioxus; nor on basal jawless vertebrates, such as the hagfish; nor on the earliest fossil vertebrate or invertebrate chordates (Janvier, 1996; Shu *et al.*, 1999; Chen *et al.*, 1999). The first appendages to appear in the vertebrate lineages were unpaired fore and aft median fins as at (a) in Fig. 5.4C, followed by paired pectoral fins as at (b). Only the jawed vertebrates (gnathostomes) have two sets of paired appendages (Janvier, 1996). As Fig. 5.4C indicates, a cooption of *hox*

FIGURE 5.4   Evolutionary cooption of posterior *hox* genes for development of tetrapod limbs.   (A) Phases of expression of posterior genes of the *hoxa* and *hoxd* clusters in chick wing and leg buds, based on *in situ* hybridization data. Distal is to right, and five successive developmental stages are shown in each row. In phase 1 *hoxd9* and *hoxd10* are expressed throughout the distal region of both wing and leg buds. In phase 2 a nested set of expressions is established, oriented from posterior to anterior in a colinear manner with respect to the 5′ to 3′ chromosomal sequence of the genes, such that the *hoxd11* and *hoxd12* genes are expressed most posteriorly and the *hoxd9* gene most anteriorly to the bud. In phase 3 a reverse order of expression appears: *hoxa13* and *hoxd13* are now expressed more anteriorly than are *hoxd10* and *hoxd11*. Phase 3 is similar in wing and leg, while phase 2 is not. [(A) From Nelson *et al.* (1996) *Development* **122**, 1449–1466 and The Company of Biologists, Ltd.] (B) Expression of *hoxd11* in tetrapod (mouse) forelimb bud (B1–3) and in teleost (zebrafish) pectoral fin bud (B4–6); anterior left; distal up. (B1) *hoxd11* is expressed posterior to the proximodistal axis of the mouse limb bud (dashed line between arrowheads). (B2) At day 12 expression extends across the anterior region of the bud and with reference to the proximodistal axis, the axis bounding the expression domain is now curved (dashed line). (B3) The terminal domain of expression shown in (B2) is where the autopod (hand), including the digits, will form as indicated by cartilage condensations (green stain). (B4) Expression of *hoxd11* in a posterior region of zebrafish fin bud. (B5) Elongating fin bud, in which *hoxd11* expression has decreased, but remains posterior. The expression domain continues to be bounded by the straight proximodistal axis (dashed line). (B6) The straight axis is reflected in the organization of cartilage condensations in the fin. [(B) From Sordino *et al.* (1995) *Nature* **375**, 678–681, copyright Macmillan Magazines Ltd.] (C) Summary of evolutionary development of paired appendages in vertebrates; inferred states of *hox* cluster expression during limb development and structural organization of the autopod indicated in blue. As shown at (a), jawless fish such as the Silurian *Jamoytius* have median elongate fins; or, as at (b), the armored, jawless osteostracans display paired pectoral fins (Janvier, 1996). Two pairs of appendages emerging from the trunk body wall appear only in the jawed vertebrates (gnathostomes); in modern gnathostomes (c) the posterior *hox* genes are expressed in the buds from which anterior as well as posterior appendages develop [cf. (A)]. The teleosts, represented in (B4–6) by the zebrafish, are ray finned fishes (actinopterygians), and as shown in (B6), their fin axes are straight. This character is retained by the sister group of the tetrapods, the extinct osteolepids, represented here by *Eusthenopteron* (d), while even primitive tetrapods such as the Devonian *Icthyostega* (e) display the curved axis seen in the mouse in (B5, 6). Development of the autopod is inferred to depend on the curved phase 3 pattern of *hox* cluster expression (A3); the zeugopod, adjoining the autopod, denotes the major structural component of the lower limb (i.e., forearm or calf of leg), and its development depends on phase 2 expression (cf. A2). [(C) Modified slightly from Shubin *et al.* (1997) *Nature* **388**, 639–648, copyright Macmillan Magazines Ltd.] (D) Expression of *lacz* construct driven by a zebrafish *hoxd11* enhancer in a transgenic mouse embryo. The arrow indicates expression in the forelimb bud. Expression occurs also in the caudal trunk region, the genital eminence, and the hindlimb bud. [(D) From Beckers *et al.* (1996) *Dev. Biol.* **180**, 543–553.]

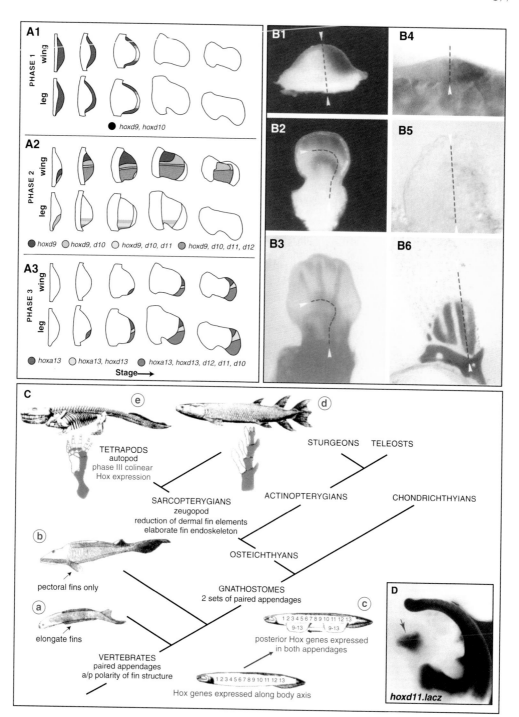

A1 PHASE 1 — wing / leg — ● hoxd9, hoxd10

A2 PHASE 2 — wing / leg — ● hoxd9 ○ hoxd9, d10 ○ hoxd9, d10, d11 ● hoxd9, d10, d11, d12

A3 PHASE 3 — wing / leg — ● hoxa13 ○ hoxa13, hoxd13 ● hoxa13, hoxd13, d12, d11, d10

Stage→

B1 B4
B2 B5
B3 B6

C

e

d

TETRAPODS
autopod
phase III colinear
Hox expression

STURGEONS   TELEOSTS

SARCOPTERYGIANS
zeugopod
reduction of dermal fin elements
elaborate fin endoskeleton

ACTINOPTERYGIANS

CHONDRICHTHYIANS

b

pectoral fins only

OSTEICHTHYANS

a

elongate fins

GNATHOSTOMES
2 sets of paired appendages

c

1 2 3 4 5 6 7 8 9 10 11 12 13
9-13   9-13

posterior Hox genes expressed
in both appendages

D

VERTEBRATES
paired appendages
a/p polarity of fin structure

1 2 3 4 5 6 7 8 9 10 11 12 13

Hox genes expressed along body axis

hoxd11.lacz

cluster activity must have been involved at this step; i.e., at (c) in Fig. 5.4C. Since forelimb and hindlimb buds develop into structurally homologous appendages, and both express sets of posterior *hox* genes, this expression denotes a discrete patterning cassette. The forelimb buds in fact utilize the posterior *hox* genes "out of register" with their use in the main body axis. Furthermore, a gnathostome common ancestor of teleosts and tetrapods had already coopted *tbx4* and *tbx5* genes to specification of forelimb and hindlimb (cf. Fig. 4.3), since this usage is also a shared property of amniotes and fish (Ruvinsky *et al.*, 2000). Figure 5.4C shows that the phase 3 expression of the posterior *hox* genes and the development of the curved axis of the autopod in part e are indeed evolutionarily novel features of the limbed tetrapods, additional evolutionary cooptions. This follows from the fact that the extinct close relatives of the tetrapods such as that shown at (d) in Fig. 5.4C (as well as living sarcopterygian relatives such as lungfish) retain the straight axis that is also retained in teleosts (Fig. 5.4B). Therefore the common gnathostome ancestor generated a straight axis of *hox* cluster expression in its limb buds as well, underlying the bony structures of its fins (Coates, 1995; Shubin *et al.*, 1997).

The cooption event which put the posterior *hox* genes to work in paired limb buds occurred no later than the Silurian (~440–410 mya), when the first jawless fish with paired appendages and then the first gnathostomes appear (Janvier, 1996). A "trans-vertebrate" gene transfer experiment suggests how this cooption might have happened. Figure 5.4D shows a mouse embryo expressing a zebrafish *hoxd11 lacz* transgene (Beckers *et al.*, 1996). This teleost regulatory element generates expression in the hindlimb and forelimb buds, and also in both the caudal tail and genital regions. The mouse *hoxd11* gene is expressed in the same regions (Dollé *et al.*, 1991; Beckers *et al.*, 1996). The inputs needed to animate the *hoxd11 cis*-regulatory system apparently survive in both evolutionary branches of the gnathostomes, i.e., teleosts and tetrapods. The implication is that these inputs were originally present at the caudal end of more basal vertebrate forms that did not have paired appendages. So the cooption would have been effected by genomic changes that caused the genes encoding the relevant transcription factors to be expressed as well in the limb bud domains. In mice some mutations that perturb posterior *hoxd* gene function coordinately affect both limb and genital development. The original cooption could have mobilized elements of a genital patterning system to the limb buds (Peichel *et al.*, 1997; Kondo *et al.*, 1997).

Colinear Expression of *hox* Genes in the Somatocoel of a Sea Urchin Larva

The example of colinear *hox* gene expression most remote from axial A/P patterning in chordates and arthropods is illustrated in Fig. 5.5 (Arenas-Mena *et al.*, 2000). Here we see a sequential set of *hox* gene expression domains that traverse the somatocoels of a microscopic larval sea urchin. The paired somatocoels are the posterior coeloms, which at the stages shown surround the larval gut, and from which various mesenteric elements of the adult body plan will eventually be formed (Arenas-Mena *et al.*, 2000; Pearse and Cameron, 1991). The fivefold radially symmetric structures of the adult echinoderm body plan form within the rudiment (cf. Fig. 5.1), which as Fig. 5.5 illustrates, overlies the left somatocoelar sac. There is only a single *hox* gene cluster in this sea urchin genome (Martinez *et al.*, 1999), and the five, 5'-most *hox* genes known (those related in sequence to vertebrate *hox7–13*) are all expressed in the somatocoels during early and middle stages of postembryonic development (e.g., Fig. 5.5C). The pattern of *hox* gene expression is affected by the rudiment, so that as development proceeds it becomes different on the left as compared to the right somatocoel (Fig. 5.5D1–4). Though the developmental significance of these somatocoelar *hox* cluster expression patterns are unknown, they have some remarkable features.

The first point is that the expression pattern is colinear with *hox* gene order in the cluster. The colinear pattern describes a curved axis through the somatocoels (white arrows in Fig. 5.5D1–2), but this conforms to the axes of neither adult nor larval body plans. The axes of the larval body plan are shown in Fig. 5.5B. Though it is folded up upon itself, in the adult echinoid the primordial A/P axis runs inward from the mouth (anterior, face down in sea urchins as they crawl along the substrate; Peterson *et al.*, 2000c). In the diagram of Fig. 5.5D3 the future adult axis would run from the oral surface of the rudiment which faces the viewer down into the page. Second, though the body plan of the adult is radially symmetric, the *hox* gene expression patterns in the somatocoels are bilateral and also somewhat asymmetric, as are the fates of the right and left somatocoels (Peterson *et al.*, 2000c). The colinear somatocoelar expression patterns may well have descended from a straight colinear pattern of A/P expression that was parallel to the gut in a remote bilaterian ancestor of echinoderms plus hemichordates, since the most posterior (i.e., 5') of the genes, *hox11/13b*, is expressed around the anus. That is, the ancestral A/P pattern of coelomic mesoderm expression probably obtained its curvature in the course of echinoderm evolution together with the gut, which the coeloms enclose. So though it looks different, this aspect of the colinear somatocoelar expression pattern can be considered conserved. The somatocoelar *hox* gene expression pattern (Fig. 5.3D) obviously has some downstream functional significance, or it could not have survived. The somatocoels give rise to a unique set of structures, the stacked mesenteries of the adult echinoid body plan (Peterson *et al.*, 2000c). Perhaps the posterior *hox* genes have been coopted for some patterning function required for development of these structures. Third, note the microscopic scale on which this *hox* cluster expression pattern is formulated: the whole somatocoel is only a couple of hundred microns across at these stages. The pattern is also dynamic and transient (Arenas-Mena *et al.*, 2000). This example delivers the warning that there probably remain to be discovered many other colinear *hox* gene cooptions in bilaterians, special uses of the *hox* gene cluster that are different from those with which we are familiar along the A/P body axes of arthropods and chordates.

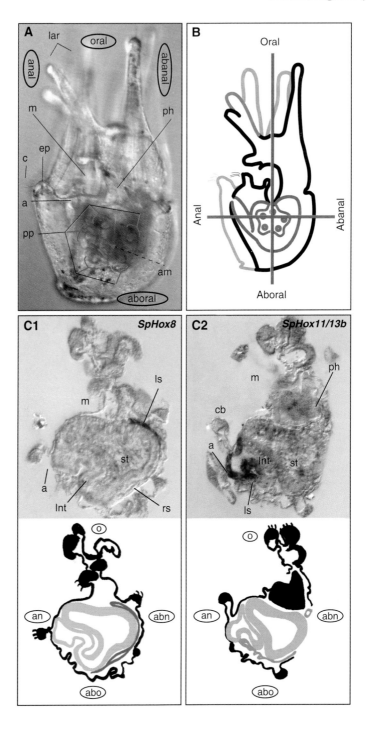

FIGURE 5.5 *(Continued)*

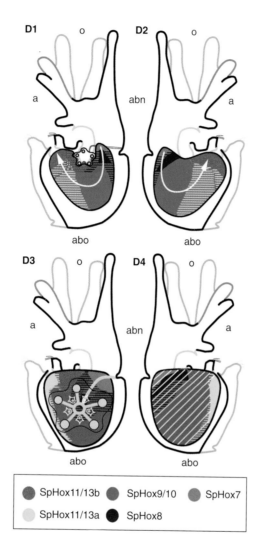

The feature of *hox* cluster usage in the bilaterians that emerges is its remarkable ability to be coopted for diverse patterning functions. The *hox* gene cluster has the unusual property of generating a set of transcriptional domains that form an oriented spatial sequence; as remarked at the outset, this has the abstract property of a vectorial patterning system. The individual transcription factors the cluster encodes are useful in multiple contexts, perhaps because they function combinatorially with diverse cofactors (e.g., Li *et al.*, 1999). These transcription factors provide inputs to *cis*-regulatory elements that operate at all levels of pattern formation systems. The *hox* cluster system is linked into an amazing variety of

---

FIGURE 5.5   Domains of expression of posterior *hox* genes in the larval somatocoels of a sea urchin. (A) Left lateral view of an eight-armed larva of *S. purpuratus*: m, mouth; ph, pharynx; a, anus; lar, larval arms; ep, epaulettes; c, cilia; pp, the primary podia of the adult rudiment; am, adult mouth (red arrowhead), which will form in the center of the rudiment. At this stage the rudiment has grown to occupy a large region of the larval body alongside the stomach, which lies beneath the rudiment in this view. The pentameral organization of the rudiment is clearly evident. (B) Axial designations for the larva itself; the axis of the rudiment (and adult) runs from the mouth into the page. (C) Examples of somatocoelar *hox* gene expression patterns in sections oriented as in (B); below, somatocoels (green), larval ectoderm (not included in juvenile at metamorphosis, black); gut (yellow). In these drawings, o, oral; an, anal; abo, aboral; abn, abanal. In sections, ls, left somatocoel; rs, right somatocoel; st, stomach; int, intestine; cb, ciliary band; other abbreviations as in (A). (C1) Expression of *hox8* gene; (C2) Expression of *hox11/13b* gene (nomenclature of *hox* genes as in Martinez *et al.*, 1999). (D) Summary of patterns of expression of five "posterior" *hox* genes in the larval somatocoels. Early stages are represented in (D1) and (D2); and later stages in (D3) and (D4). The color code (bottom) indicates the domains of expression of individual genes; abbreviations as in (C). Domains of overlapping gene expression are represented as bicolored stripes. (D1), (D3) left views, where pentameral rudiment can be seen; (D2), (D4) right views. The most 5′ gene of the cluster so far known, *hox11/13b*, is expressed in the anal region and consecutively more "anterior" genes are expressed in a sequential set of domains describing a curved pattern along the somatocoels (white arrows). Except for *hox11/13a* the gene expression domains are expanded or are stronger on the left side near the rudiment as development progresses; compare (D3) with (D4). In (D3) the pentameral water vascular system is indicated within the outline of the rudiment in gray, and the dental sacs in white. In (D4) the patterns of expression in the right somatocoel are similar to those in (D2), except for the absence of *hox11/13b* expression around the anus; the transcript levels are also lower (indicated by oblique white lines), compared to those in the left somatocoel (D3), except for *hox11/13a*, which stains similarly in both somatocoels. In (D3) expression of *hox7* and *hox8* is seen around the canal leading from the center of the rudiment to the outside. None of these five genes is expressed in a pentameral pattern. [From Arenas-Mena *et al.* (2000) *Development*, in press.]

morphogenetic functions, some which are not aspects of A/P patterning at all. And even for those which are, a close look at the patterning gene networks in which *hox* genes participate shows that each specific usage is a novel evolutionary cooption (as in the examples of the *Drosophila* wing and vertebrate hindbrain considered in this and in Chapter 4). Utilization of the *hox* complex in bilaterian evolution has included a great many nonconservative changes.

Nor in this light is it so obvious what were the original functions for which the vectorial *hox* cluster patterning system was utilized. It was indeed a key A/P patterning device in the bilaterian common ancestor, but it could have had many other uses in the precursors and cousins of that animal. The case of the sea urchin larval somatocoel shows that *hox* cluster genes could have evolved to pattern a variety of structures in a microscopically scaled stem group fauna, one branch of which gave rise to the bilaterians (Davidson *et al.*, 1995; Peterson and Davidson, 2000; Runnegar, 2000). This chapter is about evolutionary change in patterning systems, and the *hox* gene system provides some of the most dramatic examples of cooptive change in the evolutionary history of the Bilateria.

SMALL CHANGES

There is an interesting consequence to the theory that the evolutionary basis of diversity within the bilaterians is cooptive change in developmental gene regulatory networks. This is that the morphological effect of an evolutionary change in network linkages depends entirely on where in the network the linkage is, rather than on other things; for example, the effect of the change cannot be predicted by the identity of the transcription factor or signaling system affected by it. Given transcription factors and signaling systems participate in multiple developmental regulatory networks, which are required to build different parts of the body plan. But often they also participate at different levels within the same network. The *hox* genes again provide an example that is right to the point. Though their more famous functions are in the early-mid stages of the development of given body parts, they are also involved in numerous terminal developmental programs that control small details of morphogenesis.

Some examples are reproduced in Fig. 5.6 that illustrate the role of *hox* genes in terminal aspects of development. Evolutionary changes that affect these particular *hox* gene functions would cause small morphological or physiological changes, of the sort that distinguish species from one another. For instance, in Fig. 5.6A we see that the *ubx* gene, the same which globally represses the wing patterning network in the haltere imaginal disc (Fig. 5.2), and that carries out so many other upstream developmental processes, also determines the fine spatial distribution of hairs (trichomes) on the 2nd femur of the *Drosophila* leg (Stern, 1998). In *D. melanogaster* there is a small patch of naked cuticle lacking trichomes on the posterior side of the femur (Fig. 5.6A1); in *D. simulans* the naked patch is much larger; in *D. virilis* it is not there at all (Fig. 5.6A4). The naked patch depends on localized

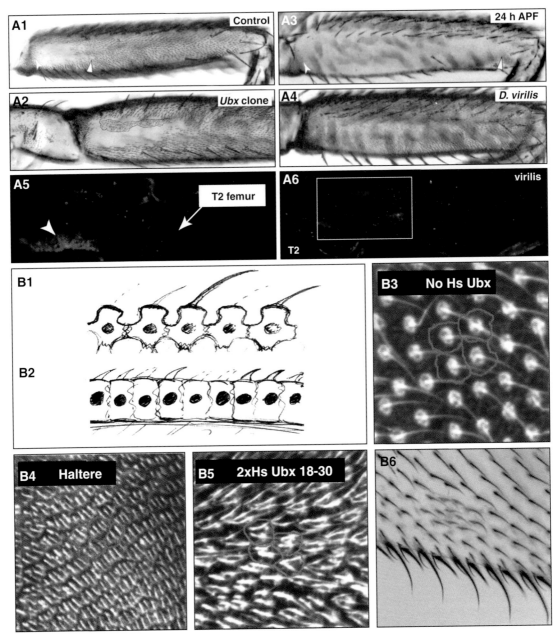

FIGURE 5.6 **Some effects of *hox* genes on species-specific, terminal mor-phogenetic functions.** (A) Species-specific trichome pattern in *Drosophila* depends on *ubx* expression. (A1) Patch of naked cuticle (between arrow and arrowhead) on posterior side of the 2nd femur of the 2nd leg of *D. melanogaster.* Trichomes cover the

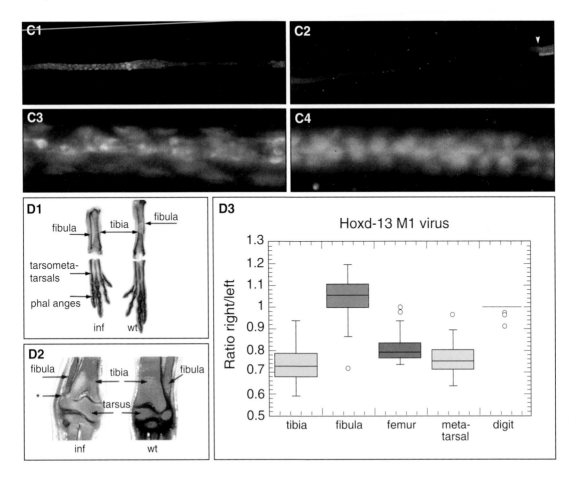

remainder of the cuticle. (A2) Trichomes appear within a *ubx⁻* clone (outlined in black) induced in the naked cuticle region. (A3) Expansion of naked cuticle domain to almost the whole of the posterior side of the 2nd femur, following ectopic production of Ubx under control of a heat responsive *cis*-regulatory system; heat shock 24 h after puparium formation. (A4) Normal pattern of trichomes on 2nd femur of *D. virilis*, which lacks a naked cuticle domain. (A5) Ubx protein, displayed by immunocytology in 2nd femur of *D. melanogaster*. Arrowhead marks Ubx accumulation in the future naked cuticle domain. (A6) Ubx protein is not present at equivalent levels in the homologous region of *D. virilis* femur (box). [(A) From Stern (1998) *Nature* **396**, 463–466, copyright Macmillan Magazines Ltd.] (B) Ubx control of morphological detail in epidermal cells of the *Drosophila* haltere. (B1) Drawing of flattened wing intervein cells, each of which mounts a single actin hair. (B2) Drawing of cuboidal haltere cells, each of which carries one to four short hairs. The cells do not flatten as in the wing, and the hairs are spaced

(Continues)

expression of the *ubx* gene late in development (Fig. 5.6A5), i.e., many hours after puparium formation. Trichomes appear ectopically in a clone lacking *ubx* function (Fig. 5.6A2), while ectopic *ubx* expression causes the naked patch to expand (Fig. 5.6A3). The *D. virilis* phenotype is evidently due to the absence of *ubx* expression in the 2nd femur (Fig 5.6A6). The *ubx* gene also controls the detailed cellular morphology of the epithelium in the *Drosophila* haltere (Roch and Akam, 2000). In the haltere the cells are cuboidal and they mount closely spaced small hairs, while the intervein cells of the wing epithelium are flattened, and each mounts only a single hair (Fig. 5.6B1, B2). Forced expression of *ubx* in the wing epithelium just before puparium formation results in a haltere-like cellular morphology and a close spacing of hairs (Fig. 5.6B3–B6).

As a third example, a terminal role of the *labial* (*lab; box1*) gene is illustrated in Fig. 5.6C. This is to control differentiation of a specific differentiated cell type, the "copper cells" of the *Drosophila* midgut (Hoppler and Bienz, 1994). These distinctive cells fluoresce orange when larvae are fed copper, and they may endow the larva with particular absorptive capabilities for ingested substances. In *labial*

---

much more closely. (B3) Normal wing surface, confocal image using phalloidin stain. (B4) Normal haltere surface. (B5) Transformation of wing cells to haltere epidermal morphology, induced by expression of Ubx under control of the heat shock regulatory system; heat shock at 18–30 h after puparium formation. (B6) Clone of wing cells producing Ubx protein marked also by GFP expression (green); multiple short hairs have formed on each cell, just as in the haltere epidermis. [(B) From Roch and Akam (2000) *Development* 127, 97–107 and The Company of Biologists Ltd..] (C) Expression of *labial* (*lab*) controls presence of copper cells in *Drosophila* midgut. (C1) Normal midgut, copper cells displayed by their orange fluorescence following larval ingestion of copper; blue is background fluorescence. (C2) Absence of copper cells in *lab⁻* midgut. (C3) Elongate copper cells in wild-type midgut, at higher magnification than in (C1) surrounding other cells (blue) of the gut lumen. (C4) All cells of the midgut differentiate like copper cells after ectopic *lab* expression driven by the heat shock regulatory system. [(C) From Hoppler and Bienz (1994) *Cell* 76, 689–702, copyright Cell Press.] (D) Control of distal bone length by the *hoxd13* gene in chick hindlimb. (D1) Comparison of alcian stained bones from hindlimbs of 10-day chick: those on right are normal; those on left show the effect of ectopic *hoxd13* expression. The limb bud from which these skeletal elements derived had been injected with a viral expression vector expressing *hoxd13* late in development (infected, inf). (D2) Ankle region: on ectopic *hoxd13* expression the fibula forms a new junction with the ankle, indicated by the arrow with askerisk; wt, contralateral uninfected ankle. (D3) Relative lengths of indicated bones, comparing hindlimbs formed after ectopic *hoxd13* expression to contralateral normal limbs. A ratio of 1 (ordinate) indicates no effect of ectopic Hoxd13 protein, as for the fibula. [(D) From Goff and Tabin (1997) *Development* 124, 627–636 and The Company of Biologists Ltd.]

mutants no copper cells form at all (Fig. 5.6C2), while *lab* overexpression converts all of the cells of the midgut into copper cells (Fig. 5.6C4). Just as *ubx* is required early in the patterning network for haltere specification, and also at its terminus, so *lab* is required early on for midgut specification (Panganiban *et al.*, 1990; reviewed by Bienz, 1994), as well as for the terminal regulatory function illustrated here. Another example in *Drosophila* is control by *ubx* and *abd-A* genes of the segment-specific patterning of denticles on the ventral surface of the larva. The morphology of these hook-like structures, which are secreted by the epithelial cells, depends on EGF and Serrate signaling systems, and a key spatial aspect is the localized secretion of Serrate (a Notch ligand): however, expression of the *serrate* gene in the abdominal epidermis depends on *ubx* and *abd-A* activity (Wiellette and McGinnis, 1999).

There are many examples as well in vertebrates. For instance, in the chicken some of the middle *hox* genes are expressed in the epidermal feather placodes (Chuong *et al.*, 1996; Kanzler *et al.*, 1997), while the *hoxd13* gene is expressed in scale-forming domains of the foot. Retinoic acid downregulates *hoxd13* (and of course probably affects other genes as well) and its application causes scale primordia to express feather keratins and to develop feather structures (Kanzler *et al.*, 1997). The *hoxd13* gene has evidently been coopted to serve several different roles late in hindlimb development. Figure 5.6D provides another, and different example. Here we see that the length of the distal bones of the leg, i.e., the amount of growth of the bone-forming domains, is limited by late *hoxd13* expression (Goff and Tabin, 1997). Shorter bones are formed by limb buds injected with a virus expressing *hoxd13*.

The *hox* genes are here just convenient examples, and it may be that every gene which participates in upstream aspects of pattern formation is somehow also utilized in the specification or execution of terminal developmental programs which control the detailed properties of a body part. There are numerous examples not involving *hox* genes, for instance, the use of the *engrailed/hedgehog/cubitus interruptus* transcription control system in butterfly wings. This apparatus is used for primary A/P patterning in butterfly as in *Drosophila* wing imaginal discs. But it is also used later to determine the location of the pigmented "eye spots" in butterfly wings (Keys *et al.*, 1999). These are species-specific foci that organize pigment cell functions around them (Nijhout, 1994). Terminal specification of individual muscles provides another remarkable example from *Drosophila*, an example that includes a number of genes encoding transcription factors which also operate elsewhere, within different pattern formation networks. In each abdominal hemisegment of the larva there are 30 muscles. They are individually distinct in location, points of insertion, and structure (reviewed by Baylies *et al.*, 1998; Bate, 1993). Each muscle descends from a founder cell that establishes its identity and then fuses with a pool of surrounding myoblasts. Remarkably, the individual founder cells are distinguished by expression of specific genes encoding transcriptional regulators, all familiar from their roles in embryonic or imaginal disc patterning: *evenskipped* and *ladybird* are expressed in single individual muscle founder cells; *krüppel* (*kr*), *vestigial, apterous,* and *s59* (which encodes a homeodomain factor) are each expressed in specific subgroups of muscle founder cells. *kr* expression has been proved directly to be required for establishment of the identity of those muscle founder cells in which this gene is activated (Ruiz-Gómez *et al.*, 1997). We have met all of these genes (but *s59*) earlier, doing other things; here they are contributing to the final details of the functional morphology of the larva.

Small evolutionary changes in terminal developmental mechanism produce species-specific phenotypes, which immediately confront the selective forces of the environment. These changes must happen continuously in evolutionary time because speciation occurs continuously. The morphological differences among congeneric species depend largely on such small changes. At the DNA level these are changes in the *cis*-regulatory control systems of genes that act to set in place specific differentiation programs; that encode signals which affect such genes; or that themselves run differentiation programs in specific contexts. These kinds of change necessarily occur more easily than do more upstream changes in pattern formation networks. As we have seen many times in Chapters 3 and 4, self-reinforcing or feedback loops are common in network elements that control the initial stages of specification of cell types or of transcriptional domains. Other common features are lateral connections amongst regulatory genes, and multiple linkages from each such gene to other genes. Changes in these regions of the network will have many consequences, but at the periphery of each network the streets are all one-way, and changes that occur there affect only terminal properties. This is surely a key aspect of the frequency of small changes, which is to say, of the "evolvability" of terminal properties. Conversely, this argument also explains the relative (but only relative) evolutionary stability of more upstream pattern formation processes that work earlier in development.

So, the same regulatory genes and signal transduction systems can be used to effect small changes as to effect great changes. But the polarity of the evolutionary processes by which new body parts have appeared in the Bilateria is thereby up for grabs. In the following we explore a view which is exactly the opposite of the traditional idea that ontogeny is a guide to phylogeny.

EVOLUTIONARY ORIGINS OF BODY PARTS

Polarity in Body Part Evolution

Differentiation gene batteries are clearly older than are the monophyletic Bilateria (Davidson *et al.*, 1995). Indeed, as noted earlier, differentiated cell types are present in nonbilaterian metazoans, e.g., in cnidarians. In the most primitive form of bilaterian development, which is still widely used in invertebrates (i.e., Type 1 embryogenesis), the main job to be done is installing the activity of given differentiation gene batteries in the appropriate cell lineages, i.e., the job of direct cell type specification which was the subject of Chapter 3 of this book. But, as summarized in Chapter 4, formation of adult bilaterian body parts requires a more complex, multistep process of transcriptional pattern formation, on top of which the differentiation gene batteries are installed. It follows that evolution of body parts has consisted of the erection of novel gene networks for generation of spatial transcriptional patterns, and only then transfer of control of the differentiation gene batteries needed in each part to the terminal apparatus of these networks. This simple argument has major consequences. Among other things it provides a possible solution to a paradox that has arisen from comparing body part development in diverse animals, and also a way of thinking about how network evolution occurs.

The paradox is as follows. When we survey the diverse Bilateria we recognize major body parts that subjectively seem analogous to us, such as heads, hearts, appendages, brains, eyes, tail ends, axial central nervous systems. But objectively we see that the anatomy and morphology of these parts are very different, comparing distant clades, and so in detail are the underlying developmental processes by which they are formed during embryogenesis. Yet, surprisingly, over and over the same transcriptional regulators are found to be used for what appear at least externally to be similar purposes: the same regulatory genes are needed for development of insect and vertebrate brains, insect and vertebrate hearts, appendages, eyes, and so forth. In response to this paradox, an almost automatic response has been that though they may look different these body parts are actually homologous; that there are basic and still hidden pattern formation processes underlying their development, and that these were already present in the bilaterian common ancestor. Conservation must be the reason the same genes are used in the development of each part in diverse bilaterians, so this argument goes. The conclusion that these developmental processes are conserved features seems, according to this view, to be required by the observation that orthologous genes are used in the development of the "same" body part.

But if we sum all the assertions of this sort we produce an impossible and illogical image of the bilaterian common ancestor. It would have been equipped with brain, seeing eyes, moving appendages, beating heart, etc. Something is very wrong with this picture because the way these body parts develop in diverse branches of the Bilateria actually share little in the details of their respective pattern formation processes. If the common ancestor had appendages, for example, it could not easily be ancestral to insect and mouse appendages both, because the structures and processes through which these develop are completely different. The pattern formation process for appendage development cannot depend to any large extent on a broadly conserved regulatory network. So there must be another explanation for the fact that the same genes are used to make body parts that we call by the same name in very different kinds of animal; and indeed there is.

This explanation is that the only things which are actually held in common between the analogous body parts of diverse animals, and are truly conserved features, are the differentiation gene batteries used to accomplish analogous functions; not the morphologies of these body parts nor their developmental programs. However they are structured, brains must deploy neuronal differentiation programs, hearts need certain kinds of contractile cells; eyes need photoreceptor cells; guts need digestive and absorptive cells, and so forth. So a possible solution to our paradox is that the regulatory genes which we find employed in patterning analogous body parts originally ran the differentiation gene batteries required for their respective functions, and since these genes were expressed in the right place they could be coopted during evolution to produce successively more elaborate pattern formation functions, differently in each clade. As we shall see, such genes still run versions of these "body part-specific" differentiation gene batteries. The bilaterian common ancestor is now relieved of the responsibility of trying to mount advanced body parts that are ancestral to incompatible offspring, or of harboring pattern formation processes conserved in all of its descendants, and it can be returned to the comfort of a much simpler format. A corollary is that some regulatory genes will very often be found operating at several different levels of the developmental process leading to body part formation, from initial specification to terminal installation of particular cell types.

To sharpen the matter a bit the cartoon in Fig. 5.7 shows the stages by which the developmental use of a differentiation gene battery might evolve into a pattern formation process. In this cartoon the same gene battery, and the same regulatory gene(s), are used at each stage. This is an arbitrary construct, not intended to represent any particular gene or body part, but the regulatory elements of which it is composed will all be too familiar from earlier chapters. As described in the legend, the figure indicates three evolutionary stages (A, B, C). In the first, spatial specification cues in the embryo are used to turn on a small regulatory network, causing transcriptional activity of a differentiation gene battery, as in all the systems discussed in Chapter 3. We focus on one of these regulatory genes (the red gene): in the second evolutionary stage (B) the two sets of ancestral linkages in (A) are still preserved, i.e., those by which the red gene is turned on at some particular spatial address in the embryo; and those by which the red gene activates its differentiation gene battery. But this battery is now mounted on some morphological structure that was not present in the ancestor at (A). Since it was already running in the right embryonic domain, the red gene has now been used for setting up the transcriptional pattern underlying the morphogenesis of this structure within that same domain. A growth routine has been called into play in this domain as well. Another transcriptional regulator has also been coopted (the purple gene), and in order to respond to it in turn, the red gene has acquired a second *cis*-regulatory module. Through this module, the late phase of expression of the red gene is controlled by the purple gene product and by negative signaling from an adjacent domain, so that it now activates its target battery of differentiation genes only in the appropriate spatial component of the structure. In the third stage (C) the pattern formation process has become more elaborate; there are now two different gene batteries expressed in the structure, and the red gene has acquired an additional regulatory function (control of the blue gene). But the red gene still sets the address of the embryonic progenitor field, and it is still used to operate one of the differentiation gene batteries mounted on the structure; and in addition it participates in the evolutionarily novel pattern formation process.

This exercise illustrates the course of the process of cooption. Cooption of a regulatory gene occurs when a useful prior network linkage is conserved and another linkage is changed. For instance the prior upstream linkage which activates transcription of the gene in a given spatial domain of the developing animal may be conserved, but the downstream linkage changes so that the gene activates different (or additional) targets in that domain. Or the reverse could happen, and the upstream linkage changes so the gene is expressed in response to different inputs, while retaining its downstream targets. Both are illustrated for the red gene in Fig. 5.7. So cooption, the major mechanism of evolutionary change at the gene network level, is like walking: one linkage, upstream or downstream, stays where it was last put and bears functional weight, while the other moves; and then if its move is useful, it may serve as the functional anchor while the first changes. After a few such "steps" all the linkages surrounding a given phase of activity of a regulatory gene may be different from the ancestral stage, and we get examples such as shown in Fig. 5.1.

This view of the mechanisms by which body parts evolve has some useful aspects. First, it suggests a reasonable evolutionary explanation for the surprising multiplicity of *cis*-regulatory modules in the control systems of genes that carry out diverse pattern formation functions within the same network. Second, it indicates why we so often see the same regulator operating at multiple levels of a developmental process of body part formation. Third, it gives us a way of approaching the paradox of similar gene use in analogous but morphologically dissimilar body parts. Fourth, it implies a selectively useful form of the body part at every evolutionary stage, since its key functionality is due to its terminal differentiation gene batteries, which are expressed at every evolutionary stage (as in Fig. 5.7).

This last is of course why the old saw about ontogeny and phylogeny fails in respect to the evolution of body parts. In ontogeny the differentiation gene batteries are loaded into the operating system last, after all the pattern formation steps. But in evolution they were there first, and continued to be used in order for the body part or its antecedents to be useful and to survive. As comparative anatomy has indicated all along, it is the pattern formation processes that in evolution have been loaded in later, in successive steps.

The Case of *pax6* and Some Other Amazing Examples

The *pax6* gene is a member of a huge family of genes encoding transcriptional regulators that are defined by their "paired domain" DNA recognition sequences. The genes of this family are utilized for many diverse purposes in bilaterian development. But though *pax6* always has other functions as well, in every animal examined that has eyes this gene proves to be involved in eye development (reviewed by Gehring and Ikeo, 1999; Jean *et al.*, 1998). Furthermore it is necessary:

FIGURE 5.7   "Bottom-up" evolution of networks for body part formation. Abstract representations of three stages of evolution of an imaginery pattern formation system are shown. Colored boxes indicate transcriptional domains, the state of which are dependent (in part) on the presence of the product of the gene of the same color, represented in a prior box as a thick horizontal line. (A) Initial evolutionary state, similar to that found in Type I embryonic systems. A gene battery, at left (a–g), encodes proteins used for some differentiated cell type. Each *cis*-regulatory element has multiple inputs, two of which are indicated in red and orange; "x" denotes various other inputs, not necessarily the same for each gene. The battery is activated by a small network of genes (red, green, orange horizontal lines in box at right) encoding transcription factors. These genes perform the job of transducing spatial embryonic cues (thick upward green arrow) into activation of the transcriptional regulators of this differentiation gene battery. The initial response gene (green) activates two other genes (orange and red) and the orange gene also cross-regulates the red gene (see Chapter 3 for examples of this kind of apparatus). In (B) and (C) green and orange genes are omitted, but they still function in the initial stage of the developmental process. Attention is mainly focused on the red gene and its functions. In (A), at the initial stage of evolution portrayed, a single *cis*-regulatory module of the red gene, which responds to the products of the green and orange genes, is responsible for the activity of the red gene (solid green box). (B) A later evolutionary stage. The same differentiation gene battery is now mounted in a pattern formation system. The boxes interposed between the initial and final portions of the process represent multicellular spatial transcriptional domains. Though only the red gene and its effects are explicit, several interlocking genes are of course likely to be involved in each function. The initial embryonic transduction system mediated by the green and red genes persists, but the red gene output is now used to activate a new regulatory gene (purple). A growth circuit is also called into play. The red gene has acquired a second *cis*-regulatory module, enabling it either to be activated by the purple gene product or repressed by a signal (S) from the underlying spatial domain. Two spatial transcriptional domains are set up, in only one of which is the red gene expressed. Activation occurs by *cis*-regulatory interactions at the purple *cis*-regulatory module of the red gene (purple box). The two purple transcriptional domains can be imagined as different parts of a morphogenetic structure, with consequent activation of the red gene battery in one of the domains. (C) Further evolutionary elaboration of the pattern formation regulatory apparatus. There are now two differentiation gene batteries, the red battery and the blue battery. Each is expressed in a different domain of the structure, due to the indicated circuitry: a new regulatory gene, the blue gene, which is a controller of the blue battery, has been coopted by the introduction into its *cis*-regulatory system of an element that responds to the purple transcription factor (purple solid box). The result is that the structure has more complexity and expresses diverse gene batteries, but the association of the red regulatory gene with its pathway of differentiation persists from the earliest evolutionary stage (A), and this gene is also still used to start development of the structure in the embryo, i.e., to define the embryonic progenitor field for the structure. In addition it executes intermediate pattern formation functions.

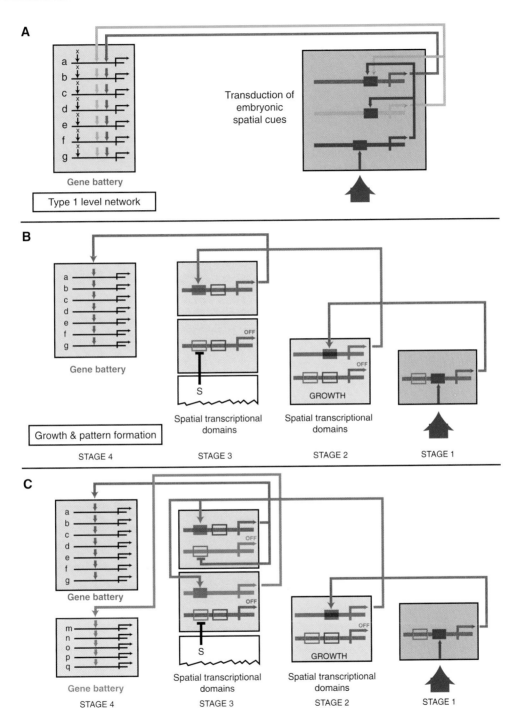

A

Gene battery

Transduction of embryonic spatial cues

Type 1 level network

B

Gene battery

Growth & pattern formation

Spatial transcriptional domains

Spatial transcriptional domains

OFF

S

GROWTH

OFF

STAGE 4   STAGE 3   STAGE 2   STAGE 1

C

Gene battery

Gene battery

Spatial transcriptional domains

Spatial transcriptional domains

OFF

OFF

S

GROWTH

OFF

STAGE 4   STAGE 3   STAGE 2   STAGE 1

in both mice (Hill *et al.*, 1991; Grindley *et al.*, 1995; Fig. 4.1D) and in flies (Quiring *et al.*, 1994) *pax6* mutations severely affect development of eyes. *Drosophila* has two closely related *pax6* genes, *eyeless* (*ey*) and *twin of eyeless* (*toy*). Ectopic expression of either of these genes in antennal, leg, and wing imaginal discs causes the appearance of reasonably well formed ectopic eyes in the corresponding appendages (Halder *et al.*, 1995; Czerny *et al.*, 1999). In *Xenopus* embryos ectopic injection of Pax6 mRNA into animal blastomeres also causes the appearance of ectopic eye-like structures (Chow *et al.*, 1999; Fig. 5.8E). Therefore *pax6* is not only required but also is capable of setting in train the whole program of eye development. So here is our paradox. There is very little in the anatomical organization of eyes or (where it is known) in the process of eye morphogenesis that is common to all the animals in which *pax6* appears to function as a regulator of eye development. Aside from vertebrates and arthropods, these include invertebrate chordates such as amphioxus (Glardon *et al.*, 1998), a cephalopod mollusc (Tomarev *et al.*, 1997), a nemertean worm (Kmita-Cunisse *et al.*, 1998), and even a flatworm (Pineda *et al.*, 2000). How can this one gene be dedicated to control of development of such diverse eyes, which even though their function is (to some extent) analogous, are so very different in structure and organization?

The functions of *pax6* in eye development have now been looked at relatively closely, and the results indicate a solution to the paradox along the lines summarized above. Some key observations are collected in Fig. 5.8. We begin with the *cis*-regulatory system of the mouse *pax6* gene. Figure 5.8A, B shows that this gene has a strikingly modular control system (Kammandel *et al.*, 1999). There is not just a single "eye module"; instead expression in the different parts of the eye, i.e., in the retina, and in the cornea (plus lens, tear gland, and conjunctiva), is controlled by widely separated enhancers. There may be additional elements finely controlling temporal and spatial expression within the developing eye (Xu *et al.*, 1998). Furthermore, fish and mice share interchangeable retinal expression elements (Fig. 5.8B1, B2). These domain-specific *cis*-regulatory modules denote vertebrate cooptions, and they suggest that the regulatory network elements underlying development of the vertebrate eye were assembled separately.

The *pax6* gene clearly plays upstream roles in the eye development network, as demonstrated most impressively by its ability to promote development of ectopic eyes, as just mentioned, and also by the nature of some of its explicit target genes (see below). But *pax6* functions at the downstream end of the network as well, in controlling terminal differentiation gene batteries that encode eye functions. This essential point is illustrated in Fig.5.8C, D. *cis*-Regulatory target sites for the Pax6 transcription factor that are of demonstrated functional importance are shown for three lens crystallin genes in Fig. 5.8C (among other such genes; Cvekl and Piatigorsky, 1996). Crystallin proteins have been coopted in evolution from prior uses as enzymes, and they have in common only that in high concentration they process the requisite optical properties. In itself this is a fascinating example of evolutionary flexibility in gene use (reviewed by Piatigorsky, 1992). The αA crystallins are found in all vertebrate lenses. They are there as a result of the evolutionary recruitment to an eye differentiation gene battery of genes encoding proteins of the heat shock/chaperone family. The chicken lens δ1-crystallin is similarly the product of a gene encoding an arginosuccinate lyase enzyme that was recruited for function within the eye lens, but only in the bird/reptile branch of the vertebrates. These cooptions to eye lens differentiation gene batteries must have been potentiated by the prior presence of Pax6 in the appropriate domain of the eye anlage during development.

The Pax6 transcription factor, the product of the *ey* gene, also turns out to function as a key transcriptional regulator of the *Drosophila rhodopsin* genes, and of other photoreceptor genes (Sheng *et al.*, 1997; Fortini and Rubin, 1990; Mismer and Rubin, 1987). Expression of a *rhodopsin.lacz* construct in the *Drosophila* photoreceptor cells is shown in Fig. 5.8D1, and the dependence of this expression on its Pax6 target sites in Fig. 5.8D2. In a sense the observation that Pax6 controls photoreceptor differentiation genes is the key that unlocks the *pax6* conundrum. All eyes include photoreceptor cells and use photoreceptor gene batteries, and the fact that *pax6* is a key regulator thereof essentially explains why it is always associated with eye development (Zuker, 1994; Sheng *et al.*, 1997; Gehring and Ikeo, 1999). The photoreceptors and their differentiation genes are the shared features descendant from the bilaterian common ancestor. Indeed, the *pax6 cis*-regulatory system that drives expression in photoreceptor cells is widely conserved: an *ey* photoreceptor enhancer from *Drosophila* produces expression in photoreceptor cells of the late embryonic mouse retina (and in immediately surrounding cells) and a similar enhancer active in differentiating retinal photoreceptor cells is also present in the quail *pax6* gene (Xu *et al.*, 1998; Plaza *et al.*, 1995).

The upstream functions of *pax6* in the eye development network of *Drosophila* are equally explicit. Figure 5.8F shows that the *toy* gene activates the *ey* gene, and this in turn activates the *sine oculis* (*so*) and *eyes absent* (*eya*) genes. Evidence for these relationships is given in Fig. 5.8G and legend. The *eya* gene generates a nuclear protein which physically interacts with the So protein, a homeodomain transcription factor, to form a transcriptional regulatory complex required for various steps in eye patterning (Pignoni *et al.*, 1997). Amongst their targets is *dachshund*, which encodes another nuclear protein that also interacts with Eya (Chen *et al.*, 1997). All three of these proteins, particularly in concert, can generate ectopic eyes if they are expressed ectopically, and they are necessary for eye formation; therefore they operate in upstream (as well as other) portions of the eye patterning network (Pignoni *et al*, 1997; Chen *et al.*, 1997 Czerny *et al.*, 1999). Once activated, they in turn stimulate *ey* expression, thus establishing a reinforcing regulatory loop (reviewed by Czerny *et al.*, 1999).

In summary, Fig. 5.8D–G show that the *pax6* genes control both upstream and terminal functions in the gene network for eye development. The argument is that the upstream (i.e., morphogenetic) functions were added later in evolution: since there was a prior assignment of *pax6* to photoreceptor regulatory functions, this gene could subsequently be coopted to various patterning functions in the diverse eye development programs of different clades (the "intercalary evolution" of

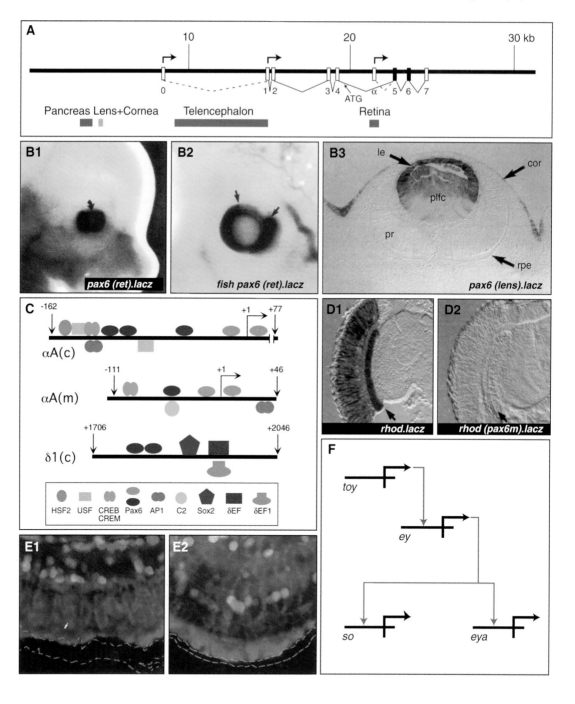

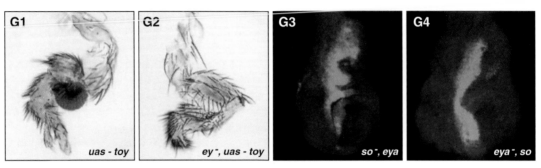

FIGURE 5.8   Regulatory transactivations of Pax6 in vertebrate and Droso-phila eyes.   (A) Modular organization of *cis*-regulatory elements controlling mouse *pax6* expression during development. Two modules control expression in the eye, an upstream element that mediates expression in lens, cornea, tear gland, and conjunctiva (tan); and an intron enhancer responsible for expression in the retina (red). A complex pattern of expression in the brain and spinal cord is activated by elements in another intron (telencephalon enhancer; blue); and expression in the pancreas is controlled by an upstream element (green) near the cornea/lens enhancer module. (B) Function of mouse and fish *pax6* eye enhancers: (B1) embryonic expression of *lacz* construct driven by mouse retina enhancer, 13.5 dpc. (B2) Same, except the enhancer driving the construct is a similar DNA sequence from the zebrafish *pax6* gene (the curved arrow in B1 and the two arrows in B2 denote the dorsal edge of the retinal domain where these constructs do not express as robustly as does the endogenous gene). [(A, B1, B2) From Kammandel *et al.* (1999) *Dev. Biol.* **205**, 79–97.] (B3) Expression of *lacz* construct driven by mouse *pax6* lens/cornea enhancer, 12.5 dpc. Lacz staining in section is shown: le, lens; plfc, primary lens fiber cells; cor, cornea; pr, prospective retina; rpe, presumptive retinal pigmented epithilium. This module produces expression in cornea plus lens, not in retina. [(B3) From Williams *et al.* (1998) *Mech. Dev.* **73**, 225–229, copyright Elsevier Science.] (C) *cis*-Regulatory elements of some lens crystallin genes. Known transcription factors are indicated, including *pax6* (dark blue, regulatory significance of binding interactions shown experimentally; light blue, observation that Pax6 paired domain binds *in vitro*). αA(c), chick αA crystallin; αA(m), mouse αA crystallin; δ1(c), chick δ1 crystallin. Transcription factors which serve as activators are shown above line representing the DNA; those below the line function as repressors. [(C) Simplified from Cvekl and Piatigorsky (1996) *BioEssays* **18**, 621–630, copyright Wiley-Liss, Inc., a subsidiary of John Wiley & Sons, Inc.] with information from Ilagan *et al.* (1999) added (see these sources for further details and additional factors which regulate these and many other lens crystallin genes). (D) Control of *rhodopsin1* gene of *Drosophila* by Pax6, the product of the *eyeless* (*ey*) gene. (D1) Expression of *rhodopsin.lacz* construct in section of fly head; arrow indicates lamina. Staining is confined to photoreceptor cells (R1–R6) and their projections to the lamina. (D2) This construct is not expressed when the Pax6 binding region is mutated. [(D)

(Continues)

From Sheng et al. (1997) Genes Dev. 11, 1122–1131.] (E) Expression of retinal markers in ectopic eye-like structures of Xenopus tadpoles, produced by injection of Pax6 RNA into animal blastomeres at 16-cell stage. (E1) Normal eye, to display markers, here revealed by immunocytology: red, Islet-1 (transcription factor), present in retinal cell nuclei; green, rhodopsin of photoreceptors; blue, Müller cells (identified by an antibody to glutamine synthetase) which are closely associated with neural cells of the inner retinal layer. (E2) Same markers in ectopic eye. [(E) From Chow et al. (1999) Development 126, 4213–4222 and The Company of Biologists Ltd.] (F) Interactions among Drosophila genes for which evidence is shown in (G). The four genes included in the diagram are all involved in pattern formation for eye development (reviewed by Czerny et al., 1999): twin of eyeless (toy), and ey, encode largely similar though not identical Pax6 factors, but have different cis-regulatory systems; sine oculis (so), encodes a homeodomain nuclear factor required for eye development; and eyes absent (eya), encodes a nuclear protein which physically interacts with the so gene product (Pignoni et al. 1997). (G) Epistasis experiments. (G1, G2) toy is upstream of ey. (G1) Ectopic eye (indicated by red eye pigment) appearing on a malformed leg of a fly in which toy expression was forced to occur in the leg imaginal disc under control of a dpp-Gal4/UAS-toy expression system; (G2) same experiment, but carried out in an ey mutant fly. Though the leg is still malformed, no ectopic eye forms, because ey is required for eye development and toy works through ey [cf.(F)]. Other evidence that toy is upstream of ey (Czerny et al., 1999) is that the toy gene is expressed normally in ey mutants; that ectopic toy expression causes ectopic ey expression but not vice versa; that ectopic ey expression causes appearance of ectopic eyes (Halder et al., 1995) in toy mutants, but (as illustrated here) not vice versa; and that the Toy protein binds directly to ey cis-regulatory sites essential for ey expression. [(G1, G2) From Czerny et al. (1999) Mol. Cell 3, 297–307, copyright Cell Press.] (G3, G4) ey is upstream of so and eya: (G3) third instar wing disc of so mutant, ectopically expressing Eya, as detected by immunocytology (orange fluorescence). Expression occurs in a region along the A/P boundary where ey expression has been forced by use of dpp-Gal4/UAS-ey. This does not cause ectopic eye formation, because the so gene product is required for eye development. (G4) Converse experiment: expression of ey forced along A/P axis of wing disc just as in (G3), but in an eya mutant; expression of so gene product is here displayed (orange fluorescence). Ectopic eyes do not form because the eya gene product is also required for eye development. [(G3, G4) From Halder et al. (1998) Development 125, 2181–2191 and The Company of Biologists Ltd.] Other evidence for the relation between ey and so shown in (F) is that the Ey protein binds directly to so cis-regulatory elements and activates this gene (Niimi et al., 1999). Furthermore, ey is expressed earlier than are so and eya, and ey expression is required for expression of both eya and so in the eye disc, but not vice versa. Ectopic expression of eya, so, and also dachshund (another gene downstream of eya and so in the eye patterning network) produces ectopic eyes, more so when they are expressed in combination (Pignoni et al., 1997; Chen et al., 1997; Czerny et al., 1999).

Gehring and Ikeo, 1999). This looks to be a canonical example for the kind of evolutionary process envisioned in Fig. 5.7.

Other cases come to mind as well. Heart development is one: *tinman* (*tin*) and its vertebrate orthologues, the *nkx2.5* and *nkx2.3* genes, are alike required for upstream pattern formation processes in the development of the heart in *Drosophila* and in vertebrates, respectively (Chapter 4). But these factors also exert distinct *cis*-regulatory control over terminal heart cell differentiation genes, e.g., in amniotes, genes encoding cardiac atrial natriuretic peptide (Lee *et al.*, 1998; Durocher and Nemer, 1998) and cardiac α-actin (Sepulveda *et al.*, 1998). Furthermore, activation of these differentiation genes requires the participation of Gata factors, and Gata regulators control several other terminal differentiation genes in the heart in addition to the above (Wang *et al.*, 1998; Grepin *et al.*, 1995; Molkentin *et al.*, 1998). On the other hand, *gata* genes too are required in various upstream pattern formation processes in heart development, in both mouse and *Drosophila* (Chapter 4). The morphogenetic features of *Drosophila* and vertebrate hearts are (of course) different: the *nkx2.5* gene participates in multiple vertebrate-specific functions beyond initial specification of the progenitor field, *viz,* formation of the heart tube, specification of its asymmetric chamber domains, and completion of its complex morphology (Lin *et al.*, 1997; see above). Different interactions are required of Nkx2.5 than of the Tin factor and if introduced into the genome *nkx2.5* cannot rescue heart formation in *Drosophila tin* mutants, unless it is equipped with a piece of the *tin* coding sequence (Ranganayakulu *et al.*, 1998). Here again control of terminal heart cell differentiation processes may be the shared ancestral property, and the pattern formation processes mediated by both *gata* and *tin/nkx2.5* genes may have been inserted stepwise into developmental networks for heart formation, differently in the evolution of different animal clades.

Perhaps these steps are indicated by the modular *cis*-regulatory organization of the *Drosophila tin* (Fig. 4.8E) and the mouse *nkx2.5* genes (reviewed by Schwartz and Olson, 1999). The vertebrate heart has a modular anatomical construction (Fishman and Olson, 1997), and expression of the *nkx2.5* gene in the various subdomains of the developing heart turns out to be controlled by many distinct *cis*-regulatory elements. These operate in the different spatial domains of the forming heart, at different stages, e.g., when the heart is a simple linear tube, or when it is forming left and right ventricles, or later in development. Each enhancer responds to different inputs, so that *lacz* constructs which are made from them are each expressed differently (for references, see Schwartz and Olson, 1999). As suggested in Fig. 5.7, modularity in the control systems of genes that function during pattern formation for a body part may provide a living trail of the evolutionary stages through which the patterning network was assembled.

How generally could the concept of body part evolution here on the table be applied? There are a great many examples of the paradox with which this section begins. Some are more trivial than others. For example, it is improbable that any particular use of BMP signaling pathways could reliably indicate survival of a dedicated ancestral bilaterian pathway for making a given structure, since BMP family members are deployed at some point in practically every structure in the body (reviewed by Kingsley, 1994). An interesting case, on the other hand, is that of the *caudal* (*cad*) genes. These encode a family of homeodomain regulators that throughout Bilateria seem to be used in the embryonic specification of the hind end of the animal, including the posterior gut (Marom *et al.*, 1997; Gamer and Wright, 1993; Katsuyama *et al.*, 1999). The *cad* genes clearly execute upstream pattern formation functions. For example, in *Drosophila cad* activates *forkhead* and *wg* genes needed for posterior gut formation in the embryo (Wu and Lengyel, 1998). The *cad* gene is necessary and sufficient for specification of the most posterior adult body segment, in which process it activates the *brachyenteron* and *evenskipped* genes in the hindgut, and the *distal-less* gene in the forming anal plates (Moreno and Morata, 1999). In mouse the *cdx caudal* orthologues provide direct transcriptional input into *hox* genes (Charité *et al.*, 1998) another upstream function; and in *Xenopus cad* overexpression or loss of function disturbs embryonic A/P patterning (Epstein *et al.*, 1997). But once again *cad/cdx* genes may have terminal differentiation roles as well. The *Drosophila cad* gene controls expression of the *short gastrulation* gene, which encodes a secreted molecule required for cell movement during hindgut invagination (Wu and Lengyel, 1998). In chick and mouse the *cdx* genes are expressed even in adult intestine (James *et al.*, 1994; Marom *et al.*, 1997). The terminal roles of *cad* genes, perhaps control of differentiation gene batteries in the posterior gut, may have been their ancestral evolutionary function. Because these genes were activated at the posterior end, cooption could have led during evolution to their inclusion in morphogenetic patterning networks for posterior structures. A similar situation obtains at the opposite end of the animal. Transcription factors encoded by the *orthodenticle* gene of *Drosophila* and the *otx* genes of vertebrates participate in the pattern formation networks underlying brain development in both (for *Drosophila*, reviewed by Cohen and Jürgens, 1991; Finkelstein and Boncinelli, 1994; Younossi-Hartenstein *et al.*, 1997; for *Xenopus*, Pannese *et al.*, 1995; for mouse, Fig. 4.1 and references in Chapter 4 related thereto). But the *otx* genes are also expressed in specific neurons, and in specific cell types of olfactory, ocular, and acoustic sense organs, which are affected in *otx* knockouts (Acampora *et al.*, 1996). Perhaps the original role of *otx* genes was control of sensory neuron differentiation, which necessarily was installed at the anterior end of the bilaterian ancestor. These genes could then have been coopted by inclusion in diverse patterning gene networks during the evolution of brains in the different bilaterian clades.

The argument made in this section boils down to the proposition that what is actually shared amongst all bilaterians in respect to particular body parts is only the basal repertoire of differentiation gene batteries required in each, including their specific controllers. The pattern formation process underlying the morphogenesis of analogous body parts in different clades are the outcome of a long series of cooptions and reorganizations, that have retained selective value at each step due to expression of their cell type-specific functions. This is a rather different view of the evolutionary process from that in which the shared use of upstream regulators is held as evidence for conservation of a genomic program for pattern formation. That view leads to the position that seeing is not believing; that though analogous adult body parts of diverse clades look different, beneath the surface there lies hidden a common, unchanging morphogenetic deep structure inherited from the bilaterian common ancestor. But if there is one impression that all the figures and examples in this chapter should convey, it would be of the mutability of pattern formation networks, and the ubiquity and prevalence of evolutionary cooption of regulatory genes in these networks.

CONCLUDING COMMENT: CONCEIVING EVOLUTION AS A PROCESS OF CHANGE IN REGULATORY GENE NETWORKS

For anyone interested in mechanism, there is in fact no other way to conceive of the basis of evolutionary change in bilaterian form than by change in the underlying developmental gene regulatory networks. This of course means change in the *cis*-regulatory DNA linkages that determine the functional architecture of all such networks. A glance toward the horizon suggests that some of the most intriguing concepts and problems in evolutionary biology will in some measure undergo redefinition as comparative knowledge of gene regulatory networks expands.

Take for example the concept of homology as applied to usage of genes. Whole symposia, treatises, and a good deal of argument are devoted to the meaning and implications of homology in evolution (Meyer, 1999). But with reference to the genetic basis of evolutionary change, we can ask the precise question "what is a homologous use of a given gene in a comparison of diverse organisms," and we can look forward to a precise answer. This emerges from the structure of gene networks. A gene performs a homologous function in two animals if at least some of its upstream or its downstream linkages (or both) remain the same in the two genomes, and the function it performs is descendant from their common ancestor. Otherwise homologous use cannot be supported.

Or consider the appearance of sudden "bursts" of evolutionary change in morphology, followed by periods of stability, which is suggested by some aspects of the bilaterian fossil record. This is somewhat a fuzzy proposition from the outset, since everything depends on what level of evolutionary change is meant. Considered in terms of regulatory gene networks, small changes at the level of development of terminal differentiation gene batteries are likely to be going on continuously (as in the bristle patterns of Fig. 5.6). These changes affect species-specific and intraspecific characters, and they are fixed according to population parameters, selective factors and so forth. But the architecture of the gene networks that control more momentous aspects of form requires that some changes will have major effects, and others smaller effects. For instance, large morphological effects will follow relocation or duplication of progenitor fields for given body parts (Chapter 4), as clearly occurred in the evolution of vertebrate appendages (Fig. 5.4C). The result of a successful upstream change of this kind is to create a new set of possibilities downstream, a "morphospace" to fill: this indeed happened, for example, when paired limbs evolved in diverse ways in different gnathostome clades. Given that gene networks have upstream and downstream polarities, major evolutionary change in developmental process must be discontinuous in rate.

The fields of development and evolution are convergent because both subjects are rooted in the genomic regulatory programs for body part formation. And the convergence is going to go much further than it has. Very soon we will be able to understand exactly how the *cis*-regulatory systems of given gene networks function to generate the spatial aspects of development. By comparative reference to other phylogenetically informative genomes, we will thereby learn just what happened at the DNA level which led in evolution to various forms of the body part. This will perhaps require something like what comparative linguists do when they reconstruct extinct proto-languages from a family of extant descendants: i.e., reconstruction of ancestral-gene regulatory networks from those of related modern animals. But first we need to obtain and to understand in detail some of the important developmental regulatory networks in the genome.

This leads to a final comment. Though the ancestors of modern animals are extinct the evidence of how they worked is not. The evidence is swimming, walking, and flying around outside, in the DNA of modern bilaterians. What happened in evolution will emerge from knowledge of the regulatory pathways of development, the subject of most of this book; and from application of computational tools for tracing, analyzing, and conceiving regulatory network architecture. Comparative sets of genomic network "wiring" diagrams will emerge. Strange as it may seem, it is these that will tell us what the Bilateria are and where they came from.

REFERENCES

Aamodt, E. J., Chung, M. A., and McGhee, J. D. (1991). Spatial control of gut-specific gene expression during *Caenorhabditis elegans* development. *Science* **252**, 579–252.

Abzhanov, A., and Kaufman, T. C. (1999). Homeotic genes and the arthropod head: Expression patterns of *labial*, *proboscipedia*, and *Deformed* genes in crustaceans and insects. *Proc. Natl. Acad. Sci. USA* **96**, 10224–10229, 267.

Abzhanov, A., and Kaufman, T. C. (2000). Crustacean (malacostracan) *Hox* genes and the evolution of the arthropod trunk. *Development* **127**, 2239–2249.

Abzhanov, A., Popadic, A., and Kaufman, T. C. (1999). Chelicerate *Hox* genes and the homology of arthropod segments. *Evol. Dev.* **1**, 77–89.

Acampora, D., Mazan, S., Avantaggiato, V., Barone, P., Tuorto, F., Lallemand, Y., Brûlet, P., and Simeone, A. (1996). Epilepsy and brain abnormalities in mice lacking the *Otx1* gene. *Nature Genet.* **14**, 218–222.

Adams, M. D. *et al.* (2000). The genome sequence of *Drosophila melanogaster*. *Science* **287**, 2185–2195.

Adoutte, A., Balavoine, G., Lartillot, N., Lespinet, O., and Prud'homme, B. (2000). The new animal phylogeny: Reliability and implications. *Proc. Natl. Acad. Sci. USA* **97**, 4453–4456.

Aguinaldo, A. M. A., Turbeville, J. M., Linford, L. S., Rivera, M. C., Garey, J. R., Raff, R. A., and Lake, J. A. (1997). Evidence for a clade of nematodes, arthropods and other moulting animals. *Nature* **387**, 489–493.

Ahlgren, U., Jonsson, J., and Edlund, H. (1996). The morphogenesis of the pancreatic mesenchyme is uncoupled from that of the pancreatic epithelium in IPF1/PDX1-deficient mice. *Development* **122**, 1409–1416.

Akam, M. (1995). *Hox* genes and the evolution of diverse body plans. *Phil. Trans. R. Soc. Lond. B.* **349**, 313–319.

Akam, M. (2000). Arthropods: Developmental diversity within a (super) phylum. *Proc. Natl. Acad. Sci. USA* **97**, 4438–4441.

Akasaka, K., Ueda, T., Higashinakagawa, T., Yamada, K., and Shimada, H. (1990). Spatial patterns of arylsulfatase mRNA expression in sea urchin embryo. *Dev. Growth Differ.* **32**, 9–13.

Altabef, M., Logan, C., Tickle, C., and Lumsden, A. (2000). *Engrailed-1* misexpression in chick embryos prevents apical ridge formation but preserves segregation of dorsal and ventral ectodermal compartments. *Dev. Biol.* **222**, 307–316.

Andersen, B., and Rosenfeld, M. G. (1994). *Pit-1* determines cell types during development of the anterior pituitary gland. A model for transcriptional regulation of cell phenotypes in mammalian organogenesis. *J. Biol. Chem.* **269**, 29335–29338.

Anderson, K. V., Jurgens, G., and Nüsslein-Volhard, C. (1985). Establishment of dorsal-ventral polarity in the *Drosophila* embryo: Genetic studies on the *toll* gene product. *Cell* **42**, 779–789.

Anderson, M. G., Perkins, G. L., Chittick, P., Shrigley, R. J., and Johnson, W. A. (1995). *drifter*, a *Drosophila* POU-domain transcription factor, is required for correct differentiation and migration of tracheal cells and midline glia. *Genes Dev.* **9**, 123–137.

Anderson, M. G., Certel, S. J., Certel, K., Lee, T., Montell, D. J., and Johnson, W. A. (1996). Function of the *Drosophila* POU domain transcription factor Drifter as an upstream regulatory of Breathless receptor tyrosine kinase expression in developing trachea. *Development* **122**, 4169–4178.

Angerer, L. M., and Angerer, R. C. (2000). Animal-vegetal axis patterning mechanisms in the early sea urchin embryo. *Development* **218**, 1–12.

Angerer, L. M., Dolecki, G. J., Gagnon, M., Lum, R., Wang, G., Yang, Q., Humphreys, T., and Angerer, R. C. (1989). Progressively restricted expression of homeo box gene within the aboral ectoderm of developing sea urchin embryos. *Genes Dev.* **3**, 370–383.

Araki, I., and Satoh, N. (1996). *Cis*-regulatory elements conserved in the proximal promoter region of an ascidian embryonic muscle myosin heavy-chain gene. *Dev. Genes Evol.* **206**, 54–63.

Arenas-Mena, C., Martinez, P., Cameron, R. A., and Davidson, E. H. (1998). Expression of the *Hox* gene complex in the indirect development of a sea urchin. *Proc. Natl. Acad. Sci. USA* **95**, 13062–13067.

Arenas-Mena, C., Cameron, R. A., and Davidson, E. H. (2000). Spatial expression of *Hox* cluster genes in the ontogeny of a sea urchin. *Development,* **127**, 4631–4643.

Arendt, D., and Nübler-Jung, K. (1994). Inversion of dorsoventral axis? *Nature* **371**, 26.

Arnone, M., and Davidson, E. H. (1997). The hardwiring of development: Organization and function of genomic regulatory systems. *Development* **124**, 1851–1864.

Arnone, M. I., Bogarad, L. D., Collazo, A., Kirchhamer, C. V., Cameron, R. A., Rast, J. P., Gregorians, A., and Davidson, E. H. (1997). Green fluorescent protein in the sea urchin: New experimental approaches to transcriptional regulatory analysis in embryos and larvae. *Development* **124**, 4649–4659.

Arnone, M. I., Martin, E. L., and Davidson, E. H. (1998). *Cis*-regulation downstream of cell type specification: A single compact element controls the complex expression of the *CyIIa* gene in sea urchin embryos. *Development* **125**, 1381–1395.

Arnosti, D. N., Barolo, S., Levine, M., and Small, S. (1996). The *eve* stripe 2 enhancer employs multiple modes of transcriptional synergy. *Development* **122**, 205–214.

Artavanis-Tsakonas, S., Rand, M. D., and Lake, R. J. (1999). Notch signaling: Cell fate control and signal integration in development. *Science* **284**, 770–776.

Averof, M., and Akam, M. (1995). Insect-crustacean relationships: Insights from comparative developmental and molecular studies. *Phil. Trans. R. Soc. Lond. B* **347**, 293–303.

Averof, M., and Patel, N. H. (1997). Crustacean appendage evolution associated with changes in *Hox* gene expression. *Nature* **388**, 682–686.

Azpiazu, N., and Frasch, M. (1993). *tinman* and *bagpipe*: Two homeobox genes that determine cell fates in the dorsal mesoderm of *Drosophila*. *Genes Dev.* **7**, 1325–1340.

Bailey, A. M., and Posakony, J. W. (1995). Suppressor of Hairless directly activates transcription of *Enhancer of split* Complex genes in response to Notch receptor activity. *Genes Dev.* **9**, 2609–2622.

Balavoine, G. (1997). The early emergence of platyhelminths is contradicted by the agreement between 18S rRNA and *Hox* genes data. *C. R. Acad. Sci., Paris* **329**, 83–94.

Balavoine, G., and Adoutte, A. (1998). One or three Cambrian radiations? *Science* **280**, 397–398.

Bally-Cuif, L., and Boncinelli, E. (1997). Transcription factors and head formation in vertebrates. *BioEssays* **19**, 127–135.

Bao, Z.-Z., Bruneau, B. G., Seidman, J. G., Seidman, C. E., and Cepko, C. L. (1999). Regulation of chamber-specific gene expression in the developing heart by *Irx4*. *Science* **283**, 1161–1164.

Baonza, A., de Celis, J. F., and García-Bellido, A. (2000) Relationships between *extramacrochaetae* and *Notch* signalling in *Drosophila* wing development. *Development* **27**, 2383–2393.

Barrow, J. R., Stadler, H. S., and Capecchi, M. R. (2000). Roles of *Hoxa1* and *Hoxa2* in patterning the early hindbrain of the mouse. *Development* **127**, 933–944.

Bassham, S., and Postlethwait, J. (2000). *Brachyury* (*T*) expression in embryos of a larvacean urochordate, *Oikopleura dioica*, and the ancestral role of *T*. *Dev. Biol.* **220**, 322–332.

Batchelder, C., Dunn, M. A., Choy, B., Suh, Y., Cassie, C., Shim, E. Y., Shin, T. H., Mello, C., Seydoux, G., and Blackwell, T. K. (1999). Transcriptional repression by the *Caenorhabditis elegans* germ-line protein PIE-1. *Genes Dev.* **13**, 202–212.

Bate, M. (1993). The mesoderm and its derivatives. *In* "The Development of *Drosophila melanogaster*." (M. Bates and A. Martinez Arias, Eds.), pp. 1013–1090. Cold Spring Harbor Laboratory Press, New York.

Baumgartner, S., and Noll, M. (1991). Network of interactions among pair-rule genes regulating paired expression during primordial segmentation of *Drosophila*. *Mech. Dev.* **33**, 1–18.

Baylies, M. K., Bate, M., and Ruiz Gomez, M. (1998). Myogenesis: A view from *Drosophila*. *Cell* **93**, 921–927.

Beck, F., Tata, F., and Chawengsaksophak, K. (2000). Homeobox genes and gut development. *BioEssays* **22**, 431–441.

Beckendorf, S. K., and Kafatos, F. C. (1976). Differentiation in the salivary glands of *Drosophila melanogaster*: Characterization of the glue proteins and their developmental appearance. *Cell* **9**, 365–373.

Beckers, J., Gérard, M., and Duboule, D. (1996). Transgenic analysis of a potential *Hoxd-11* limb regulatory element present in tetrapods and fish. *Dev. Biol.* **180**, 543–553.

Bender, W., Akam, M., Karch, F., Beachy, P. A., Peifer, M., Spierer, P., Lewis, E. B., and Hogness, D. S. (1983). Molecular genetics of the Bithorax Complex in *Drosophila melanogaster*. *Science* **221**, 23–29.

Benson, S. C., Sucov, H. M., Stephens, L., Davidson, E. H., and Wilt, F. (1987). A lineage-specific gene encoding a major matrix protein of the sea urchin embryo spicule. I. Authentication of the cloned gene and its developmental expression. *Dev. Biol.* **120**, 499–506.

Biben, C., and Harvey, R. P. (1997). Homeodomain factor Nkx2–5 controls left/right asymmetric expression of bHLH gene *eHand* during murine heart development. *Genes Dev.* **11**, 1357–1369.

Biehs, B., Sturtevant, M. A., and Bier, E. (1998). Boundaries in the *Drosophila* wing imaginal disc organize vein-specific genetic programs. *Development* **125**, 4245–4257.

Bienz, M. (1994). Homeotic genes and positional signalling in the *Drosophila* viscera. *Trends Genet.* **10**, 22–26.

Bier, E. ((2000) Drawing lines in the *Drosophila* wing: Initiation of wing vein development. *Curr. Opin. Genet. Dev.* 10, 393–398.

Bilder, D., Graba, Y., and Scott, M. P. (1998). Wnt and TGFβ signals subdivide the AbdA Hox domain during *Drosophila* mesoderm patterning. *Development* **125**, 1781–1790.

Bodmer, R. (1993). The gene *tinman* is required for specification of the heart and visceral muscles in *Drosophila*. *Development* **118**, 719–729.

Bodmer, R., and Frasch, M. (1999). Genetic determination of *Drosophila* heart development. *In* "Heart Development." (R. P. Harvey and N. Rosenthal, Eds.), pp. 65–90. Academic Press, San Diego.

Bogarad, L. D., Arnone, M. I., Chieh, C., and Davidson, E. H. (1998). TKO, a general approach to functionally knock-out transcription factors *in vivo*, using a novel genetic expression vector. *Proc. Natl. Acad. Sci. USA* **95**, 14827–14832.

Boncinelli, E. (1999). *Otx* and *Emx* homeobox genes in brain development. *The Neuroscientist* **5**, 164–172.

Borycki, A.-G., and Emerson, C. P. (1997). Muscle determination: Another key player in myogenesis? *Curr. Biol.* **7**, R620–R623.

Boube, M., Llimargas, M., and Casanova, J. (2000). Cross-regulatory interactions among tracheal genes support a co-operative model for the induction of tracheal fates in the *Drosophila* embryo. *Mech. Dev.* **91**, 271–278.

Bowerman, B. (1998). Maternal control of pattern formation in early *Caenorhabditis elegans* embryos. *Curr. Topics Dev. Biol.* **39**, 73–117.

Bowerman, B., Eaton, B. A., and Priess, J. R. (1992). *skn-1*, a maternally expressed gene required to specify the fate of ventral blastomeres in the early *C. elegans* embryo. *Cell* **68**, 1061–1075.

Bowerman, B., Draper, B. W., Mello, C. C., and Priess, J. R. (1993). The maternal gene *skn-1* encodes a protein that is distributed unequally in early *C. elegans* embryos. *Cell* **74**, 443–452.

Bowerman, B., Ingram, M. K., and Hunter, C. P. (1997). The maternal *par* genes and the segregation of cell fate specification activities in early *Caenorhabditis elegans* embryos. *Development* **124**, 3815–3826.

Brannon, M., Gomperts, M., Sumoy, L., Moon, R. T., and Kimelman, D. (1997). A β-catenin/XTcf-3 complex binds to the *siamois* promoter to regulate dorsal axis specification in *Xenopus*. *Genes Dev.* **11**, 2359–2370.

Brenner, S., Elgar, G., Sandford, R., Macrae, A., Venkatesh, B., and Aparcio, S. (1993). Characterization of the pufferfish (*Fugu*) genome as a compact model vertebrate genome. *Nature* **366**, 265–268.

Britten, R. J. (1997). Mobile elements inserted in the distant past have taken on important functions. *Gene* **205**, 177–182.

Britten, R. J., and Davidson, E. H. (1969). Gene regulation for higher cells: A theory. *Science* **165**, 349–358.

Britten, R. J., and Davidson, E. H. (1971). Repetitive and non-repetitive DNA sequences and a speculation on the origins of evolutionary novelty. *Quart. Rev. Biol.* **46**, 111–138.

Bruneau, B. G., Bao, Z.-Z., Tanaka, M., Schott, J.-J., Izumo, S., Cepko, C. L., Seidman, J. G., and Seidman, C. E. (2000). Cardiac expression of the ventricle-specific homeobox gene *Irx4* is modulated by Nkx2–5 and dHand. *Dev. Biol.* **217**, 266–277.

Brunschwig, K., Wittmann, C., Schnabel, R., Bürglin, T. R., Tobler, H., and Müller, F. (1999). Anterior organization of the *Caenorhabditis elegans* embryo by the *labial*-like *Hox* gene *ceh-13*. *Development* **126**, 1537–1546.

Budd, G. E. (1996). The morphology of *Opabinia regalis* and the reconstruction of the arthropod stem-group. *Lethaia* **29**, 1–14.

Burz, D. S., Rivera-Pomar, R., Jäckle, H., and Hanes, S. D. (1998). Cooperative DNA-binding by Bicoid provides a mechanism for threshold-dependent gene activation in the *Drosophila* embryo. *EMBO J.* **17**, 5998–6009.

The *C. elegans* Sequencing Consortium (1998). Genome sequence of the nematode *C. elegans*: A platform for investigating biology. *Science* **282**, 2012–2018.

Cai, H. N., Arnosti, D. N., and Levine, M. (1996). Long-range repression in the *Drosophila* embryo. *Proc. Natl. Acad. Sci. USA* **93**, 9309–9314.

Calzone, F. J., Thézé, N., Thiebaud, P., Hill, R. L., Britten, R. J., and Davidson, E. H. (1988). Developmental appearance of factors that bind specifically to *cis*-regulatory sequences of a gene expressed in the sea urchin embryo. *Genes Dev.* **2**, 1074–1088.

Cameron, C. B., Garey, J. R., and Swalla, B. J. (2000). Evolution of the chordate body plan: New insights from phylogenetic analyses of deuterostome phyla. *Proc. Natl. Acad. Sci. USA* **97**, 4469–4474.

Cameron, R. A., and Davidson, E. H. (1997). LiCl perturbs ectodermal *veg*₁ lineage allocations in *Strongylocentrotus purpuratus* embryos. *Dev. Biol.* **187**, 236–239.

Cameron, R. A., Hough-Evans, B. R., Britten, R. J., and Davidson, E. H. (1987). Lineage and fate of each blastomere of the eight-cell sea urchin embryo. *Genes Dev.* **1**, 75–85.

Cameron, R. A., Fraser, S. E., Britten, R. J., and Davidson, E. H. (1989). The oral-aboral axis of a sea urchin embryo is specified by first cleavage. *Development* **106**, 641–647.

Cameron, R. A., Britten, R. J., and Davidson, E. H. (1993). The embryonic ciliated band of the sea urchin, *Strongylocentrotus purpuratus*, derives from both oral and aboral ectoderm territories. *Dev. Biol.* **160**, 369–376.

Cameron, R. A. *et al.* (2000). A sea urchin genome project: Sequence scan, virtual map, and additional resources. *Proc. Natl. Acad. Sci. USA* **97**, 9514–9518.

Capovilla, M., Brandt, M., and Botas, J. (1994). Direct regulation of *decapentaplegic* by *Ultrabithorax* and its role in *Drosophila* midgut morphogenesis. *Cell* **76**, 461–475.

Carnac, G., Kodjabachian, L., Gurdon, J. B., and Lemaire, P. (1996). The homeobox gene *Siamois* is a target of the Wnt dorsalisation pathway and triggers organiser activity in the absence of mesoderm. *Development* **122**, 3055–3065.

Carroll, S. B. (1995). Homeotic genes and the evolution of arthropods and chordates. *Nature* **376**, 479–485.

Casares, F., and Mann, R. S. (2000). A dual role for homothorax in inhibiting wing blade development and specifying proximal wing identities in *Drosophila*. *Development* **127**, 1499–1508.

Castresana, J., Feldmaier-Fuchs, G., Yokobori, S., Satoh, N., and Paabo, S. (1998). The mitochondrial genome of the hemichordate *Balanoglossus carnosus* and the evolution of deuterostome mitochondria. *Genetics* **150**, 1115–1123.

Cavallo, R. A., Cox, R. T., Moline, M. M., Roose, J., Polevoy, G. A., Clevers, H., Peifer, M., and Bejsovec, A. (1998). *Drosophila* Tcf and Groucho interact to repress Wingless signalling activity. *Nature* **395**, 604–608.

Charité, J., de Graaff, W., Consten, D., Reijnen, M. J., Korving, J., and Deschamps, J. (1998). Transducing positional information to the *Hox* genes: Critical interaction of *cdx* gene products with position-sensitive regulatory elements. *Development* **125**, 4349–4358.

Chen, C.-K., Kühnlein, R. P., Eulenberg, K. G., Vincent, S., Affolter, M., and Schuh, R. (1998). The transcription factors KNIRPS and KNIRPS RELATED control cell migration and branch morphogenesis during *Drosophila* tracheal development. *Development* **125**, 4959–4968.

Chen, J. Y., Huang, D. Y., and Li, C. W. (1999). An early Cambrian craniate-like chordate. *Nature* **402**, 518–522.

Chen, R., Amoui, M., Zhang, Z., and Mardon, G. (1997). Dachshund and eyes absent proteins form a complex and function synergistically to induce ectopic eye development in *Drosophila. Cell* **91**, 893–903.

Chow, R. L., Altmann, C. R., Lang, R. A., and Hemmati-Brivanlou, A. (1999). *Pax6* induces ectopic eyes in a vertebrate. *Development* **126**, 4213–4222.

Christoffels, V. M., Habets, P.E.M.H., Franco, D., Campione, M., de Jong, F., Lamers, W. H., Harvey, R. P., and Moorman, A. F. M. (2000). Chamber formation and morphogenesis in the developing mammalian heart. *Dev. Biol.* **223**, 266–278.

Chuang, C.-K., Wikramanayake, A. H., Mao, C.-A., Li, X, and Klein, W. H. (1996). Transient appearance of *Strongylocentrotus purpuratus* Otx in micromere nuclei: Cytoplasmic retention of SpOtx possibly mediated through an α-actinin interaction. *Dev. Genet.* **19**, 231–237.

Chuong, C. M., Widelitz, R. B., Ting-Berreth, S., and Jiang, T. X. (1996). Early events during avian skin appendage regeneration: Dependence on epithelial-mesenchymal interaction and order of molecular reappearance. *J. Invest. Dermatol.* **107**, 639–646.

Coates, M. I. (1995). Fish fins or tetrapod limbs—a simple twist of fate? *Curr. Biol.* **5**, 844–848.

Coffman, J. A., and Davidson, E. H. (2000). Oral-aboral axis specification in the sea urchin embryo. I. Axis entrainment by respiratory asymmetry. *Dev. Biol.*, in press.

Coffman, J. A., Kirchhamer, C. V., Harrington, M. G., and Davidson, E. H. (1996). SpRunt-1, a new member of the Runt-domain family of transcription factors, is a positive regulator of the aboral ectoderm-specific *CyIIIa* gene in sea urchin embryos. *Dev. Biol.* **174**, 43–54.

Coffman, J. A., Kirchhamer, C. V., Harrington, M. G., and Davidson, E. H. (1997). SpMyb functions as an intramodular repressor to regulate spatial expression of *CyIIIa* in sea urchin embryos. *Development* **124**, 4717–4727.

Cohen, B., Simcox, A. A., and Cohen, S. M. (1993). Allocation of the thoracic imaginal primordia in the *Drosophila* embryo. *Development* **117**, 597–608.

Cohen, B. L., Gawthrop, A., and Cavalier-Smith, T. (1998). Molecular phylogeny of brachiopods and phoronids based on nuclear-encoded small subunit ribosomal RNA gene sequences. *Phil. Trans. R. Soc. Lond. Ser. B. Biol. Sci.* **353**, 2039–2061.

Cohen, S., and Jürgens, G. (1991). *Drosophila* headlines. *Trends Genet.* **7**, 267–272.

Cohen, S. M. (1993). Imaginal disc development. *In* "The Development of Drosophila melanogaster." (M. Bate and A. Martinez Arias, Eds.), pp. 747–841. Cold Spring Harbor Press, NY.

Cohen, S. M., and Jürgens, G. (1989). Proximal–distal pattern formation in *Drosophila*: Cell autonomous requirement for *Distal-less* gene activity in limb development. *EMBO J.* **8**, 2045–2055.

Cohn, M. J., Izpisùa-Belmonte, J. C., Abud, H., Heath, J. K., and Tickle, C. (1995). Fibroblast growth factors induce additional limb development from the flank of chick embryos. *Cell* **80**, 739–746.

Cohn, M. J., Patel, K., Krumlauf, R., Wilkinson, D. G., Clarke, J. D. W., and Tickle, C. (1997). *Hox9* genes and vertebrate limb specification. *Nature* **387**, 97–101.

Collins, F. S. (1995). Ahead of schedule and under budget: The genome project passes its fifth birthday. *Proc. Natl. Acad. Sci. USA* **92**, 10821–10823.

Conklin, E. G. (1905). The organization and cell lineage of the ascidian egg. *J. Acad. Nat. Sci. (Philadelphia)* **13**, 1–119.

Cooper, M. T. D., Tyler, D. M., Furriols, M., Chalkiadaki, A., Delidakis, C., and Bray, S. (2000). Spatially restricted factors cooperate with Notch in the regulation of *Enhancer of split* genes. *Dev. Biol.* **221**, 390–403.

Corbo, J. C., Levine, M., and Zeller, R. W. (1997). Characterization of a notochord-specific enhancer from the *Brachyury* promoter region of the ascidian, *Ciona intestinalis*. *Development* **124**, 589–602.

Corbo, J. C., Fujiwara, S., Levine, M., and Di Gregorio, A. (1998). Suppressor of Hairless activates *Brachyury* expression in the *Ciona* embryo. *Dev. Biol.* **203**, 358–368.

Cordes, S. P., and Barsh, G. S. (1994). The mouse segmentation gene *kr* encodes a novel basic domain-leucine zipper transcription factor. *Cell* **79**, 1025–1034.

Cripps, R. M., Black, B. L., Zhao, B., Lien, C.-L., Schulz, R. A., and Olson, E. N. (1998). The myogenic regulatory gene *Mef2* is a direct target for transcriptional activation by Twist during *Drosophila* myogenesis. *Genes Dev.* **12**, 422–434.

Cripps, R. M., Zhao, B., and Olson, E. N. (1999). Transcription of the myogenic regulatory gene *Mef2* in cardiac, somatic, and visceral muscle cell lineages is regulated by a *tinman*-dependent core enhancer. *Dev. Biol.* **215**, 420–430.

Cserjesi, P., Brown, D., Lyons, G. E., and Olson, E. N. (1995). Expression of the novel basic helix-loop-helix gene *eHAND* in neural crest derivatives and extra-embryonic membranes during mosue development. *Dev. Biol.* **170**, 664–678.

Cubadda, Y., Heitzler, P., Ray, R. P., Bourouis, M., Ramain, P., Gelbart, W., Simpson, P., and Haenlin, M. (1997). *u-shaped* encodes a zinc finger protein that regulates the proneural genes *achaete* and *scute* during the formation of bristles in *Drosophila*. *Genes Dev.* **11**, 3083–3095.

Cubas, P., de Celis, J.-F., Campuzano, S., and Modolell, J. (1991). Proneural clusters of *achaete-scute* expression and the generation of sensory organs in the *Drosophila* imaginal wing disc. *Genes Dev.* **5**, 996–1008.

Cvekl, A., and Piatigorsky, J. (1996). Lens development and crystallin gene expression: Many roles for Pax-6. *BioEssays* **18**, 621–630.

Czerny, T., Halder, G., Kloter, U., Souabni, A., Gehring, W. J., and Busslinger, M. (1999). *twin of eyeless*, a second *Pax-6* gene of *Drosophila*, acts upstream of *eyeless* in the control of eye development. *Mol. Cell* **3**, 297–307.

Czihak, G. (1963). Entwicklungsphysiologische Untersuchungen an Echiniden (Verteilung and Bedeutung der Cytochromoxydase). *Wilhelm Roux' Arch. EntwMech Org.* **154**, 272–292.

Damen, W. G. M., Hausdorf, M., Seyfarth, E.-A., and Tautz, D. (1998). A conserved mode of head segmentation in arthropods revealed by the expression pattern of *Hox* genes in a spider. *Proc. Natl. Acad. Sci. USA* **95**, 10665–10670.

Dasen, J. S., and Rosenfeld, M. G. (1999). Combinatorial codes in signaling and synergy: Lessons from pituitary development. *Curr. Opin. Genet. Dev.* **9**, 566–574.

Dasen, J. S., O'Connell, S. M., Flynn, S. E., Treier, M., Gleiberman, A. S., Szeto, D. P., Hooshmand, F., Aggarwal, A. K., and Rosenfeld, M. G. (1999). Reciprocal interactions of Pit1 and GATA2 mediate signaling gradient-induced determination of pituitary cell types. *Cell* **97**, 587–598.

Davidson, E. H. (1968). "Gene Activity in Early Development." Academic Press, New York.

Davidson, E. H. (1986). "Gene Activity in Early Development." Academic Press, Orlando, Florida.

Davidson, E. H. (1989). Lineage-specific gene expression and the regulative capacities of the sea urchin embryo: A proposed mechanism. *Development* **105**, 421–445.

Davidson, E. H. (1990). How embryos work: A comparative view of diverse modes of cell fate specification. *Development* **108**, 365–389.

Davidson, E. H. (1991). Spatial mechanisms of gene regulation in metazoan embryos. *Development* **113**, 1–26.

Davidson, E. H. (1993). Later embryogenesis: Regulatory circuitry in morphogenetic fields. *Development* **118**, 665–690.

Davidson, E. H., and Britten, R. J. (1971). Note on the control of gene expression during development. *J. Theor. Biol.* **32**, 123–130.

Davidson, E. H., Peterson, K., and Cameron, R. A. (1995). Origin of the adult bilaterian body plans: Evolution of developmental regulatory mechanisms. *Science* **270**, 1319–1325.

Davidson, E. H., Cameron, R. A., and Ransick, A. (1998). Specification of cell fate in the sea urchin embryo: Summary and some proposed mechanisms. *Development* **125**, 3269–3290.

de Celis, J. F. (1998). Positioning and differentiation of veins in the *Drosophila* wing. *Int. J. Dev. Biol.* **42**, 335–343.

de Celis, J. F., and Barrio, R. (2000). Function of the *spalt/spalt-related* gene complex in positioning the veins in the *Drosophila* wing. *Mech. Dev.* **91**, 31–41.

de Celis, J. F., Llimargas, M., and Casanova, J. (1995). Ventral veinless, the gene encoding the Cf1A transcription factor, links positional information and cell differentiation during embryonic and imaginal development in *Drosophila melanogaster. Development* **121**, 3405–3416.

de Rosa, R., Grenier, J. K., Andreeva, T., Cook, C. E., Adoutte, A., Akam, M., Carroll, S. B., and Balavoine, G. (1999). *Hox* genes in brachiopods and priapulids and protostome evolution. *Nature* **399**, 772–776.

Di Bernardo, M., Russo, R., Oliveri, P., Melfi, R., and Spinelli, G. (1995). Homeobox-containing gene transiently expressed in a spatially restricted pattern in the early sea urchin embryo. *Proc. Natl. Acad. Sci. USA* **92**, 8180–8184.

Di Gregorio, A., and Levine, M. (1999). Regulation of *Ci-tropomyosin-like*, a *Brachyury* target gene in the ascidian, *Ciona intestinalis*. *Development* **126**, 5599–5609.

DiMattia, G. E., Rhodes, S. J., Krones, A., Carrière, C., O'Connell, S., Kalla, K., Arias, C., Sawchenko, P., and Rosenfeld, M. G. (1997). The PIT-1 gene is regulated by distinct early and late pituitary-specific enhancers. *Dev. Biol.* **182**, 180–190.

Dollé, P., Izpisúa-Belmonte, J. C., Brown, J. M., Tickle, C., and Duboule, D. (1991). *Hox-4* genes and the morphogenesis of mammalian genitalia. *Genes Dev.* **5**, 1767–1776.

Donoviel, D. B., Shield, M. A., Buskin, J. N., Haugen, H. S., Clegg, C. H., and Hauschka, S. D. (1996). Analysis of muscle creatine kinase gene regulatory elements in skeletal and cardiac muscles of transgenic mice. *Mol. Cell. Biol.* **16**, 1649–1658.

Dover, G. (1982). Molecular drive: A cohesive mode of species evolution. *Nature* **299**, 111–117.

Dover, G. A. (1987). DNA turnover and the molecular clock. *J. Mol. Evol.* **26**, 47–58.

Driever, W., and Nüsslein-Volhard, C. (1988). A gradient of *bicoid* protein in *Drosophila* embryos. *Cell* **54**, 83–93.

Dubnicoff, T., Valentine, S. A., Chen, G. Q., Shi, T., Lengyel, J. A., Paroush, Z., and Courey, A. J. (1997). Conversion of Dorsal from an activator to a repressor by the global corepressor Groucho. *Genes Dev.* **11**, 2952–2957.

Dudley, A. T., and Tabin, C. J. (2000) Constructive antagonism in limb development. *Curr. Opin. Genet. Dev.* **10**, 387–392.

Dupé, V., Davenne, M., Brocard, J., Dollé, P., Mark, M., Dierich, A., Chambon, P., and Rijli, F. M. (1997). *In vivo* functional analysis of the *Hoxa1* 3′ retinoic acid response element (3′RARE). *Development* **124**, 399–410.

Durocher, D., and Nemer, M. (1998). Combinatorial interactions regulating cardiac transcription. *Dev. Genet.* **22**, 250–262.

Edgar, L. G., and McGhee, J. D. (1986). Embryonic expression of a gut-specific esterase in *Caenorhabditis elegans*. *Dev. Biol.* **114**, 109–118.

Edgar, L. G., Wolf, N., and Wood, W. B. (1994). Early transcription in *Caenorhabditis elegans* embryos. *Development* **120**, 443–451.

Eernisse, D. J. (1997). "Arthropod and Annelid Relationships Re-Examined." Chapman & Hall, London.

Egan, C. R., Chung, M. A., Allen, F. L., Heschl, M. F. P., Van Buskirk, C. L., and McGhee, J. D. (1995). A gut-to-pharynx/tail switch in embryonic expression of the *Caenorhabditis elegans ges-1* gene centers on two GATA sequences. *Dev. Biol.* **170**, 397–419.

Emily-Fenouil, F., Ghiglione, C., Lhomond, G., Lepage, T., and Gache, C. (1998). GSK3β/shaggy mediates patterning along the animal-vegetal axis of the sea urchin embryo. *Development* **125**, 2489–2498.

Epstein, M., Pillemer, G., Yelin, R., Yisraeli, J. K., and Fainsod, A. (1997). Patterning of the embryo along the anterior-posterior axis: The role of the *caudal* genes. *Development* **124**, 3805–3814.

Ericson, J., Norlin, S., Jessell, T. M., and Edlund, T. (1998). Integrated FGF and BMP signaling controls the progression of progenitor cell differentiation and the emergence of pattern in the embryonic anterior pituitary. *Development* **125**, 1005–1015.

Erives, A., and Levine, M. (2000a). Characterization of a maternal T-box gene in *Ciona intestinalis. Dev. Biol.*, **225**, 169–178.

Erives, A., and Levine, M. (2000b). *cis*-Regulation of the *Ciona* prototypical chordate collagen, *CiCOL1* in the ascidian tail. Submitted for publication.

Erives, A., Corbo, J. C., and Levine, M. (1998). Lineage-specific regulation of the *Ciona snail* gene in the embryonic mesoderm and neuroectoderm. *Dev. Biol.* **194**, 213–225.

Ernst, S. G., Hough-Evans, B. R., Britten, R. J., and Davidson, E. H. (1980). Limited complexity of the RNA in micromeres of sixteen-cell sea urchin embryos. *Dev. Biol.* **79**, 119–127.

Ettensohn, C. A. (1992). Cell interactions and mesodermal cell fates in the sea urchin embryo. *Development* **116S**, 43–51.

Evans, S. M. (1999). Vertebrate *tinman* homologues and cardiac differentiation. *Semin. Cell Dev. Biol.* **10**, 73–83.

Evans, T., Felsenfeld, G., and Reitman, M. (1990). Control of globin gene transcription. *Annu. Rev. Cell Biol.* **6**, 95–124.

Evans, T. C., Crittenden, S. L., Kodoyianni, V., and Kimble, J. (1994). Translational control of maternal *glp-1* mRNA establishes an asymmetry in the *C. elegans* embryo. *Cell* **77**, 183–194.

Fabre-Suver, C., and Hauschka, S. D. (1996). A novel site in the muscle creatine kinase enhancer is required for expression in skeletal but not cardiac muscle. *J. Biol. Chem.* **271**, 4646–4652.

Ferretti, E., Marshall, H., Pöpperl, H., Maconochie, M., Krumlauf, R., and Blasi, F. (2000). Segmental expression of *Hoxb2* in r4 requires two separate sites that integrate cooperative interactions between Prep1, Pbx and Hox proteins. *Development* **127**, 155–166.

Finkelstein, R., and Boncinelli, E. (1994). From fly head to mammalian forebrain: The story of *otd* and *Otx. Trends Genet.* **10**, 310–315.

Finnerty, J. R. (1998). Homeoboxes in sea anemones and other nonbilaterian animals: Implications for the evolution of the Hox cluster and the zootype. *Curr. Topics Dev. Biol.* **40**, 211–254.

Finnerty, J. R., and Martindale, M. Q. (1998). The evolution of the *Hox* cluster: Insights from outgroups. *Curr. Opin. Genet. Dev.* **8**, 681–687.

Finnerty, J. R., and Martindale, M. Q. (1999). Ancient origins of axial patterning genes: *Hox* genes and *ParaHox* genes in the Cnidaria. *Evol. Dev.* **1**, 16–23.

Firulli, A. B., and Olson, E. N. (1997). Modular regulation of muscle gene transcription: A mechanism for muscle cell diversity. *Trends Genet.* **13**, 364–369.

Fisher, A., and Caudy, M. (1998). The function of hairy-related bHLH repressor proteins in cell fate decisions. *BioEssays* **20**, 298–306.

Fisher, A. L., Ohsako, S., and Caudy, M. (1996). The WRPW motif of the hairy-related basic helix-loop-helix repressor proteins acts as a 4–amino-acid transcription repression and protein-protein interaction domain. *Mol. Cell. Biol.* **16**, 2670–2677.

Fishman, M. C., and Chien, K. R. (1997) Fashioning the vertebrate heart: Earliest embryonic decisions. *Development* **124**, 2099–2117.

Fishman, M. C., and Olson, E. N. (1997). Parsing the heart: Genetic modules for organ assembly. *Cell* **91**, 153–156.

Florence, B., Guichet, A., Ephrussi, A., and Laughon, A. (1997). Ftz-F1 is a cofactor in Ftz activation of the *Drosophila engrailed* gene. *Development* **124**, 839–847.

Fortini, M. E., and Rubin, G. M. (1990). Analysis of *cis*-acting requirements of the *Rh3* and *Rh4* genes reveals a bipartite organization to rhodopsin promoters in *Drosophila melanogaster. Genes Dev.* **4**, 444–463.

Fraidenraich, D., Lang, R., and Basilico, C. (1998). Distinct regulatory elements govern *Fgf4* gene expression in the mouse blastocyst, myotomes, and developing limb. *Dev. Biol.* **204**, 197–209.

Frise, E., Knoblich, J. A., Younger-Shepherd, S., Jan. L. Y., and Jan, Y. N. (1996). The *Drosophila* Numb protein inhibits signaling of the Notch receptor during cell-cell interaction in sensory organ lineage. *Proc. Natl. Acad. Sci. USA* **93**, 11925–11932.

Fu, Y., Yan, W., Mohun, T. J., and Evans, S. M. (1998). Vertebrate *tinman* homologues *XNkx2–3* and *XNkx2–5* are required for heart formation in a functionally redundant manner. *Development* **125**, 4439–4449.

Fujioka, M., Emi-Sarker, Y., Yusibova, G. L., Goto, T., and Jaynes, J. B. (1999). Analysis of an *even-skipped* rescue transgene reveals both composite and discrete neuronal and early blastoderm enhancers, and multi-stripe positioning by gap gene repressor gradients. *Development* **126**, 2527–2538.

Fujiwara, S., Corbo, J. C., and Levine, M. (1998). The Snail repressor establishes a muscle/notochord boundary in the *Ciona* embryo. *Development* **125**, 2511–2520.

Fukushige, T., Schroeder, D. F., Allen, F. L., Goszczynski, B., and McGhee, J. D. (1996). Modulation of gene expression in the embryonic digestive tract of *C. elegans. Dev. Biol.* **178**, 276–288.

Fukushige, T., Hawkins, M. G., and McGhee, J. D. (1998). The GATA-factor *elt-2* is essential for formation of the *Caenorhabditis elegans* intestine. *Dev. Biol.* **198**, 286–302.

Fukushige, T., Hendzel, M. J., Bazett-Jones, D. P., and McGhee, J. D. (1999). Direct visualization of the *elt-2* gut-specific GATA factor binding to a target promoter inside the living *Caenorhabditis elegans* embryo. *Proc. Natl. Acad. Sci. USA* **96**, 11883–11888.

Gajewski, K., Kim, Y., Lee, Y. M., Olson, E. N., and Schulz, R. A. (1997). *D-mef2* is a target for Tinman activation during *Drosophila* heart development. *EMBO J.* **16**, 515–522.

Gajewski, K., Kim, Y., Choi, C. Y., and Schulz, R. A. (1998). Combinatorial control of *Drosophila mef2* gene expression in cardiac and somatic muscle cell lineages. *Dev. Genes Evol.* **208**, 382–392.

Gajewski, K., Fossett, N., Molkentin, J. D., and Schulz, R. A. (1999). The zinc finger proteins Pannier and GATA4 function as cardiogenic factors in *Drosophila*. *Development* **126**, 5679–5688.

Gamer, L. W., and Wright, C. V. E. (1993). Murine *Cdx-4* bears striking similarities to the *Drosophila caudal* gene in its homeodomain sequence and early expression pattern. *Mech. Dev.* **43**, 71–81.

Gan, L., Wessel, G. M., and Klein, W. H. (1990). Regulatory elements from the related *Spec* genes of *Strongylocentrotus purpuratus* yield different spatial patterns with a lacZ reporter gene. *Dev. Biol.* **142**, 346–359.

Garcia-Bellido, A. (1981). From the gene to the pattern: Chaete differentiation. *In* "Cellular Controls on Differentiation." (C. D. Lloyd and D. A. Rees, Eds.), pp. 281–301. Academic Press, New York.

Garcia-Bellido, A., and de Celis, J. F. (1992). Developmental genetics of the venation pattern of *Drosophila. Annu. Rev. Genet.* **26**, 275–302.

Garcia-Fernandez, J., and Holland, P. W. H. (1994). Archetypal organization of the amphioxus *Hox* gene cluster. *Nature* **370**, 563–566.

García-García, M. J., Ramain, P., Simpson, P., and Modolell, J. (1999). Different contributions of *pannier* and *wingless* to the patterning of the dorsal mesothorax of *Drosophila. Development* **126**, 3523–3532.

Garey, J. R., Near, T. J., Nonnemacher, M. R., and Nadler, S. A. (1996). Molecular evidence for Acanthocephala as a subtaxon of Rotifera. *J. Mol. Evol.* **43**, 287–292.

Garrell, J., and Modolell, J. (1990). The *Drosophila extramacrochaetae* locus, an antagonist of proneural genes that, like these genes, encodes a helix-loop-helix protein. *Cell* **61**, 39–48.

Garrity, P. A., Chen, D., Rothenberg, E. V., and Wold, B. J. (1994). Interleukin-2 transcription is regulated *in vivo* at the level of coordinated binding of both constitutive and regulated factors. *Mol. Cell. Biol.* **14**, 2159–2169.

Gavalas, A., and Krumlauf, R. (2000) Retinoid signalling and hindbrain patterning. *Curr. Opin. Genet. Dev.* **10**, 380–386.

Gavis, E. R., and Lehmann, R. (1994). Translational regulation of *nanos* by RNA localization. *Nature* **369**, 315–318.

Gehring, W. J., and Ikeo, K. (1999). *Pax6*: Mastering eye morphogenesis and eye evolution. *Trends Genet.* **15**, 371–377.

Gellon, G., and McGinnis, W. (1998). Shaping animal body plans in development and evolution by modulation of *Hox* expression patterns. *BioEssays* **20**, 116–125.

Gho, M., Bellaïche, Y., and Schweisguth, F. (1999) Revisiting the *Drosophila* microchaete lineage: A novel intrinsically asymmetric cell division generates a glial cell. *Development* **126**, 3573–3584

Gibson-Brown, J. J., Agulnik, S. I., Silver, L. M., Niswander, L., and Papaioannou, V. E. (1998). Involvement of T-box genes *Tbx2–Tbx5* in vertebrate limb specification and development. *Development* **125**, 2499–2509.

Giribet, G., and Ribera, C. (1998). The position of arthropods in the animal kingdom: A search for a reliable outgroup for internal arthropod phylogeny. *Mol. Phylogenet. Evol.* **9**, 481–488.

Gisselbrecht, S., Skeath, J. B., Doe, C. Q., and Michelson, A. M. (1996). *heartless* encodes a fibroblast growth factor receptor (DFR1/DRGF-R2) involved in the directional migration of early mesodermal cells in the *Drosophila* embryo. *Genes Dev.* **10**, 3003–3017.

Glardon, S., Holland, L. Z., Gehring, W. J., and Holland, N. D. (1998). Isolation and developmental expression of the amphioxus *Pax-6* gene (*AmphiPax-6*): Insights into eye and photoreceptor evolution. *Development* **125**, 2701–2710.

Godin, R. E., Urry, L. A., and Ernst, S. G. (1996). Alternative splicing of the *Endo16* transcript produces differentially expressed mRNAs during sea urchin gastrulation. *Dev. Biol.* **179**, 148–159.

Goff, D. J., and Tabin, C. J. (1997). Analysis of *Hoxd-13* and *Hoxd-11* misexpression in chick limb buds reveals that *Hox* genes affect both bone condensation and growth. *Development* **124**, 627–636.

Goldstein, B. (1992). Induction of gut in *Caenorhabditis elegans* embryos. *Nature* **357**, 255–257.

Goldstein, B. (1993). Establishment of gut fate in the E lineage of *C. elegans*: The roles of lineage-dependent mechanisms and cell interactions. *Development* **118**, 1267–1277.

Goldstein, B. (1995). An analysis of the response to gut induction in the *C. elegans* embryo. *Development* **121**, 1227–1236.

Goldstein, B., and Hird, S. N. (1996). Specification of the anteroposterior axis in *Caenorhabditis elegans*. *Development* **122**, 1467–1474.

Goldstein, R. E., Jiménez, G., Cook, O., Gur, D., and Paroush, Z. (1999). Huckebein repressor activity in *Drosophila* terminal patterning is mediated by Groucho. *Development* **126**, 3747–3755.

Gómez-Skarmeta, J. L., and Modolell, J. (1996). *araucan* and *caupolican* provide a link between compartment subdivisions and patterning of sensory organs and veins in the *Drosophila* wing. *Genes Dev.* **10**, 2935–2945.

Gómez-Skarmeta, J. L., Rodríguez, I., Martínez, C., Culí, J., Ferrés-Marcó, D., Beamonte, D., and Modolell, J. (1995). *cis*-Regulation of *achaete* and *scute*: Shared enhancer-like elements drive their coexpression in proneural clusters of the imaginal discs. *Genes Dev.* **9**, 1869–1882.

Gómez-Skarmeta, J.-L., Diez del Corral, R., de la Calle-Mustienes, E., Ferrés-Marcó, D., and Modolell, J. (1996). *araucan* and *caupolican*, two members of the novel iroquois complex, encode homeoproteins that control proneural and vein-forming genes. *Cell* **85**, 95–105.

González-Crespo, S., and Morata, G. (1996). Genetic evidence for the subdivision of the arthropod limb into coxopodite and telopodite. *Development* **122**, 3921–3928.

Gorfinkiel, N., Morata, G., and Guerrero, I. (1997). The homeobox gene *Distal-less* induces ventral appendage development in *Drosophila*. *Genes Dev.* **11**, 2259–2271.

Goto, S., and Hayashi, S. (1997). Specification of the embryonic limb primordium by graded activity of Decapentoplegic. *Development* **124**, 125–132.

Gould, A., Itasaki, N., and Krumlauf, R. (1998). Initiation of rhombomeric *Hoxb4* expression requires induction by somites and a retinoid pathway. *Neuron* **21**, 39–51.

Grapin-Botton, A., and Melton, D. A. (2000). Endoderm development—from patterning to organogenesis. *Trends Genet.* **16**, 124–130.

Gray, S., and Levine, M. (1996). Short-range transcriptional repressors mediate both quenching and direct repression within complex loci in *Drosophila*. *Genes Dev.* **10**, 700–710.

Gray, S., Szymanski, P., and Levine, M. (1994). Short-range repression permits multiple enhancers to function autonomously within a complex promoter. *Genes Dev.* **8**, 1829–1838.

Grayson, J., Bassel-Duby, R., and Williams, R. S. (1998). Collaborative interactions between MEF-2 and Sp1 in muscle-specific gene regulation. *J. Cell. Biochem.* **70**, 366–375.

Grenier, J. K., Garber, T. L., Warren, R., Whitington, P. M., and Carroll, S. (1997). Evolution of the entire arthropod *Hox* gene set predated the origin and radiation of the onychophoran/arthropod clade. *Curr. Biol.* **7**, 547–553.

Grépin, C., Robitaille, L., Antakly, T., and Nemer, M. (1995). Inhibition of transcription factor GATA-4 expression blocks *in vitro* cardiac muscle differentiation. *Mol. Cell. Biol.* **15**, 4095–4102.

Grindley, J. C., Davidson, D. R., and Hill, R. E. (1995). The role of *Pax-6* in eye and nasal development. *Development* **121**, 1433–1442.

Grow, M. W., and Krieg, P. A. (1998). Tinman function is essential for vertebrate heart development: Elimination of cardiac differentiation by dominant inhibitory mutants of the *tinman*-related genes, *XNkx2–3* and *XNkx2–5*. *Dev. Biol.* **204**, 187–196.

Gupta, M. P., Zak, R., Libermann, T. A., and Gupta, M. P. (1998). Tissue-restricted expression of the cardiac α-myosin heavy chain gene is controlled by a downstream repressor element containing a palindrome of two Ets-binding sites. *Mol. Cell. Biol.* **18**, 7243–7258.

Gurdon, J. B., Lemaire, P., and Kato, K. (1993). Community effects and related phenomena in development. *Cell* **75**, 831–834.

Guss, K. A., and Ettensohn, C. A. (1997). Skeletal morphogenesis in the sea urchin embryo: Regulation of primary mesenchyme gene expression and skeletal rod grown by ectoderm-derived cues. *Development* **124**, 1899–1908.

Gutjahr, T., Vanario-Alonso, C. E., Pick, L., and Noll, M. (1994). Multiple regulatory elements direct the complex expression pattern of the *Drosophila* segmentation gene *paired*. *Mech. Dev.* **48**, 119–128.

Häder, T., La Rosée, A., Ziebold, U., Busch, M., Taubert, H., Jäckle, H., and Rivera-Pomar, R. (1998). Activation of posterior pair-rule stripe expression in response to maternal *caudal* and zygotic *knirps* activities. *Mech. Dev.* **71**, 177–186.

Haenlin, M., Cubadda, Y., Blondeau, F., Heitzler, P., Lutz, Y., Simpson, P., and Ramain, P. (1997). Transcriptional activity of Pannier is regulated negatively by heterodimerization of the GATA DNA-binding domain with a cofactor encoded by the *u-shaped* gene of *Drosophila*. *Genes Dev.* **11**, 3096–3108.

Halanych, K. M. (1995). The phylogenetic position of the pterobranch hemichordates based on 18S rDNA sequence data. *Mol. Phylogenet. Evol.* **4**, 72–76.

Halanych, K. M. (1996). Testing hypotheses of chaetognath origins: Long branches revealed by 18S ribosomal DNA. *Syst. Biol.* **45**, 223–246.

Halder, G., Callaerts, P., and Gehring, W. J. (1995). Induction of ectopic eyes by targeted expression of the *eyeless* gene in *Drosophila*. *Science* **267**, 1788–1792.

Halder, G., Callaerts, P., Flister, S., Walldorf, U., Kloter, U., and Gehring, W. J. (1998). Eyeless initiates the expression of both *sine oculis* and *eyes absent* during *Drosophila* compound eye development. *Development* **125**, 2181–2191.

Halder, G., Polaczyk, P., Kraus, M. E., Hudson, A., Kim, J., Laughon, A., and Carroll, S. (1998). The Vestigial and Scalloped proteins act together to directly regulate wing-specific gene expression in *Drosophila*. *Genes Dev.* **12**, 3900–3909.

Harada, Y., Yasuo, H., and Sato, N. (1995). A sea urchin homologue of the chordate *Brachyury* (*T*) gene is expressed in the secondary mesenchyme founder cells. *Development* **121**, 2747–2754.

Hardin, S. H., Carpenter, C. D., Hardin, P. E., Bruskin, A. M., and Klein, W. H. (1985). Structure of the *Spec1* gene encoding a major calcium-binding protein in the embryonic ectoderm of the sea urchin, *Strongylocentrotus purpuratus*. *J. Mol. Biol.* **186**, 243–255.

Harland, R., and Gerhart, J. (1997). Formation and function of Spemann's organizer. *Annu. Rev. Cell Dev. Biol.* **13**, 611–667.

Harvey, R. P. (1996). *NK-2* homeobox genes and heart development. *Dev. Biol.* **178**, 203–216.

Henry, J. J., Klueg, K. M., and Raff, R. A. (1992). Evolutionary dissociation between cleavage, cell lineage and embryonic axes in sea urchin embryos. *Development* **114**, 931–938.

Hepker, J., Blackman, R. K., and Holmgren, R. (1999). Cubitus interruptus is necessary but not sufficient for direct activation of a wing-specific *decapentaplegic* enhancer. *Development* **126**, 3669–3677.

Herrmann, B. G., and Kispert, A. (1994). The *T* genes in embryogenesis. *Trends Genet.* **10**, 280–286.

Herschbach, B. M., Arnaud, M. B., and Johnson, A. D. (1994). Transcriptional repression directed by the yeast alpha-2 protein *in vitro*. *Nature* **370**, 309–311.

Hill, R. E., Favor, J., Hogan, B. L. M., Ton, C. C. T., Saunders, G. F., Hanson, I. M., Prosser, J., Jordan, T., Hastie, N. D., and van Heyningen, V. (1991). Mouse *Small eye* results from mutations in a paired-like homeobox-containing gene. *Nature* **354**, 522–525.

Hinegardner, R. (1974). Cellular DNA content of the Echinodermata. *Comp. Biochem. Physiol.* **49B**, 219–226.

Hirano, T., and Nishida, H. (1997). Developmental fates of larval tissues after metamorphosis in ascidian *Halocynthia roretzi.* I. Origin of mesodermal tissues of the juvenile. *Dev. Biol.* **192**, 199–210.

Hoch, M., Gerwin, N., Taubert, H., and Jäckle, H. (1992). Competition for overlapping sites in the regulatory region of the *Drosophila* gene *Krüppel. Science* **256**, 94–97.

Holland, P. W. H., Garcia-Fernandez, J., Williams, N. A., and Sidow, A. (1994). Gene duplications and the origins of vertebrate development. *Development* **120S**, 125–133.

Holland, P. W. H., Koschorz, B., Holland, L. Z., and Herrmann, B. G. (1995). Conservation of *Brachyury (T)* genes in amphioxus and vertebrates: Developmental and evolutionary implications. *Development* **121**, 4283–4291.

Hoppler, S., and Bienz, M. (1994). Specification of a single cell type by a *Drosophila* homeotic gene. *Cell* **76**, 689–702.

Hori, S., Saitoh, T., Matsumoto, M., Makabe, K. W., and Nishida, H. (1997). *Notch* homologue from *Halocynthia roretzi* is preferentially expressed in the central nervous system during ascidian embryogenesis. *Dev. Genes Evol.* **207**, 371–380.

Hörstadius, S. (1939). The mechanics of sea urchin development, studied by operative methods. *Rev. Cambridge Phil. Soc.* **14**, 132–179.

Hörstadius, S. (1973). "Experimental Embryology of Echinoderms." Clarendon Press, Oxford.

Hoskins, R. A. *et al.* (2000). A BAC-based physical map of the major autosomes of *Drosophila melanogaster. Science* **287**, 2271–2274.

Hotta, K., Takahashi, H., Erives, A., Levine, M., and Satoh, N. (1999). Temporal expression patterns of 39 *Brachyury*-downstream genes associated with notochord formation in the *Ciona intestinalis* embryo. *Dev. Growth Differ.* **41**, 657–664.

Hough-Evans, B. R., Jacob-Lorena, M., Cummings, M. R., Britten, R. J., and Davidson, E. H. (1980). Complexity of RNA in eggs of *Drosophila melanogaster* and *Musca domestica. Genetics* **95**, 81–94.

Hough-Evans, B. R., Franks, R. R., Zeller, R. W., Britten, R. J., and Davidson, E. H. (1990). Negative spatial regulation of the lineage specific *CyIIIa* actin gene in the sea urchin embryo. *Development* **110**, 41–50.

Huang, A. M., Rusch, J., and Levine, M. (1997). An anteroposterior Dorsal gradient in the *Drosophila* embryo. *Genes Dev.* **11**, 1963–1973.

Huang, J.-D., Schwyter, D. H., Shirokawa, J. M., and Courey, A. J. (1993). The interplay between multiple enhancer and silencer elements defines the pattern of *decapentaplegic* expression. *Genes Dev.* **7**, 694–704.

Huang, J.-D., Dubnicoff, T., Liaw, G.-J., Bai, Y., Valentine, S. A., Shirokawa, J. M., Lengyel, J. A., and Courey, A. J. (1995). Binding sites for transcription factor NTF-1/Elf-1 contribute to the ventral repression of *decapentaplegic*. *Genes Dev.* **9**, 3177–3189.

Huang, L., Li, X., El-Hodiri, H. M., Dayal, S., Wikramanayake, A. H., and Klein, W. H. (2000). Involvement of Tcf/Lef in establishing cell types along the animal-vegetal axis of sea urchins. *Dev. Genes Evol.* **210**, 73–81.

Hunt, J. A. (1974). Rate of synthesis and half-life of globin messenger ribonucleic acid. Rate of synthesis of globin messenger ribon ucleic acid calculated from data of cell haemoglobin content. *Biochem. J.* **138**, 499–510.

Hutter, H., and Schnabel, R. (1994). glp-1 and induction establishing embryonic axes in *C. elegans*. *Development* **120**, 2051–2064.

Hyman, L. H. (1940). "The Invertebrates: Protozoa through Ctenophora." McGraw-Hill Book Company, Inc., New York, London.

Inazawa, T., Okamura, Y., and Takahashi, K. (1998). Basic fibroblast growth factor induction of neuronal ion channel expression in ascidian ectodermal blastomeres. *J. Physiol. London* **511**, 347–359.

Ilagan, J. G., Cvekl, A., Kantorow, M., Piatigorsky, J., and Sax, C. M. (1999). Regulation of αA-crystallin gene expression. Lens specificity achieved through the differential placement of similar transcriptional control elements in mouse and chicken. *J. Biol. Chem.* **274**, 19973–19978.

Ip, Y. T., Park, R. E., Kosman, D., Yazdanbakhsh, K., and Levine, M. (1992a). dorsal-twist interactions establish snail expression in the presumptive mesoderm of the *Drosophila* embryo. *Genes Dev.* **6**, 1518–1530.

Ip, Y. T., Park, R. E., Kosman, D., Bier, E., and Levine, M. (1992b). The dorsal gradient morphogen regulates stripes of rhomboid expression in the presumptive neuroectoderm of the *Drosophila* embryo. *Genes Dev.* **6**, 1728–1739.

Irvine, K. D., Helfand, S. L., and Hogness, D. S. (1991). The large upstream control region of the *Drosophila* homeotic gene *Ultrabithorax*. *Development* **111**, 407–424.

Isaac, A., Cohn, M. J., Ashby, P., Ataliotis, P., Spicer, D. B., Cooker, J., and Tickle, C. (2000) FGF and genes encoding transcription factors in early limb specification. *Mech. Dev.* 93, 41–48.

Isaac, D. D., and Andrew, D. J. (1996). Tubulogenesis in *Drosophila*: A requirement for the *trachealess* gene product. *Genes Dev.* **10**, 103–117.

Jacobs, Y., Schnabel, C. A., and Cleary, M. L. (1999). Trimeric association of Hox and TALE homeodomain proteins mediates *Hoxb2* hindbrain enhancer activity. *Mol. Cell. Biol.* **19**, 5134–5142.

Jagla, K., Frasch, M., Jagla, T., Dretzen, G., Bellard, F., and Bellard, M. (1997). *ladybird*, a new component of the cardiogenic pathway in *Drosophila* required for diversification of heart precursors. *Development* **124**, 3471–3479.

James, R., Erler, T., and Kazenwadel, J. (1994). Structure of the murine homeobox gene *cdx-2*. *J. Biol. Chem.* **269**, 15229–15237.

Jan, Y. N., and Jan, L. Y. (1994). Genetic control of cell fate specification in *Drosophila* peripheral nervous system. *Annu. Rev. Genet.* **28**, 373–393.

Janvier, P. (1996). "Early Vertebrates." Clarendon Press, Oxford.

Jean, D., Ewan, K., and Gruss, P. (1998). Molecular regulators involved in vertebrate eye development. *Mech. Dev.* **76**, 3–18.

Jiang, J., and Levine, M. (1993). Binding affinities and cooperative interactions with bHLH activators delimit threshold responses to the dorsal gradient morphogen. *Cell* **72**, 741–752.

Jiang, J., and Struhl, G. (1996). Complementary and mutually exclusive activities of *decapentaplegic* and *wingless* organize axial patterning during *Drosophila* leg development. *Cell* **86**, 401–409.

Jiang, J., Cai, H., Zhou, Q., and Levine, M. (1993). Conversion of a *dorsal*-dependent silencer into an enhancer: Evidence for *dorsal* corepressors. *EMBO J.* **12**, 3201–3209.

Jiménez, G., Paroush, Z., and Ish-Horowicz, D. (1997). Groucho acts as a corepressor for a subset of negative regulators, including Hairy and Engrailed. *Genes Dev.* **11**, 3072–3082.

Johnson, R. L., and Tabin, C. J. (1997). Molecular models for vertebrate limb development. *Cell* **90**, 979–990.

Joyner, A. L. (1996). *Engrailed, Wnt* and *Pax* genes regulate midbrain-hindbrain development. *Trends Genet.* **12**, 15–20.

Kalb, J. M., Lau, K. K., Goszczynski, B., Fukushige, T., Moons, D., Okkema, P. G., and McGhee, J. D. (1998). *pha-4* is *Ce-fkh-1*, a *fork head*/HNF-3α,β,γ homolog that functions in organogenesis of the *C. elegans* pharynx. *Development* **125**, 2171–2180.

Kammandel, B., Chowdhury, K., Stoykova, A., Aparicio, S., Brenner, S., and Gruss, P. (1999). Distinct *cis*-Essential modules direct the time-space pattern of the *Pax6* gene activity. *Dev. Biol.* **205**, 79–97.

Kanzler, B., Prin, F., Thelu, J., and Dhouailly, D. (1997). *CHOXC-8* and *CHOXD-13* expression in embryonic chick skin and cutaneous appendage specification. *Dev. Dynam.* **210**, 274–287.

Kapoun, A. M., and Kaufman, T. C. (1995). A functional analysis of 5', intronic and promoter regions of the homeotic gene *proboscipedia* in *Drosophila melanogaster*. *Development* **121**, 2127–2141.

Kasmir, J., Zhu, J., and Rothman, J. H. (2000). Dual role of POP-1 in repression and *Wnt*-dependent activation of endoderm development. In preparation.

Katsuyama, Y., Sato, Y., Wada, S., and Saiga, H. (1999). Ascidian tail formation requires *caudal* function. *Dev. Biol.* **213**, 257–268.

Kavaler, J., Fu, W., Duan, H., Noll, M., and Posakony, J. W. (1999). An essential role for the *Drosophila Pax2* homolog in the differentiation of adult sensory organs. *Development* **126**, 2261–2272.

Kawaminani, S., and Nishida, H. (1997). Induction of trunk lateral cells, the blood cell precursors, during ascidian embryogenesis. *Dev. Biol.* **181**, 14–20.

Kehl, B. T., Cho, K.-O., and Choi, K.-W. (1998). *mirror*, a *Drosophila* homeobox gene in the *iroquois* complex, is required for sensory organ and alula formation. *Development* **125**, 1217–1227.

Kenny, A. P., Kozlowski, D., Oleksyn, D. W., Angerer, L. M., and Angerer, R. C. (1999). SpSoxB1, a maternally encoded transcription factor asymmetrically distributed among early sea urchin blastomeres. *Development* **126**, 5473–5483.

Keys, D. N., Lewis, D. L., Selegue, J. E., Pearson, B. J., Goodrich, L. V., Johnson, R. L., Gates, J., Scott, M. P., and Carroll, S. B. (1999). Recruitment of a *hedgehog* regulatory circuit in butterfly eyespot evolution. *Science* **283**, 532–534.

Khaner, O., and Wilt, F. (1990). The influence of cell interaction and tissue mass on differentiation of sea urchin mesomeres. *Development* **109**, 625–634.

Killian, C. E., and Wilt, F. H. (1989). The accumulation and translation of spicule matrix protein mRNA during sea urchin embryo development. *Dev. Biol.* **133**, 148–156.

Kim, G. J., and Nishida, H. (1999). Suppression of muscle fate by cellular interaction is required for mesenchyme formation during ascidian embryogenesis. *Dev. Biol.* **214**, 9–22.

Kim, J., Sebring, A., Esch, J. J., Kraus, M. E., Vorwerk, K., Magee, J., and Carroll, S. B. (1996). Integration of postitional signals and regulation of wing formation and identity by *Drosophila vestigial* gene. *Nature* **382**, 133–138.

Kim, J., Johnson, K., Chen, H. J., Carroll, S., and Laughon, A. (1997a). *Drosophila* Mad binds to DNA and directly mediates activation of *vestigial* by Decapentaplegic. *Nature* **388**, 304–308.

Kim, J., Magee, J., and Carroll, S. B. (1997b). Intercompartmental signaling and the regulation of *vestigial* expression at the dorsoventral boundary of the developing *Drosophila* wing. *Cold Spring Harbor Symp. Quant. Biol.* **62**, 283–291.

Kim, T. K., and Maniatis, T. (1997). The mechanism of transcriptional synergy of an *in vitro* assembled interferon-β enhanceosome. *Mol. Cell* **1**, 119–129.

Kimelman, D. (1999). Transcriptional regulation in *Xenopus*: A bright and froggy future. *Curr. Opin. Genet. Dev.* **9**, 553–558.

Kingsley, D. M. (1994). The TGF-β-superfamily: New members, new receptors, and new genetic tests of function in different organisms. *Genes Dev.* **8**, 133–146.

Kirchhamer, C. V., and Davidson, E. H. (1996). Spatial and temporal information processing in the sea urchin embryo: Modular and intramodular organization of the *CyIIIa* gene *cis*-regulatory system. *Development* **122**, 333–348.

Kirchhamer, C. V., Bogarad, L. D., and Davidson, E. H. (1996). Developmental expression of synthetic *cis*-regulatory systems composed of spatial control elements from two different genes. *Proc. Natl. Acad. Sci. USA* **93**, 13849–13854.

Kispert, A., and Herrmann, B.G. (1994). Immunohistochemical analysis of the *Brachyury* protein in wild-type and mutant mouse embryos. *Dev. Biol.* **161**, 179–193.

Kispert, A., Herrmann, B. G., Leptin, M., and Reuter, R. (1994). Homologs of the mouse *Brachyury* gene are involved in the specification of posterior terminal structures in *Drosophila, Tribolium,* and *Locusta. Genes Dev.* **8**, 2137–2150.

Klein, P. S., and Melton, D. A. (1996). A molecular mechanism for the effect of lithium on development. *Proc. Natl. Acad. Sci. USA* **93**, 8455–8459.

Kmita, M., van der Hoeven, F., Zákány, J., Krumlauf, R., and Duboule, D. (2000). Mechanisms of *Hox* gene colinearity: Transposition of the anterior *Hoxb1* gene into the posterior *HoxD* complex. *Genes Dev.* **14**, 198–211.

Kmita-Cunisse, M., Loosli, F., Bièrne, J., and Gehring, W. J. (1998). Homeobox genes in the ribbonworm *Lineus sanguineus*: Evolutionary implications. *Proc. Natl. Acad. Sci. USA* **95**, 3030–3035.

Knoepfler, P. S., and Eisenman, R. N. (1999). Sin meets NuRD and other tails of rerpression. *Cell* **99**, 447–450.

Kondo, T., Zákány, J., Innis, J. W., and Duboule, D. (1997). Of fingers, toes and penises. *Nature*, **390**, 29.

Kovacs, A. M., and Zimmer, W. E. (1998). Cell-specific transcription of the smooth muscle γ-actin gene requires both positive-and negative-acting *cis* elements. *Gene Expression* **7**, 115–129.

Kozlowski, D. J., Gagnon, M. L., Marchant, J. K., Reynolds, S. D., Angerer, L. M., and Angerer, R. C. (1996). Characterization of a SpAn promoter sufficient to mediate correct spatial regulation along the animal-vegetal axis of the sea urchin embryo. *Dev. Biol.* **176**, 95–107.

Kuchin, S., and Carlson, M. (1998). Functional relationships of Srb10-Srb11 kinase, carboxy-terminal domain kinase CTDK-I, and transcriptional corepressor Ssn6-Tup1. *Mol. Cell. Biol.* **18**, 1163–1171.

Kühnlein, R. P., and Schuh, R. (1996). Dual function of the region-specific homeotic gene *spalt* during *Drosophila* tracheal system development. *Development* **122**, 2215–2223.

Kühnlein, R. P., Brönner, G., Taubert, H., and Schuh, R. (1997). Regulation of *Drosophila spalt* gene expession. *Mech. Dev.* **66**, 107–118.

Kunisch, M., Haenlin, M., and Campos-Ortega, J. A. (1994). Lateral inhibition mediated by the *Drosophila* neurogenic gene Delta is enhanced by proneural proteins. *Proc. Natl. Acad. Sci. USA* **91**, 10139–10143.

Kusakabe, T., Hikosaka, A., and Satoh, N. (1995). Coexpression and promoter function in two muscle actin gene complexes of different structural organization in the ascidian *Halocynthia roretzi. Dev. Biol.* **169**, 461–472.

Kusch, T., and Reuter, R. (1999). Functions for *Drosophila brachyenteron* and *forkhead* mesoderm specification and cell signalling. *Development* **126**, 3991–4003.

Lacalli, T. C. (1996). Frontal eye circuitry, rostral sensory pathways and brain organization in amphioxus larvae: Evidence from 3D reconstructions. *Phil. Trans. R. Soc. London B* **351**, 243–263.

La Rosée, A., Häder, T., Taubert, H., Rivera-Pomar, R., and Jäckle, H. (1997). Mechanism and Bicoid-dependent control of *hairy* stripe 7 expression in the posterior region of the *Drosophila* embryo. *EMBO J.* **16**, 4403–4411.

Langeland, J. A., Attai, S. F., Vorwerk, K., and Carroll, S. B. (1994). Positioning adjacent pair-rule stripes in the posterior *Drosophila* embryo. *Development* **120**, 2945–2955.

Laurent, M. N., Blitz, I. L., Hashimoto, C., Rothbacher, U., and Cho, K. W.-Y. (1997). The *Xenopus* homeobox gene *twin* mediates Wnt induction of *goosecoid* in establishment of Spemann's organizer. *Development* **124**, 4905–4916.

Lawrence, P. A. (1992). "The Making of a Fly: The Genetics of Animal Design." Blackwell Scientific Publications, Oxford, Boston.

Lawrence, P. A., and Struhl, G. (1996). Morphogens, compartments, and pattern: Lessons from *Drosophila*? *Cell* **85**, 951–961.

Lecourtois, M., and Schweisguth, F. (1998). Indirect evidence for Delta-dependent intracellular processing of Notch in *Drosophila* embryos. *Curr. Biol.* **8**, 771–774.

Lecuit, T., and Cohen, S. M. (1997). Proximal-distal axis formation in the *Drosophila* leg. *Nature* **388**, 139–145.

Lee, Y., Shioi, T., Kasahara, H., Jobe, S. M., Wiese, R. J., Markham, B. E., and Izumo, S. (1998). The cardiac tissue-restricted homeobox protein Csx/Nkx2.5 physically associates with the zinc finger protein GATA4 and cooperatively activates atrial natriuretic factor gene expression. *Mol. Cell. Biol.* **18**, 3120–3129.

Lee, Y.-H., Britten, R. J., and Davidson, E. H. (1999). *SM37*, a new skeletogenic gene of the sea urchin embryo isolated by regulatory target site screening. *Dev., Growth Differ.* **41**, 303–312.

Lemaire, P., Garrett, N., and Gurdon, J. B. (1995). Expression cloning of *siamois*, a *Xenopus* homeobox gene expressed in dorsal-vegetal cells of blastulae and able to induce a complete secondary axis. *Cell* **81**, 85–94.

Lemon, B. D., and Freedman, L. P. (1999). Nuclear receptor cofactors as chromatin remodelers. *Curr. Opin. Genet. Dev.* **9**, 499–504.

Leung, B., Hermann, G. J., and Priess, J. R. (1999). Organogenesis of the *Caenorhabditis elegans* intestine. *Dev. Biol.* **216**, 114–134.

Lewis, E. B. (1978). A gene complex controlling segmentation in *Drosophila*. *Nature* **276**, 565–570.

Li, H. S., Yang, J. M., Jacobson, R. D., Pasko, D., and Sundin, O. (1994). *Pax-6* is first expressed in a region of ectoderm anterior to the early neural plate— Implications for stepwise determination of the lens. *Dev. Biol.* **162**, 181–194.

Li, X., Chuang, C. K., Mao, C. A., Angerer, L. M., and Klein, W. H. (1997). Two Otx proteins generated from multiple transcripts of a single gene in *Strongylocentrotus purpuratus*. *Dev. Biol.* **187**, 253–266.

Li, X., Veraksa, A., and McGinnis, W. (1999). A sequence motif distinct from *Hox* binding sites controls the specificity of a Hox response element. *Development* **126**, 5581–5589.

Li, X., Wikramanayake, A. H., and Klein, W. H. (1999). Requirement of SpOtx in cell fate decisions in the sea urchin embryo and possible role as a mediator of β-catenin signaling. *Dev. Biol.* **212**, 425–439.

Liaw, G.-J., Rudolph, K. M., Huang, J.-D., Dubnicoff, T., Courey, A. J., and Lengyel, J. A. (1995). The *torso* response element binds GAGA and NTF-1/Elf-1, and regulates *tailless* by relief of respression. *Genes Dev.* **9**, 3163–3176.

Ligoxygakis, P., Bray, S. J., Apidianakis, Y., and Delidakis, C. (1999). Ectopic expression of individual *E(spl)* genes has differential effects on different cell fate decisions and underscores the biphasic requirement for Notch activity in wing margin establishment in *Drosophila. Development* **126**, 2205–2214.

Lilly, B., Zhao, B., Ranganayakulu, G., Paterson, B. M., Schultz, R. A., and Olson, E. N. (1995). Requirement of MADS domain transcription factor D-MEF2 for muscle formation in *Drosophila. Science* **267**, 688–693.

Lin, Q., Srivastava, D., and Olson, E. N. (1997). A transcriptional pathway for cardiac development. *Cold Spring Harbor Symp. Quant. Biol.* **62**, 405–411.

Lin, R., Thompson, S., and Priess, J. R. (1995). *pop-1* encodes an HMG box protein required for specification of a mesoderm precursor in early *C. elegans* embryos. *Cell* **83**, 599–609.

Livant, D. L., Hough-Evans, B. R., Moore, J. G., Britten, R. J., and Davidson, E. H. (1991). Differential stability of expression of similarly specified endogenous and exogenous genes in the sea urchin embryo. *Development* **113**, 385–398.

Logan, C., Hornbruch, A., Campbell, I., and Lumsden, A., (1997). The role of *Engrailed* in establishing the dorsoventral axis of the chick limb. *Development* **124**, 2317–2324.

Logan, C. Y., and McClay, D. R. (1997). The allocation of early blastomeres to the ectoderm and endoderm is variable in the sea urchin embryo. *Development* **124**, 2213–2223.

Logan, C. Y., Miller, J. R., Ferkowicz, M. J., and McClay, D. R. (1999). Nuclear β-catenin is required to specify vegetal cell fates in the sea urchin embryo. *Development* **126**, 345–357.

Logan, M., and Tabin, C. J. (1999). Role of *Pitx1* upstream of *Tbx4* in specification of hindlimb identity. *Science* **283**, 1736–1739.

Logan, M., Simon, H.-G., and Tabin, C. (1998). Differential regulation of T-box and homeobox transcription factors suggests roles in controlling chick limb-type identity. *Development* **125**, 2825–2835.

Lowe, C. J., and Wray, G. A. (1997). Radical alterations in the roles of homeobox genes during echinoderm evolution. *Nature* **389**, 718–721.

Lu, J., Webb, R., Richardson, J. A., and Olson, E. N. (1999). MyoR: A muscle-restricted basic helix-loop-helix transcription factor that antagonizes the actions of MyoD. *Proc. Natl. Acad. Sci. USA* **96**, 552–557.

Lumsden, A., and Krumlauf, R. (1996). Patterning the vertebrate neuraxis. *Science* **274**, 1109–1115.

Lynn, D., Angerer, L., Bruskin, A., Klein, W., and Angerer, R. (1983). Localization of a family of mRNAs in a single cell type and its precursors in sea urchin embryos. *Proc. Natl. Acad. Sci. USA* **80**, 2656–2660.

MacLellan, W. R., Lee, T.-C., Schwartz, R. J., and Schneider, M. D. (1994). Transforming growth factor-β response elements of the skeletal α-actin gene. *J. Biol. Chem.* **269**, 16754–16760.

Maconochie, M. K., Nonchev, S., Studer, M., Chan, S.-K., Pöpperl, H., Sham, M. H., Mann, R. S., and Krumlauf, R. (1997). Cross-regulation in the mouse *HoxB* complex: The expression of *Hoxb2* in rhombomere 4 is regulated by *Hoxb1*. *Genes Dev.* **11**, 1885–1895.

Maduro, M., Meneghini, M. D., Bowerman, B., Broitman-Maduro, G., and Rothman, J. H. (2000). Restriction of mesendoderm to a single blastomere by the combined action of SKN-1 and GSK-3β homolog is mediated by MED-1 and -2 in *C. elegans*. Submitted for publication.

Maggert, K., Levine, M., and Frasch, M. (1995). The somatic-visceral subdivision of the embryonic mesoderm is initiated by dorsal gradient thresholds in *Drosophila*. *Development* **121**, 2107–2116.

Makabe, K. W., Kirchhamer, C. V., Britten, R. J., and Davidson, E. H. (1995). *Cis*-regulatory control of the *SM50* gene, an early marker of skeletogenic lineage specification in the sea urchin embryo. *Development* **121**, 1957–1970.

Mallamaci, A., Mercurio, S., Muzio, L., Cecchi, C., Pardini, C. L., Gruss, P., and Boncinelli, E. (2000). The lack of *Emx2* causes impairment of *Reelin* signaling and defects of neuronal migration in the developing cerebral cortex. *J. Neurosci.* **20**, 1109–1118.

Manak, J. R., Mathies, L. D., and Scott, M. P. (1995). Regulation of a *decapentaplegic* midgut enhancer by homeotic proteins. *Development* **120**, 3605–3619.

Mango, S. E., Lambie, E. J., and Kimble, J. (1994). The *pha-4* gene is required to generate the pharyngeal primordium of *Caenorhabditis elegans*. *Development* **120**, 3019–3031.

Mannervik, M., Nibu, Y., Zhang, H., and Levine, M. (1999). Transcriptional co-regulators in development. *Science* **284**, 606–609.

Manzanares, M., Cordes, S., Kwan, C.-T., Sham, M. H., Barsh, G. S., and Krumlauf, R. (1997). Segmental regulation of *Hoxb*-3 by *kreisler*. *Nature* **387**, 191–195.

Manzanares, M., Trainor, P. A., Nonchev, S., Ariza-McNaughton, L., Brodie, J., Gould, A., Marshall, H., Morrison, A., Kwan, C.-T., Sham, M.-H., Wilkinson, D. G., and Krumlauf, R. (1999a). The role of *kreisler* in segmentation during hindbrain development. *Dev. Biol.* **211**, 220–237.

Manzanares, M., Cordes, S., Ariza-McNaughton, L., Sadl, V., Maruthainar, K., Barsh, G., and Krumlauf, R. (1999b). Conserved and distinct roles of *kreisler* in regulation of the paralogous *Hoxa3* and *Hoxb3* genes. *Development* **126**, 759–769.

Mao, C.-A., Wikramanayake, A. H., Gan, L., Chuang, C.-K., Summers, R. G., and Klein, W. H. (1996). Altering cell fates in sea urchin embryos by overexpressing SpOtx, an Orthodenticle-related protein. *Development* **122**, 1489–1498.

Mardon, G., Solomon, N. M., and Rubin, G. M. (1994). *dachshund* encodes a nuclear protein required for normal eye and leg development in *Drosophila*. *Development* **120**, 3473–3486.

Marikawa, Y., Yoshida, S., and Satoh, N. (1994). Development of egg fragments of the ascidian *Ciona savignyi*: The cytoplasmic factors responsible for muscle differentiation are separated into a specific fragment. *Dev. Biol.* **162**, 134–142.

Marín, F., and Charnay, P. (2000). Positional regulation of *Krox-20* and *mafB/kr* expression in the developing hindbrain: Potentialities of prospective rhombomeres. *Dev. Biol.* **218**, 220–234.

Marom, K., Shapira, E., and Fainsod, A. (1997). The chicken *caudal* genes establish an anterior-posterior gradient by partially overlapping temporal and spatial patterns of expression. *Mech. Dev.* **64**, 41–52.

Martin, E. L., Consales, C., Davidson, E. H., and Arnone, M. I. (2000). Evidence for a pan-mesodermal embryonic regulator of the sea urchin *cyIIa* gene. Submitted for publication.

Martinez, P., Rast, J. P., Arenas-Mena, C., and Davidson, E. H. (1999). Organization of an echinoderm *Hox* gene cluster. *Proc. Natl. Acad. Sci. USA* **96**, 1469–1474.

McClay, D. R., Armstrong, N. A., and Hardin, J. (1992). Pattern formation during gastrulation in the sea urchin embryo. *Development* 1992 Supplement, 33–41.

McClay, D. R., Peterson, R. E., Range, R. C., Winter-Vann, A. M., and Ferkowitz, M. J. (2000). A micromere induction signal is activated by β-catenin and acts through Notch to initiate specification of secondary mesenchyme cells in the sea urchin embryo. *Development*, in press.

McCormick, A., Coré, N., Kerridge, S., and Scott, M. P. (1995). Homeotic response elements are tightly linked to tissue-specific elements in a transcriptional enhancer of the *teashirt* gene. *Development* **121**, 2799–2812.

McGinnis, W., and Krumlauf, R. (1992). Homeobox genes and axial patterning. *Cell* **68**, 283–302.

McKendry, R., Hsu, S.-C., Harland, R. M., and Grosschedl, R. (1997). LEF-1/TCF proteins mediate wnt-inducible transcription from the *Xenopus* nodal-related 3 promoter. *Dev. Biol.* **192**, 420–431.

Mechta-Grigoriou, F., Garel, S., and Charnay, P. (2000). Nab proteins mediate a negative feedback loop controlling *Krox-20* activity in the developing hindbrain. *Development* **127**, 119–128.

Meedel, T. H., Crowther, R. J., and Whittaker, J. R. (1987). Determinative properties of muscle lineages in ascidian embryos. *Development* **100**, 245–260.

Meneghini, M. D., Ishitani, T., Carter, J. C., Hisamoto, N., Ninomiya-Tsuji, J., Thorpe, C. J., Hamill, D. R., Matsumoto, K., and Bowerman, B. (1999). Map kinase and Wnt pathways converge to downregulate an HMG-domain repressor in *Caenorhabditis elegans*. *Nature* **399**, 793–797.

Metzger, R. J., and Krasnow, M. A. (1999). Genetic control of branching morphogenesis. *Science* **284**, 1635–1639.

Meyer, A. (1999). Homology and homoplasy: The retention of genetic programmes. *In* "Homology." (Novartis Foundation Symposium 222), pp. 141–157. John Wiley & Sons, Chichester.

Miklos, G. L. G. (1993). Emergence of organizational complexities during metazoan evolution: Perspectives from molecular biology, palaeontology and neo-Darwinism. *Memoir of the Assoc. Austral-Asian Palaeontol.* **15**, 7–41.

Milán, M., and Cohen, S. M. (2000). Subdividing cell populations in the developing limbs of *Drosophila*: Do wing veins and leg segments define units of growth control? *Dev. Biol.* **217**, 1–9.

Mismer, D., and Rubin, G. M. (1987). Analysis of the promoter of the *ninaE* opsin gene in *Drosophila melanogaster*. *Genetics* **116**, 565–587.

Mitani, Y., Takahashi, H., and Satoh, N. (1999). An ascidian T-box gene *As-T2* is related to the *Tbx6* subfamily and is associated with embryonic muscle cell differentiation. *Dev. Dynam.* **215**, 62–68.

Miya, T., Morita, K., Ueno, N., and Satoh, N. (1996). An ascidian homologue of vertebrate *BMPs-5–8* is expressed in the midline of the anterior neuroectoderm and in the midline of the ventral epidermis of the embryo. *Mech. Dev.* **57**, 181–190.

Miyabayashi, T., Palfreyman, M. T., Sluder, A. E., Slack, F., and Sengupta, P. (1999). Expression and function of members of a divergent nuclear receptor family in *Caenorhabditis elegans*. *Dev. Biol.* **215**, 314–331.

Modolell, J. (1997). Patterning of the adult peripheral nervous system of *Drosophila*. *Perspect. Dev. Neurobiol.* **4**, 285–296.

Modolell, J., and Campuzano, S. (1998). The *achaete-scute* complex as an integrating device. *Int. J. Dev. Biol.* **42**, 275–282.

Mohler, J., Seecoomar, M., Agarwal, S., Bier, E., and Hsai, J. (2000) Activation of *knot* (*kn*) specifies the 3–4 intervein region in the *Drosophila* wing. *Development* 127, 55–63.

Molkentin, J. D., and Olson, E. N. (1996). Defining the regulatory networks for muscle development. *Curr. Opin. Genet. Dev.* **6**, 445–453.

Molkentin, J. D., Lin, Q., Duncan, S. A., and Olson, E. N. (1997). Requirement of the transcription factor GATA4 for heart tube formation and ventral morphogenesis. *Genes Dev.* **11**, 1061–1072.

Molkentin, J. D., Lu, J.-R., Antos, C. L., Markham, B., Richardson, J., Robbins, J., Grant, S. R., and Olson, E. N. (1998). A calcineurin-dependent transcriptional pathway for cardiac hypertrophy. *Cell* **93**, 215–228.

Moon, R. T., and Kimelman, D. (1998). From cortical rotation to organizer gene expression: Toward a molecular explanation of axis specification in *Xenopus*. *BioEssays* **20**, 536–545.

Moreno, E., and Morata, G. (1999). *Caudal* is the *Hox* gene that specifies the most posterior *Drosophile* segment. *Nature* **400**, 873–877.

Morgan, T. H. (1934). "Embryology and Genetics." Columbia University Press, New York.

Morrison, A., Moroni, M. C., Ariza-McNaughton, L., Krumlauf, R., and Mavilio, F. (1996). *In vitro* and transgenic analysis of a human *HOXD4* retinoid-responsive enhancer. *Development* **122**, 1895–1907.

Morrison, A., Ariza-McNaughton, L., Gould, A., Featherstone, M., and Krumlauf, R. (1997). *HOXD4* and regulation of the group 4 paralog genes. *Development* **124**, 3135–3146.

Myers, E. W. *et al.* (2000). A whole-genome assembly of *Drosophila. Science* **287**, 2196–2204.

Nagel, A. C., Maier, D., and Preiss, A. (2000) Su(H)-independent activity of Hairless during mechano-sensory organ formation in *Drosophila. Mech. Dev.* 94, 3–12.

Nakatani, Y., and Nishida, H. (1994). Induction of notochord during ascidian embryogenesis. *Dev. Biol.* **166**, 289–299.

Nakatani, Y., and Nishida, H. (1999). Duration of competence and inducing capacity of blastomeres in notochord induction during ascidian embryogenesis. *Dev. Growth Differ.* **41**, 449–453.

Nakatani, Y., Yasuo, H., Satoh, N., and Nishida, H. (1996). Basic fibroblast growth factor induces notochord formation and the expression of *As-T*, a *Brachyury* homolog, during ascidian embryogenesis. *Development* **122**, 2023–2031.

Nellesen, D. T., Lai, E. C., and Posakony, J. W. (1999). Discrete enhancer elements mediate selective responsiveness of *Enhancer of split* complex genes to common transcriptional activators. *Dev. Biol.* **213**, 33–53.

Nelson, C. E., Morgan, B. A., Burke, A. C., Laufer, E., DiMambro, E., Murtaugh, L. C., Gonzales, E., Tessarollo, L., Parada, L. F., and Tabin, C. (1996). Analysis of *Hox* gene expression in the chick limb bud. *Development* **122**, 1449–1466.

Nemer, M., Thornton, R. D., Stuebing, E. W., and Harlow, P. (1991). Structure, spatial, and temporal expression of two sea urchin metallothionein genes, *SpMTB1* and *SpMTA. J. Biol. Chem.* **266**, 6586–6593.

Newman-Smith, E.D., and Rothman, J. H. (1998). The maternal-to-zygotic transition in embryonic patterning of *Caenorhabditis elegans. Curr. Opin. Genet. Dev.* **8**, 472–480.

Newman-Smith, E., and Rothman, J. H. (2000). The *end-2* gene specifies embryonic endoderm and pharynx in early *C. elegans* embryos, in preparation.

Nguyen, H. T., and Xu, X. (1998). *Drosophila mef2* expression during mesoderm development is controlled by a complex array of *cis*-acting regulatory modules. *Dev. Biol.* **204**, 550–566.

Nibu, Y., Zhang, H., and Levine, M. (1998). Interaction of short-range repressors with *Drosophila* CtBP in the embryo. *Science* **280**, 101–104.

Niederreither, K., Vermot, J., Schuhbaur, B., Chambon, P., and Dollé, P. (2000). Retinoic acid synthesis and hindbrain patterning in the mouse embryo. *Development* **127**, 75–85.

Nielsen, C. (1995). "Animal Evolution: Interrelationships of the Living Phyla." Oxford University Press, Oxford.

Nielsen, C. (1998). Origin and evolution of animal life cycles. *Biol. Rev.* **73**, 125–155.

Niimi, T., Seimiya, M., Kloter, U., Flister, S., and Gehring, W. J. (1999). Direct regulatory interaction of the *eyeless* protein with an eye-specific enhancer in the *sine oculis* gene during eye induction in *Drosophila. Development* **126**, 2253–2260.

Nijhout, H. F. (1994). Symmetry systems and compartments in Lepidopteran wings: The evolution of a patterning mechanism. *Development* **120S**, 225–233.

Nieto, M. A., Sechrist, J., Wilkinson, D. G., and Bronner-Fraser, M. (1995). Relationship between spatially restricted *Krox-20* gene expression in branchial neural crest and segmentation in the chick embryo hindbrain. *EMBO J.* **14**, 1697–1710.

Nishida, H. (1987). Cell lineage analysis in ascidian embryos by intracellular injection of a tracer enzyme. III. Up to the tissue restricted stage. *Dev. Biol.* **121**, 526–541.

Nishida, H. (1991). Induction of brain and sensory pigment cells in the ascidian embryo analyzed by experiments with isolated blastomeres. *Development* **112**, 389–395.

Nishida, H. (1992). Regionality of egg cytoplasm that promotes muscle differentiation in embryo of the ascidian, *Halocynthia roretzi. Development* **116**, 521–529.

Nishida, H. (1993). Localized regions of egg cytoplasm that promote expression of endoderm-specific alkaline phosphatase in embryos of the ascidian *Halocynthia roretzi. Development* **118**, 1–7.

Nishida, H. (1994). Localization of egg cytoplasm that promotes differentiation to epidermis in embryos of the ascidian *Halocynthia roretzi. Development* **120**, 235–243.

Nishida, H. (1997). Cell fate specification by localized cytoplasmic determinants and cell interactions in ascidian embryos. *Int. Rev. Cytol.* **176**, 245–306.

Nocente-McGrath, C., Brenner, C. A., and Ernst, S. G. (1989). *Endo16*, a lineage-specific protein of the sea urchin embryo, is first expressed just prior to gastrulation. *Dev. Biol.* **136**, 264–272.

Nonchev, S., Vesque, C., Maconochie, M., Seitanidou, T., Ariza-McNaughton, L., Frain, M., Marshall, H., Sham, M. H., Krumlauf, R., and Charnay, P. (1996). Segmental expression of *Hoxa-2* in the hindbrain is directly regulated by *Krox-20. Development* **122**, 543–554.

Offield, M. F., Jetton, T. L., Labosky, P.A., Ray, M., Stein, R. W., Magnuson, M. A., Hogan, B. L. M., and Wright, C. V. E. (1996). PDX-1 is required for pancreatic outgrowth and differentiation of the rostral duodenum. *Development* **122**, 983–995.

Ohsako, S., Hyer, J., Panganiban, G., Oliver, I., and Caudy, M. (1994). Hairy function as a DNA-binding helix-loop-helix repressor of *Drosophila* sensory organ formation. *Genes Dev.* **8**, 2743–2755.

Ohshiro, T., and Saigo, K. (1997). Transcriptional regulation of *breathless* FGF receptor gene by binding of TRACHEALESS/dARNT heterodimers to three central midline elements in *Drosophila* developing trachea. *Development* **124**, 3975–3986.

Oliveri, P. (1996). Espressione spazio-temporale di un gene con homeobox durante l'embriogenesi del riccio di mare *Paracentrotus lividus*. Dottorato di Ricerca in Biologia Cellulare, Universita di Palermo, Italy.

Olson, E. N. (1990). MyoD family: A paradigm for development. *Genes Dev.* **4**, 1454–1461.

Ortolani, G. (1955). The presumptive territory of the mesoderm in the ascidian germ. *Experientia* **11**, 445–446.

Panganiban, G. E. F., Reuter, R., Scott, M. P., and Hoffman, F. M. (1990). A *Drosophila* growth factor homolog, *decapentaplegic*, regulates homeotic gene expression within and across germ layers during midgut morphogenesis. *Development* **110**, 1041–1050.

Pannese, M., Polo, C., Andreazzoli, M., Vignali, R., Kablar, B., Barsacchi, G., and Boncinelli, E. (1995). The *Xenopus* homologue of *Otx2* is a maternal homeobox gene that demarcates and specifies anterior body regions. *Development* **121**, 707–720.

Papaioannou, V., and Silver, L. M. (1998). The T-box gene family. *BioEssays* **20**, 9–19.

Park, M., Wu, X., Golden, K., Axelrod, J. D., and Bodmer, R. (1996). The wingless signaling pathway is directly involved in *Drosophila* heart development. *Dev. Biol.* **177**, 104–116.

Paroush, Z., Finley, R. L., Jr., Kidd, T., Wainwright, S. M., Ingham, P. W., Brent, R., and Ish-Horowicz, D. (1994). Groucho is required for *Drosophila* neurogenesis, segmentation, and sex determination and interacts directly with hairy-related bHLH proteins. *Cell* **79**, 805–815.

Pata, I., Studer, M., van Doorninck, J. H., Briscoe, J., Kuuse, S., Engel, J. D., Grosveld, F., and Karis, A. (1999). The transcription factor GATA3 is a downstream effector of *Hoxb1* specification in rhombomere 4. *Development* **126**, 5523–5531.

Pearse, J. S., and Cameron, R. A. (1991). "Reproduction of Marine Invertebrates," Vol. VI (A. C. Giese, J. S. Pearse and V. B. Pearse, eds.). The Boxwood Press, Pacific Grove, CA.

Peichel, C. L., Prabhakaran, B., and Vogt, T. F. (1997). The mouse *Ulnaless* mutation deregulates posterior *HoxD* gene expression and alters appendicular patterning. *Development* **124**, 3481–3492.

Peterson, K. J., and Davidson, E. H. (2000). Regulatory evolution and the origin of the bilaterians. *Proc. Natl. Acad. Sci. USA* **97**, 4430–4433.

Peterson, K. P., and Eernisse, D. J. (2000). Animal phylogeny and the ancestry of bilaterians: Inferences from morphology and 18S rDNA gene sequences. *Evol. Dev.*, submitted for publication.

Peterson, K. J., Cameron, R. A., and Davidson, E. H. (1997). Set-aside cells in maximal indirect development: Evolutionary and development: Evolutionary and developmental significance. *BioEssays* **19**, 623–631.

Peterson, K. J., Cameron, R. A., Tagawa, K., Satoh, N., and Davidson, E. H. (1999a). A comparative molecular approach to mesodermal patterning in basal deuterostomes: The expression pattern of *Brachyury* in the enteropneust hemichordate *Ptychodera flava. Development* **126**, 85–95.

Peterson, K. J., Harada, Y., Cameron, R. A., and Davidson, E. H. (1999b). Expression pattern of *Brachyury* and *Not* in the sea urchin: Comparative implications for the origins of mesoderm in the basal deuterostomes. *Dev. Biol.* **207**, 419–431.

Peterson, K. J., Irvine, S. Q., Cameron, R. A., and Davidson, E. H. (2000a). Quantitative assessment of *Hox* complex expression in the indirect development of the polychaete annelid *Chaetopterus sp. Proc. Natl. Acad. Sci. US*A **97**, 4487–4492.

Peterson, K. J., Cameron, R. A., and Davidson, E. H. (2000b). Bilaterian origins: Significance of new experimental obervations. *Dev. Biol.* **219**, 1–17.

Peterson, K. J., Arenas-Mena, C., and Davidson, E. H. (2000c). The A/P axis in echinoderm ontogeny and evolution: Evidence from fossils and molecules. *Evol. Dev.* **2**, 93–101.

Piatigorsky, J. (1992). Lens crystallins. *J. Biol. Chem.* **267**, 4277–4280.

Pignoni, F., Hu, B., Zavitz, K. H., Xiao, J., Garrity, P. A., and Zipursky, S. L. (1997). The eye-specification proteins So and Eya form a complex and regulate multiple steps in *Drosophila* eye development. *Cell* **91**, 881–891.

Pineda, D., Gonzalez, J., Callaerts, P., Ikeo, K., Gehring, W. J., and Salo, E. (2000). Searching for the prototypic eye genetic network: *Sine oculis* is essential for eye regeneration in planarians. *Proc. Natl. Acad. Sci. USA* **97**, 4525–4529.

Plaza, S., Dozier, C., Langlois, M.-C., and Saule, S. (1995). Identification and characterization of a neuroretina-specific enhancer element in the quail *Pax-6* (*Pax-QNR*) gene. *Mol. Cell. Biol.* **15**, 892–903.

Pöpperl, H., Bienz, M., Studer, M., Chan, S.-K., Aparicio, S., Brenner, S., Mann, R. S., and Krumlauf, R. (1995). Segmental expression of *Hoxb-1* is controlled by a highly conserved autoregulatory loop dependent upon *exd/pbx. Cell* **81**, 1031–1042.

Posakony, J. W. (1994). Nature versus nurture: Asymmetric cell divisions in *Drosophila* bristle development. *Cell* **76**, 415–418.

Qian, S., Capovilla, M., and Pirrotta, V. (1991). The *bx* region enhancer, a distant *cis*-control element of the *Drosophila Ubx* gene and its regulation by *hunchback* and other segmentation genes. *EMBO J.* **10**, 1415–1425.

Quiring, R., Walldorf, U., Kloter, U., and Gehring, W. J. (1994). Homology of the *eyeless* gene of *Drosophila* to the *Small eye* gene in mice and *Aniridia* in humans. *Science* **265**, 785–789.

Ranganayakulu, G., Elliott, D. A., Harvey, R. P., and Olson, E. N. (1998). Divergent roles for *NK-2* class homeobox genes in cardiogenesis in flies and mice. *Development* **125**, 3037–3048.

Ransick, A., and Davidson, E. H. (1993). A complete second gut induced by transplanted micromeres in the sea urchin embryo. *Science* **259**, 1134–1138.

Ransick, A., and Davidson, E. H. (1995). Micromeres are required for normal vegetal plate specification in sea urchin embryos. *Development* **121**, 3215–3222.

Ransick, A., and Davidson, E. H. (1998). Late specification of *veg₁* lineages to endodermal fate in the sea urchin embryo. *Dev. Biol.* **195**, 38–48.

Ransick, A., Ernst, S., Britten, R. J., and Davidson, E. H. (1993). Whole mount *in situ* hybridization shows *Endo-16* to be a marker for the vegetal plate territory in sea urchin embryos. *Mech. Dev.* **42**, 117–124.

Rao, M. V., Donoghue, M. J., Merlie, J. P., and Sanes, J. R. (1996). Distinct regulatory elements control muscle-specific, fiber-type-selective, and axially graded expression of a myosin light-chain gene in transgenic mice. *Mol. Cell. Biol.* **16**, 3909–3922.

Reecy, J. M., Li, X., Yamada, M., DeMayo, F. J., Newman, C. S., Harvey, R. P., and Schwartz, R. J. (1999). Identification of upstream regulatory regions in the heart-expressed homeobox gene *Nkx2-5*. *Development* **126**, 839–849.

Reichmann, V., Irion, U., Wilson, R., Grosskortenhaus, R., and Leptin, M. (with appendix by M. Bate and M. Frasch). (1997). Control of cell fates and segmentation in the *Drosophila* mesoderm. *Development* **124**, 2915–2922.

Reverberi, G., and Minganti, A. (1947). La distribuzione delle potenze nel germe di Ascidie allo stadio di otto blastomeri, analizzata mediante le combinazioni e i trapianti di blastomeri. *Pubbl. Stn. Zool. Napoli* **21**, 1–35.

Reynolds, S. D., Angerer, L. M., Palis, J., Nasir, A., and Angerer, R. C. (1992). Early mRNAs, spatially restricted along the animal-vegetal axis of sea urchin embryos, include one encoding a protein related to tolloid and BMP-1. *Development* **114**, 769–786.

Rhodes, S. J., Chen, R., DiMattia, G. E., Scully, K. M., Kalla, K. A., Lin, S.-C., Yu, V. C., and Rosenfeld, M. G. (1993). A tissue-specific enhancer confers Pit-1–dependent morphogen inducibility and autoregulation on the *pit-1* gene. *Genes Dev.* **7**, 913–932.

Riddle, R. D., Ensini, M., Nelson, C., Tsuchida, T., Jessell, T. M., and Tabin, C. (1995). Induction of the LIM homeobox gene *Lmx1* by WNT7a establishes dorsoventral pattern in the vertebrate limb. *Cell* **83**, 631–640.

Rieger, R. M., and Tyler, S. (1995). Sister-group relationship of Gnathostomulida and Rotifera-Ancanthocephala. *Invert. Biol.* **114**, 186–188.

Ristoratore, F., Spagnuolo, A., Aniello, F., Branno, M., Fabbrini, F., and Di Lauro, R. (1999). Expression and functional analysis of *Cititf1*, an ascidian NK-2 Class gene, suggest its role in endoderm development. *Development* **126**, 5149–5159.

Rivera-Pomar, R., and Jäckle, H. (1996). From gradients to stripes in *Drosophila* embryogenesis: Filling in the gaps. *Trends Genet.* **12**, 478–483.

Rivera-Pomar, R., Lu, X., Perrimon, N., Taubert, H., and Jäckle, H. (1995). Activation of posterior gap gene expression in the *Drosophila* blastoderm. *Nature* **376**, 253–256.

Roch, F., and Akam, M. (2000). *Ultrabithorax* and the control of cell morphology in *Drosophila* halteres. *Development* **127**, 97–107.

Rocheleau, C. E., Downs, W. D., Lin, R., Wittmann, C., Bei, Y., Cha, Y.-H., Ali, M., Priess, J. R., and Mello, C. C. (1997). Wnt signaling and an APC-related gene specify endoderm in early *C. elegans* embryos. *Cell* **90**, 707–716.

Rodriguez-Esteban, C., Tsukui, T., Yonei, S., Magallon, J., Tamura, K., and Izpisua Belmonte, J. C. (1999). The T-box genes *Tbx4* and *Tbx5* regulate limb outgrowth and identity. *Nature* **398**, 814–818.

Roose, J., Molenaar, M., Peterson, J., Hurenkamp, J., Brantjes, H., Moerer, P., van de Wetering, M., Destre, O., and Clevers, H. (1998). The *Xenopus* Wnt effector XTcf-3 interacts with Groucho-related transcriptional repressors. *Nature* **395**, 608–612.

Rose, S. M. (1939). Embryonic induction in the Ascidia. *Biol. Bull.* **76**, 216–232.

Rosenthal, N., Wentworth, B., Engert, J., Grieshammer, U., Berglund, E., and Gong, X. (1992). The myosin light-chain 1/3 locus: A model for developmental control of skeletal muscle differentiation. *In* "Neuromuscular Development and Disease." (A. M. Kelly and H. M. Blau, Eds.), pp. 131–144. Raven Press, Ltd., New York.

Rossel, M., and Capecchi, M. R. (1999). Mice mutant for both *Hoxa1* and *Hoxb1* show extensive remodeling of the hindbrain and defects in craniofacial development. *Development* **126**, 5027–5040.

Rubin, G. M. *et al.* (2000). Comparative genomics of the eukaryotes. *Science* **287**, 2204–2215.

Ruddle, F. H., Bentley, K. L., Murtha, M. T., and Risch, N. (1994). Gene loss and gain in the evolution of the vertebrates. *Development* **120S** 155–161.

Rudolph, K. M., Liaw, G.-J., Daniel, A., Green, P., Courey, A. J., Hartenstein, V., and Lengyel, J. A. (1997). Complex regulatory region mediating *tailless* expression in early embryonic patterning and brain development. *Development* **124**, 4297–4308.

Ruffins, S. W., and Ettensohn, C. A. (1993). A clonal analysis of secondary mesenchyme cell fates in the sea urchin embryo. *Dev. Biol.* **160**, 285–288.

Ruffins, S. W., and Ettensohn, C. A. (1996). A fate map of the vegetal plate of the sea urchin (*Lytechinus variegatus*) mesenchyme blastula. *Development* **122**, 253–263.

Ruiz-Gómez, M., Romani, S., Hartmann, C., Jäckle, H., and Bate, M. (1997). Specific muscle identities are regulated by *Krüppel* during *Drosophila* embryogenesis. *Development* **124**, 3407–3414.

Runnegar, B. (2000). Loophole for snowball Earth. *Nature* **405**, 403–404.

Ruppert, E. E., and Barnes, R. D. (1994). "Invertebrate Zoology." Saunders College Publishing, Fort Worth, TX.

Rusch, J., and Levine, M. (1996). Threshold responses to the dorsal regulatory gradient and the subdivision of primary tissue territories in the *Drosophila* embryo. *Curr. Opin. Genet. Dev.* **6**, 416–423.

Rushlow, C. A., Han, K., Manley, J. L., and Levine, M. (1989). The graded distribution of the dorsal morphogen involves selective nuclear transport in *Drosophila*. *Cell* **59**, 1165–1177.

Ruvinsky, I., Oates, A., Silver, L. M., and Ho, R. (2000). The evolution of paired appendages in vertebrates: T-box genes in the zebrafish. *Dev. Genes Evol.* 210, 82–91.

Ruvkun, G., and Hobert, O. (1998). The taxonomy of developmental control in *Caenorhabditis elegans. Science* **282**, 2033–2041.

Salser, S. J., and Kenyon, C. (1994). Patterning *C. elegans*: Homeotic cluster genes, cell fates and cell migrations. *Trends Genet.* **10**, 159–164

Satoh, N. (1994). "Developmental Biology of Ascidians." Cambridge University Press, Cambridge, UK.

Satoh, N., Araki, I., and Satou, Y. (1996a). An intrinsic genetic program for autonomous differentiation of muscle cells in the ascidian embryo. *Proc. Natl. Acad. Sci. USA* **93**, 9315–9321.

Satoh, N., Makabe, K. W., Katsuyama, Y., Wada, S., and Saiga, H. (1996b). The ascidian embryo: An experimental system for studying genetic circuitry for embryonic cell specification and morphogenesis. *Dev. Growth Differ.* **38**, 325–340.

Satou, Y., and Satoh, N. (1996). Two *cis*-regulatory elements are essential for the muscle-specific expression of an actin gene in the ascidian embryo. *Dev. Growth Differ.* **38**, 565–573.

Satou, Y., Kusakabe, T., Araki, I., and Satoh, N. (1995). Timing of initiation of muscle-specific gene expression in the ascidian embryo precedes that of developmental fate restriction in lineage cells. *Dev. Growth Differ.* **37**, 319–327.

Schaeper, U., Boyd, J. M., Verma, S., Uhlmann, E., Subramanian, T., and Chinnadurai, G. (1995). Molecular cloning and characterization of a cellular phosphoprotein that interacts with conserved C-terminal domain of adenovirus E1A involved in negative modulation of oncogenic transformation. *Proc. Natl. Acad. Sci. USA* **92**, 10467–10471.

Schmidt-Rhaesa, A., Bartolomaeus, T., Lembuerg, C., Ehlers, U., and Garey, J. (1998). The position of the Arthropoda in the phylogenetic system. *J. Morphol.* **138**, 263–285.

Schnabel, R., and Priess, J. R. (1997). Specification of cell fates in the early embryo. In *"C. Elegans* II." (D. L. Riddle, T. Blumenthal, B. J. Meyer and J. R. Priess, Eds.), pp. 361–382. Cold Spring Harbor Laboratory Press, NY.

Schneider-Maunoury, S., Topilko, P., Seitanidou, T., Levi, G., Cohen-Tannoudji, M., Pournin, S., Babinet, C., and Charnay, P. (1993). Disruption of *Krox-20* results in alteration of rhombomere-3 and rhombomere-5 in the developing hindbrain. *Cell* **75**, 1199–1214.

Schneider-Maunoury, S., Seitanidou, T., Charnay, P., and Lumsden, A. (1997). Segmental and neuronal architecture of the hindbrain of *Krox-20* mouse mutants. *Development* **124**, 1215–1226.

Schroeder, D. F., and McGhee, J. D. (1998). Anterior-posterior patterning within the *Caenorhabditis elegans* endoderm. *Development* **125**, 4877–4887.

Schwartz, C., Locke, J., Nishida, C., and Kornberg, T. B. (1995). Analysis of *cubitus interruptus* regulation in *Drosophila* embryos and imaginal disks. *Development* **121**, 1625–1635.

Schwartz, R. J., and Olson, E. N. (1999). Building the heart piece by piece: Modularity of *cis*-elements regulating *Nkx2–5* transcription. *Development* **126**, 4187–4192.

Sepulveda, J. L., Belaguli, N., Nigam, V., Chen, C.-Y., Nemer, M., and Schwartz, R. J. (1998). GATA-4 and Nkx-2.5 coactivate *Nkx-2* DNA binding targets: Role for regulating early cardiac gene expression. *Mol. Cell. Biol.* **18**, 3405–3415.

Seydoux, G., and Dunn, M. A. (1997). Transcriptionally repressed germ cells lack a subpopulation of phosphorylated RNA polymerase II in early embryos of *Caenorhabditis elegans* and *Drosophila melanogaster*. *Development* **124**, 2191–2201.

Seydoux, G., and Fire, A. (1994). Soma-germline asymmetry in the distribution of embryonic RNAs in *C. elegans*. *Development* **120**, 2823–2834.

Seydoux, G., Mello, C. C., Pettitt, J., Wood, W. B., Priess, J. R., and Fire, A. (1996). Repression of gene expression in the embryonic germ lineage of *C. elegans*. *Nature* **382**, 713–716.

Sham, M. H., Vesque, C., Nonchev, S., Marshall, H., Frain, M., Gupta, R. D., Whiting, J., Wilkinson, D., Charnay, P., and Krumlauf, R. (1993). The zinc finger gene *Krox20* regulates *HoxB2* (*Hox2.8*) during hindbrain segmentation. *Cell* **72**, 183–196.

Shankland, M., and Seaver, E. C. (2000). Evolution of the bilaterian body plan: What have we learned from annelids? *Proc. Natl. Acad. Sci. USA* **97**, 4434–4437.

Sheng, G., Thouvenot, E., Schmucker, D., Wilson, D. S., and Desplan, C. (1997). Direct regulation of *rhodopsin 1* by Pax-6/eyeless in *Drosophila*: Evidence for a conserved function in photoreceptors. *Genes Dev.* **11**, 1122–1131.

Sherwood, D. R., and McClay, D. R. (1997). Identification and localization of a sea urchin Notch homologue: Insights into vegetal plate regionalization and Notch receptor regulation. *Development* **124**, 3363–3374.

Sherwood, D. R., and McClay, D. R. (1999). LvNotch signaling mediates secondary mesenchyme specification in the sea urchin embryo. *Development* **126**, 1703–1713.

Shimamura, K., Hartigan, D. J., Martinez, S., Puelles, L., and Rubenstein, J. L. R. (1995). Longitudinal organization of the anterior neural plate and neural tube. *Development* **121**, 3923–3933.

Shin, T. H., Yasuda, J., Rocheleau, C. E., Lin, R. L., Soto, M., Bei, Y. X., Davis, R. J., and Mello, C. C. (1999). MOM-4, a MAP kinase kinase kinase-related protein, activates WRM-1/LIT-1 kinase to transduce anterior/posterior polarity signals in *C. elegans*. *Mol. Cell* **4**, 275–280.

Shu, D.-G., Luo, H.-L., Conway Morris, S., Zhang, X.-L., Hu, S.-X., Chen, L., Han, J., Zhu, M., Li, Y., and Chen. L.-Z. (1999). Lower Cambrian vertebrates from south China. *Nature* **402**, 42–46.

Shubin, N., Tabin, C., and Carroll, S. (1997). Fossils, genes and the evolution of animal limbs. *Nature* **388**, 639–648.

Simeone, A., Acampora, D., Gulisano, M., Stornaiuolo, A., and Boncinelli, E. (1992). Nested expression domains of four homeobox genes in developing rostral brain. *Nature* **358**, 687–690.

Simmen, M. W., Leitgeb, S., Clark, V. H., Jones, S. J. M., and Bird, A. (1998). Gene number in an invertebrate chordate, *Ciona intestinalis*. *Proc. Natl. Acad. Sci. USA* **95**, 4437–4440.

Simmons, D. M., Voss, J. W., Ingraham, H. A., Holloway, J. M., Broide, R. S., Rosenfeld, M. G., and Swanson, L. W. (1990). Pituitary cell phenotypes involve cell-specific Pit-1 mRNA translation and synergistic interactions with other classes of transcription factors. *Genes Dev.* **4**, 695–711.

Simpson, P. (1990). Lateral inhibition and the development of the sensory bristles of the adult peripheral nervous system of *Drosophila*. *Development* **109**, 509–519.

Singer, J. B., Harbecke, R., Kusch, T., Reuter, R., and Lengyel, J. A. (1996). *Drosophila brachyenteron* regulates gene activity and morphogenesis in the gut. *Development* **122**, 3707–3718.

Small, S., Kraut, R., Hoey, T., Warrior, R., and Levine, M. (1991). Transcriptional regulation of a pair-rule stripe in *Drosophila*. *Genes Dev.* **5**, 827–839.

Small, S., Blair, A., and Levine, M. (1992). Regulation of *even-skipped* stripe 2 in the *Drosophila* embryo. *EMBO J.* **11**, 4047–4057.

Small, S., Blair, A., and Levine, M. (1996). Regulation of two pair-rule stripes by a single enhancer in the *Drosophila* embryo. *Dev. Biol.* **175**, 314–324.

Smith, J. (1997). *Brachyury* and the T-box genes. *Curr. Opin. Genet. Dev.* **7**, 474–480.

Smith Fernandez, A., Pieau, C., Repérant, J., Boncinelli, E., and Wassef, M. (1998). Expression of the *Emx-1* and *Dlx-1* homeobox genes define three molecularly distinct domains in the telencephalon of mouse, chick, turtle and frog embryos: Implications for the evolution of telencephalic subdivisions in amniotes. *Development* **125**, 2099–2111.

Sordino, P., van der Hoeven, F., and Duboule, D. (1995). *Hox* gene expression in teleost fins and the origin of vertebrate digits. *Nature* **375**, 678–681.

Srivastava, D., Thomas, T., Lin, Q., Kirby, M. L., Brown, D., and Olson, E. N. (1997). Regulation of cardiac mesodermal and neural crest development by the bHLH transcription factor, dHAND. *Nature Genet.* **16**, 154–160; 410 (*erratum*).

Stein, D., and Nüsslein-Volhard, C. (1992). Multiple extracellular activities in *Drosophila* egg perivitelline fluid are required for establishment of embryonic dorsal-ventral polarity. *Cell* **68**, 429–440.

Stern, D. L. (1998). A role of *Ultrabithorax* in morphological differences between *Drosophila* species. *Nature* **396**, 463–466.

Steward, R. (1989). Relocalization of the dorsal protein from the cytoplasm to the nucleus correlates with its function. *Cell* **59**, 1179–1188.

Steward, R., and Govind, S. (1993). Dorsal-ventral polarity in the *Drosophila* embryo. *Curr. Opin. Gen. Dev.* **3**, 556–561.

Struhl, G., and Adachi, A. (1998). Nuclear access and action of Notch *in vivo*. *Cell* **93**, 649–660.

Studer, M., Pöpperl, H., Marshall, H., Kuroiwa, A., and Krumlauf, R. (1994). Role of a conserved retinoic acid response element in rhombomere restriction of *Hoxb-1*. *Science* **265**, 1728–1732.

Studer, M., Gavalas, A., Marshall, H., Ariza-McNaughton, L., Rijli, F. M., Chambon, P., and Krumlauf, R. (1998). Genetic interactions between *Hoxa1* and *Hoxb1* reveal new roles in regulation of early hindbrain patterning. *Development* **125**, 1025–1036.

Sturtevant, M. A., Roark, M., and Bier, E. (1993). The *Drosophila*-rhomboid gene mediates the localized formation of wing veins and interacts genetically with components of the EGF-R signaling pathway. *Genes Dev.* **7**, 961–973.

Sturtevant, M. A., Biehs, B., Marin, E., and Bier, E. (1997). The *spalt* gene links the A/P compartment boundary to a linear adult structure in the *Drosophila* wing. *Development* **124**, 21–32.

Su, M.-T., Fujioka, M., Goto, T., and Bodmer, R. (1999). The *Drosophila* homeobox genes *zfh-1* and *even-skipped* are required for cardiac-specific differentiation of a *numb*-dependent lineage decision. *Development* **126**, 3241–3251.

Sucov, H. M., Hough-Evans, B. R., Franks, R. R., Britten, R. J., and Davidson, E. H. (1988). A regulatory domain that directs lineage-specific expression of a skeletal matrix protein gene in the sea urchin embryo. *Genes Dev.* **2**, 1238–1250.

Sulston, J. E., Schierenberg, E., White, J. G., and Thomson, J. N. (1983). The embryonic cell lineage of the nematode *Caenorhabditis elegans*. *Dev. Biol.* **100**, 64–119.

Sutherland, D., Samakovlis, C., and Krasnow, M. A. (1996). *Branchless* encodes a *Drosophila* FGF homolog that controls tracheal cell migration and the pattern of branching. *Cell* **87**, 1091–1101.

Sweet, H. C., Hodor, P. G., and Ettensohn, C. A. (1999). The role of micromere signaling in Notch activation and mesoderm specification during sea urchin embryogenesis. *Development* **126**, 5255–5265.

Szeto, D. P., Ryan, A. K., O'Connell, S. M., and Rosenfeld, M. G. (1996). P-OTX: A PIT-1-interacting homeodomain factor expressed during anterior pituitary gland development. *Proc. Natl. Acad. Sci. USA* **93**, 7706–7710.

Tagawa, K., Humphreys, T., and Satoh, N. (1998). Novel pattern of *Brachyury* gene expression in hemichordate embryos. *Mech. Dev.* **75**, 139–143.

Takahashi, H., Hotta, K., Erives, A., Di Gregorio, A., Zeller, R. W., Levine, M., and Satoh, N. (1999a). *Brachyury* downstream notochord differentiation in the ascidian embryo. *Genes Dev.* **13**, 1519–1523.

Takahashi, H., Mitani, Y., Satoh, G., and Satoh, N. (1999b). Evolutionary alterations of the minimal promoter for notochord-specific *Brachyury* expression in ascidian embryos. *Development* **126**, 3725–3734.

Takeuchi, J. K., Koshiba-Takeuchi, K., Matsumoto, K., Vogel-Höpker, A., Naitoh-Matsuo, M., Ogura, K., Takahashi, N., Yasuda, K., and Ogura, T. (1999). *Tbx5* and *Tbx4* genes determine the wing/leg identity of limb buds. *Nature* **398**, 810–814.

Tanaka, M., Chen, Z., Bartunkova, S., Yamasaki, N., and Izumo, S. (1999). The cardiac homeobox gene *Csx/Nkx2.5* lies genetically upstream of multiple genes essential for heart development. *Development* **126**, 1269–1280.

Technau, U., and Bode, H. R. (1999). *HyBra1*, a *Brachyury* homologue, acts during head formation in *Hydra*. *Development* **126**, 999–1010.

Theil, T., Frain, M., Gilardi-Hebenstreit, P., Flenniken, A., Charnay, P., and Wilkinson, D. G. (1998). Segmental expression of the EphA4 (Sek-1) receptor tyrosine kinase in the hindbrain is under direct transcriptional control of *Krox-20*. *Development* **125**, 443–452.

Theil, T., Alvarez-Bolado, G., Walter, A., and Rüther, U. (1999). *Gli3* is required for *Emx* gene expression during dorsal telencephalon development. *Development* **126**, 3561–3571.

Thézé, N., Calzone, F. J., Thiebaud, P., Hill, R. L., Britten, R. J., and Davidson, E. H. (1990). Sequences of the *CyIIIa* actin gene regulatory domain bound specifically by sea urchin embryo nuclear proteins. *Mol. Reprod. Dev.* **25**, 110–122.

Thorpe, C. J., Schlesinger, A., Carter, J. C., and Bowerman, B. (1997). Wnt signaling polarizes in early *C. elegans* blastomere to distinguish endoderm from mesoderm. *Cell* **90**, 695–705.

Tomarev. S. I., Callaerts, P., Kos, L., Zinovieva, R., Halder, G., Gehring, W., and Piatigorsky, J. (1997). Squid *Pax-6* and eye development. *Proc. Natl. Acad. Sci. USA* **94**, 2421–2426.

Tomlinson, C. R., and Klein, W. H. (1990). Temporal and spatial transcriptional regulation of the aboral ectoderm-specific *Spec* genes during sea urchin embryogenesis. *Mol. Reprod. Dev.* **25**, 328–338.

Topol, J., Dearolf, C. R., Prakash, K., and Parker, C. S. (1991). Synthetic oligonucleotides recreate *Drosophila fushi tarazu* zebra-stripe expression. *Genes Dev.* **5**, 855–867.

Torchia, J., Glass, C., and Rosenfeld, M. G. (1998). Co-activators and co-repressors in the integration of transcriptional responses. *Curr. Opin. Cell Biol.* **10**, 373–383.

Treier, M., and Rosenfeld, M. G. (1996). The hypothalamic-pituitary axis: Co-development of two organs. *Curr. Opin. Cell Biol.* **8**, 833–843.

Treier, M., Gleiberman, A. S., O'Connell, S. M., Szeto, D. P., McMahon, J. A., McMahon, A. P., and Rosenfeld, M. G. (1998). Multistep signaling requirements for pituitary organogenesis *in vivo*. *Genes Dev.* **12**, 1691–1704.

Turbeville, J. M., Schultz, J. R., and Raff, R. A. (1994). Deuterostome phylogeny and the sister group of the chordates: Evidence from molecules and morphology. *Mol. Biol. Evol.* **11**, 648–655.

Valentine, S. A., Chen, G., Shandala, T., Fernandez, J., Mische, S., Saint, R., and Courey, A. J. (1998). Dorsal-mediated repression requires the formation of a multiprotein repression complex at the ventral silencer. *Mol. Cell. Biol.* **18**, 6584–6594.

Van Auken, K., Weaver, D. C., Edgar, L. G., and Wood, W. B. (2000). *Caenorhabditis elegans* embryonic axial patterning requires two recently discovered posterior-group *Hox* genes. *Proc. Natl. Acad. Sci. USA* **97**, 4499–4503.

van der Hoeven, F., Sordino, P., Fraudeau, N., Izpisúla-Belmonte, J.-C., and Duboule, D. (1996). Teleost *HoxD* and *HoxA* genes: Comparison with tetrapods and functional evolution of the *HOXD* complex. *Mech. Dev.* **54**, 9–21.

Van Doren, M., Powell, P. A., Pasternak, D., Singson, A., and Posakony, J. W. (1992). Spatial regulation of proneural gene activity: Auto- and cross-activation of *achaete* is antagonized by *extramacrochaetae*. *Genes Dev.* **6**, 2592–2605.

Van Doren, M., Bailey, A. M., Esnayra, J., Ede, K., and Posakony, J. W. (1994). Negative regulation of proneural gene activity: hairy is a direct transcriptional repressor of *achaete*. *Genes Dev.* **8**, 2729–2742.

Vesque, C., Maconochie, M., Nonchev. S., Ariza-McNaughton, L., Kuroiwa, A., Charnay, P., and Krumlauf, R. (1996). *HoxB-2* transcriptional activation in rhombomere-3 and rhombomere-5 requires an evolutionarily conserved *cis*-acting element in addition to the *Krox-20* binding site. *EMBO J.* **15**, 5383–5396.

Vincent, S., Ruberte, E., Grieder, N. C., Chen, C.-K., Haerry, T., Schuh, R., and Affolter, M. (1997). DPP controls tracheal cell migration along the dorsoventral body axis of the *Drosophila* embryo. *Development* **124**, 2741–2750.

Vonica, A., Weng, W., Gumbiner, B. M., and Venuti, J. M. (2000). TCF is the nuclear effector of the β-catenin signal that patterns the sea urchin animal-vegetal axis. *Dev. Biol.* **217**, 230–243.

Voronoy, D. A., and Panchin, Y. V. (1998). Cell lineage in marine nematode *Enoplus brevis. Development* **125**, 143–150.

Wada, H., and Satoh, N. (1994). Details of the evolutionary history from invertebrates to vertebrates, as deduced from the sequences of 18S rDNA. *Proc. Natl. Acad. Sci. USA* **91**, 1801–1804.

Wada, S., and Saiga, H. (1999). Cloning and embryonic expression of *Hrsna*, a *snail* family gene of the ascidian *Halocynthia roretzi*: Implication in the origins of mechanisms for mesoderm specification and body axis formation in chordates. *Dev. Growth. Differ.* **41**, 9–18.

Wada, S., Katsuyama, Y., and Saiga, H. (1999). Anteroposterior patterning of the epidermis by inductive influences from the vegetal hemisphere cells in the ascidian embryo. *Development* **126**, 4955–4963.

Wang, D. G.-W., Kirchhamer, C. V., Britten, R. J., and Davidson, E. H. (1995). SpZ12–1, a negative regulator required for spatial control of the territory-specific *CyIIIa* gene in the sea urchin embryo. *Development* **121**, 1111–1122.

Wang, G. F., Nikovits, W., Jr., Schleinitz, M., and Stockdale, F. E. (1998). A positive GATA element and a negative vitamin D receptor-like element control atrial chamber-specific expression of a slow myosin heavy-chain gene during cardiac morphogenesis. *Mol. Cell. Biol.* **18**, 6023–6034.

Wang, W., Wikramanayake, A. H., Gonzalez-Rimbau, M., Vlahou, A., Flytzanis, C. N., and Klein, W. H. (1996). Very early and transient vegetal-plate expression of *SpKrox1*, a *Krüppel/Krox* gene from *Strongylocentrotus purpuratus. Mech. Dev.* **60**, 185–195.

Wappner, P., Gabay, L., and Shilo, B.-Z. (1997). Interactions between the EGF receptor and DPP pathways establish distinct cell fates in the tracheal placodes. *Development* **124**, 4707–4716.

Watanabe, T., Kim, S., Candia, A., Rothbacher, U., Hashimoto, C., Inoue, K., and Cho, K. W. Y. (1995). Molecular mechanisms of Spemann's organizer formation: Conserved growth factor synergy between *Xenopus* and mouse. *Genes Dev.* **9**, 3038–3050.

Weatherbee, S. D., and Carroll, S. B. (1999). Selector genes and limb identity in arthropods and vertebrates. *Cell* **97**, 283–286.

Weatherbee, S. D., Halder, G., Kim, J., Hudson, A., and Carroll, S. (1998). Ultrabithorax regulates genes at several levels of the wing-patterning hierarchy to shape the development of the *Drosophila* haltere. *Genes Dev.* **12**, 1474–1482.

Wells, J. M., and Melton, D. A. (2000). Early mouse endoderm is patterned by soluble factors from adjacent germ layers. *Development* **127**, 1563–1572.

Wheeler, W., Cartwright, P., and Hayashi, C. (1993). Arthropod phylogeny: a combined approach. *Cladistics* **9**, 1–39.

Wiegner, O., and Schierenberg, E. (1999). Regulative development in a nematode embryos: A hierarchy of cell fate transformations. *Dev. Biol.* **215**, 1–12.

Wiellette, E. L., and McGinnis, W. (1999). *Hox* genes differentially regulate *Serrate* to generate segment-specific structures. *Development* **126**, 1985–1995.

Wikramanayake, A. H., Huang, L., and Klein, W. H. (1998). β-Catenin is essential for patterning the maternally specified animal-vegetal axis in the sea urchin embryo. *Proc. Natl. Acad. Sci. USA* **95**, 9343–9348.

Wikramanayake, A. H., Kauffman, J., Peterson, R., Chen, J., McClay, D. R., and Klein, W. H. (2000). Sea urchin Wnt8 functions downstream of nuclear *β*-catenin to mediate endomesoderm specification and is heterochronically shifted during the evolution of direct development. In preparation.

Wilk, R., Weizman, I., and Shilo, B.-Z. (1996). *trachaeless* encodes a bHLH-PAS protein that is an inducer of tracheal cell fates in *Drosophila. Genes Dev.* **10**, 93–102.

Wilk, R., Reed, B. H., Tepass, U., and Lipshitz, H. D. (2000). The *hindsight* gene is required for epithelial maintenance and differentiation of the tracheal system in *Drosophila. Dev. Biol.* **219**, 183–196.

Williams, J. A., and Carroll, S. B. (1993). The origin, patterning and evolution of insect appendages. *BioEssays* **15**, 567–577.

Williams, J. A., Paddock, S. W., Vorwerk, K., and Carroll, S. B. (1994). Organization of wing formation and induction of a wing-patterning gene at the dorsal/ventral compartment boundary. *Nature* **368**, 299–305.

Williams, S. C., Altmann, C. R., Chow, R. L., Hemmati-Brivanlou, A., and Lang, R. A. (1998). A highly conserved lens transcriptional control element from the *Pax-6* gene. *Mech. Dev.* **73**, 225–229.

Wilt, F. H. (1987). Determination and morphogenesis in the sea urchin embryo. *Development* **100**, 559–575.

Winnepenninckx, B. M. H., Backeljau, T., and Kristensen, R. M. (1998). Relations of the new phylum Cycliophora. *Nature* **393**, 636–638.

Wray, G. A., and Bely, A. E. (1994). "The Evolution of Echinoderm Development is Driven by Several Distinct Factors". *Development* **120S**, 97–106.

Wu, J., and Cohen, S. M. (1999). Proximodistal axis formation in the *Drosophila* leg: Subdivision into proximal and distal domains by Homothorax and Distalless. *Development* **126**, 109–117.

Wu, L. H., and Lengyel, J. A. (1998). Role of *caudal* in hindgut specification and gastrulation suggests homology between *Drosophila* aminoproctodeal invagination and vertebrate blastopore. *Development* **125**, 2433–2442.

Wu, X., Golden, K., and Bodmer, R. (1995). Heart development in *Drosophila* requires the segment polarity gene *wingless*. *Dev. Biol.* **169**, 619–628.

Wu, X., Vakani, R., and Small, S. (1998). Two distinct mechanisms for differential positioning of gene expression borders involving the *Drosophila* gap protein giant. *Development* **125**, 3765–3774.

Wülbeck, C., and Simpson, P. (2000). Expression of *achaete-scute* homologues in discrete proneural clusters on the developing notum of the medfly *Ceratitis capitata*, suggests a common origin for the stereotyped bristle patterns of higher Diptera. *Development* **127**, 1411–1420.

Xu, P.-X., Zhang, X., Heaney, S., Yoon, A., Michelson, A. M., and Maas, R. L. (1999). Regulation of *Pax6* expression is conserved between mice and flies. *Development* **126**, 383–395.

Xu, X., Yin, Z., Hudson, J. B., Ferguson, E. L., and Frasch, M. (1998). Smad proteins act in combination with synergistic and antagonistic regulators to target Dpp responses to the *Drosophila* mesoderm. *Genes Dev.* **12**, 2354–2370.

Yasuo, H., and Satoh, N. (1994). An ascidian homolog of the mouse *Brachyury* (*T*) gene is expressed exclusively in notochord cells at the fate restricted stage. *Dev., Growth Differ.* **36**, 9–18.

Yasuo, H., and Satoh, N. (1998). Conservation of the developmental role of *Brachyury* in notochord formation in a urochordate, the ascidian *Halocynthia roretzi*. *Dev. Biol.* **200**, 158–170.

Yatskievych, T. A., Pascoe, S., and Antin, P. B. (1999). Expression of the homeobox gene *Hex* during early stages of chick embryo development. *Mech. Dev.* **80**, 107–109.

Yin, Z., Xu, X.-L., and Frasch, M. (1997). Regulation of the Twist target gene *tinman* by modular *cis*-regulatory elements during early mesoderm development. *Development* **124**, 4971–4982.

Younnossi-Hartenstein, A., Green, P., Liaw, G.-J., Rudolph, K., Lengyel, J., and Hartenstein, V. (1997). Control of early neurogenesis of the *Drosophila* brain by the head gap genes *tll*, *otd*, *ems*, and *btd*. *Dev. Biol.* **182**, 270–283.

Yu, Y., Yussa, M., Song, J. B., Hirsch, J., and Pick, L. (1999). A double interaction screen identifies positive and negative *ftz* gene regulators and ftz-interacting proteins. *Mech. Dev.* **83**, 95–105.

Yuh, C.-H., and Davidson, E. H. (1996). Modular *cis*-regulatory organization of *Endo16*, a gut-specific gene of the sea urchin embryo. *Development* **122**, 1069–1082.

Yuh, C.-H., Ransick, A., Martinez, P., Britten, R. J., and Davidson, E. H. (1994). Complexity and organization of DNA-protein interactions in the 5' regulatory region of an endoderm-specific marker gene in the sea urchin embryo. *Mech. Dev.* **47**, 165–186.

Yuh, C.-H., Moore, J. G., and Davidson, E. H. (1996). Quantitative functional interrelations within the *cis*-regulatory system of the *S. purpuratus Endo16* gene. *Development* **122**, 4045–4056.

Yuh, C.-H., Bolouri, H., and Davidson, E. H. (1998). Genomic *cis*-regulatory logic: Functional analysis and computational model of a sea urchin gene control system. *Science* **279**, 1896–1902.

Yuh, C. H., Bolouri, H., and Davidson, E. H. (2000). *cis*-Regulatory logic in the endo 16 gene: Switching from a specification to a differentiation mode of control. Submitted for publication.

Yun, K., and Wold, B. (1996). Skeletal muscle determination and differentiation: Story of a core regulatory network and its context. *Curr. Opin. Cell Biol.* **8**, 877–889.

Zákány, J., Fromental-Ramain, C., Warot, X., and Duboule, D. (1997). Regulation of number and size of digits by posterior *Hox* genes: A dose-dependent mechanism with potential evolutionary implications. *Proc. Natl. Acad. Sci. USA* **94**, 13695–13700.

Zeller, R. W., Cameron, R. A., Franks, R. R., Britten, R. J., and Davidson, E. H. (1992). Territorial expression of three different *trans*-genes in early sea urchin embryhos by a whole-mount fluorescence procedure. *Dev. Biol.* **151**, 382–390.

Zeller, R. W., Griffith, J. D., Moore, J. G., Kirchhamer, C. V., Britten, R. J., and Davidson, E. H. (1995). A multimerizing transcription factor of sea urchin embryos capable of looping DNA. *Proc. Natl. Acad. Sci. USA* **92**, 1111–1122.

Zelzer, E., and Shilo, B.-Z. (2000a). Interaction between the bHLH-PAS protein Trachealess and the POU-domain protein Drifter, specifies tracheal cell fates. *Mech. Dev.* **91**, 163–173.

Zelzer, E., and Shilo, B.-Z. (2000b). Cell fate choices in *Drosophila* tracheal morphogenesis. *BioEssays* **22**, 219–226.

Zhang, C.-C., Müller, J., Hoch, M., Jäckle, H., and Bienz, M. (1991). Target sequences for *hunchback* in a control region conferring *Ultrabithorax* expression boundaries. *Development* **113**, 1171–1179.

Zhang, H., and Levine, M. (1999). Groucho and dCtBP mediate separate pathways of transcriptional repression in the *Drosophila* embryo. *Proc. Natl. Acad. Sci. USA* **96**, 535–540.

Zhu, J., Hill, R. J., Heid, P. J., Fukuyama, M., Sugimoto, A., Priess, J. R., and Rothman, J. H. (1997). *end-1* encodes an apparent GATA factor that specifies the endoderm precursor in *Caenorhabditis elegans* embryos. *Genes Dev.* **11**, 2883–2896.

Zhu, J., Fukushige, T., McGhee, J. D., and Rothman, J. H. (1998). Reprogramming of early embryonic blastomeres into endodermal progenitors by a *Caenorhabditis elegans* GATA factor. *Genes Dev.* **12**, 3809–3814.

Zilberman, A., Miano, D. J., Olson, E. N., and Periasamy, M. (1998). Evolutionarily conserved promoter region containing CArG*-like elements is crucial for smooth muscle myosin heavy chain gene expression. *Circ. Res.* **82**, 566–575.

Zuker, C. S. (1994). On the evolution of eyes: Would you like it simple or compound? *Science* **265**, 742–743.

INDEX *

* Page numbers in italics refer to the pages on which the figure appears.

ل